1987

REAL-TIME MICROCOMPUTER SYSTEM DESIGN
An Introduction

COMPUTER ENGINEERING

Consulting Editor
Stephen W. Director, Carnegie-Mellon University

REAL-TIME MICROCOMPUTER SYSTEM DESIGN
An Introduction

Peter D. Lawrence
Konrad Mauch

Department of Electrical Engineering
University of British Columbia
Canada

McGraw-Hill Book Company

New York   St. Louis   San Francisco   Auckland   Bogotá   Hamburg
Johannesburg   London   Madrid   Mexico   Milan   Montreal   New Delhi
Panama   Paris   São Paulo   Singapore   Sydney   Tokyo   Toronto

This book was set in Times Roman.
The editor was Sanjeev Rao;
the cover was designed by Rafael Hernandez;
the production supervisor was Phil Galea.
Project supervision was done by The Total Book.
R.R. Donnelley & Sons Company was printer and binder.

REAL-TIME MICROCOMPUTER SYSTEM DESIGN
An Introduction

1234567890 DOCDOC 8943210987

ISBN 0-07-036731-0

Library of Congress Cataloging-in-Publication Data

Lawrence, Peter D. (Peter Donald)
   Real-time microcomputer system design.

   (McGraw-Hill series in electrical engineering)
   Includes bibliographies and index.
   1. Automatic control.   2. Microcomputers.
3. Real-time data processing.   4. System design.
I. Mauch, Konrad.   II. Title.   III. Series.
TJ223.M53L39   1987   629.8′95   86-15361
ISBN 0-07-036731-0
ISBN 0-07-036732-9 (solutions manual)

From Peter to Margaret, Katherine and Megan

From Konrad to my parents, Walter and Elizabeth

CONTENTS

5   Programming Languages   103

Part 3 Lower-Level Considerations

Part 6 Implementation Tools

Appendixes

PREFACE

Microcomputer-based real-time systems are used not only by electrical and computer engineers but also by engineers in many other disciplines. Mechanical engineers, for example, design robots and other automated machinery controlled by microcomputers. Plants designed by chemical engineers and mineral processing engineers are controlled and monitored by microcomputers. Civil engineers gather hydrological and other environmental data using microcomputer-based data-acquisition systems. Scientists in both the physical and biological sciences are also using microcomputer-based real-time systems to gather and record data from experiments and to control experimental apparatus.

USERS OF THIS BOOK

This book was primarily developed for engineering students in a broad range of disciplines. Practicing engineers and scientists working in the physical, biological, and applied sciences who are seeking a self-contained introduction to the design of real-time microcomputer systems should also find this book useful.

Even though this book is not aimed specifically at electrical engineering students, we believe that it has a place in the electrical engineering curriculum. Used either as a supplement or as the primary text, it provides a high-level perspective on the design process. References after every chapter are provided for readers who may want to pursue a topic in greater depth.

APPROACH

The objective of the book is to present a systems design methodology that involves all the components of a real-time computing system, including sensors, actuators, conditioning electronics, and interfaces as well as the CPU. To reflect a common practice in industry and to a great extent in scientific research, we have chosen to concentrate on the development of *board-based* real-time microcomputer systems.

We also emphasize the use of *high-level programming languages* to produce software for real-time systems. It is our experience that high-level programming languages can be used effectively in all but the most time-critical real-time systems. When assembly-language programming is required, it can often be reduced to a few short subroutines called by the high-level-language program.

Because of our emphasis on board-level hardware design and the use of high-level programming languages, and because of the nature of the audience that we are writing for, this book differs in content and approach from the standard textbooks on microcomputers used by electrical engineering students. Our approach is mostly top-down, starting with a look at the design process for the real-time system as a whole and then discussing the design of the various components of the system.

We do not discuss the design of microcomputer systems at the chip level in any detail. Our discussion of assembly-language programming is brief and oriented toward how to incorporate short assembly-language modules into a program written in a high-level language and write the necessary assembly-language components of an interrupt handler. However, we include material on important topics such as microcomputer-development systems, real-time operating systems, and transducers and actuators, which are usually not covered in the standard introductory textbooks on microcomputers.

ORGANIZATION

We use this book as a text for an introductory course on microcomputers. Our students are in the senior year of an engineering discipline other than electrical engineering. The majority are mechanical engineering students, but we also have bioresource, chemical, civil, engineering physics, geological, metallurgical, mining and mineral process engineering students. All students have had at least one high-level language programming course (either FORTRAN or Pascal) and an introductory electrical circuits course. Many of the students have also had an introductory electronics course which includes some exposure to digital logic. The material on number systems and digital logic elements in the first two appendixes is assigned to the students for self-study at the beginning of the course to ensure that they all have the required background information.

While we are covering high-level design and software design in lectures, our students are programming C language applications (using an auxiliary C language text) in the laboratory. We then cover assembly language and finally the hardware topics, calling this a top-down software-before hardware organization.

Our course is heavily laboratory-oriented, which is important for any course based on this text, since the general concepts introduced must be reinforced by practical experience with a real microcomputer system. Our microcomputer laboratory consists of a multiuser microcomputer development system (Tektronix 8560), which is linked to Motorola MC6800-based laboratory microcomputers. Each laboratory microcomputer is equipped with a standard backplane bus (STD bus) which contains boards that interface the computer to the apparatus which it monitors and controls. The use of a board-based microcomputer system in the laboratory reinforces the board-based hardware design approach of the text.

Our students develop programs for the laboratory computers on the microcomputer development system. The majority of the programs are written entirely in a high-level programming language (the language C).

The laboratory sequence starts with some programming exercises to introduce the student to the UNIX operating system used on the development system and to programming in C. Then the students develop a program which makes use of pumps, liquid-level sensors, and agitators in the laboratory apparatus to emulate a washing machine. This procedure introduces students to the use of microcomputers in simple sequential control (i.e., programmable controller applications) and shows them very early in the course how a computer can control physical systems. The next sequence of laboratory exercises involves the measurement and control of water temperature, which introduces students to analog interfaces, sensor linearizing, and control via computer. Students then develop a program for a real-time clock. This exercise involves the use of interrupts and the development of a short assembly-language interrupt service routine linked with the main C program. The last laboratory in the sequence varies, depending on the time available and the interests of the instructors and students, but it has included the control of a small robot arm.

OTHER COURSE ORGANIZATIONS

A second approach is to have a bottom-up software-before-hardware organization in which assembly language would be covered first. A suggested chapter order for this approach is: introduction (1), computer structure (2), assembly language (7), high-level languages (5), high-level software example (6), high-level design (3, 4), hardware topics (8 to 14). In the laboratory, the students could do assembly language first, high-level language next, and then programs interacting with interfaces, sensors, actuators, etc.

A third approach is to have a bottom-up hardware-before-software order. In that case the chapters could be ordered as hardware (2, 8, 9, 10 to 13), assembly language (7), high-level language (5, 6), high-level design (3, 4), system-design example (14), development systems (15), and operating systems (16). With this approach it is more difficult to have laboratories early in the course (since software is required to interact with the hardware and the software is not covered until later in the sequence). It might be used when there are no laboratories in the first semester of a two-semester sequence.

Although our laboratory is centered on a commercial microcomputer development system, many other arrangements are possible. A mainframe computer or minicomputer could be used to develop the software for the laboratory computers. Both Tektronix and Hewlett-Packard, for example, support Pascal and C language cross-compilers and assemblers for many microprocessors. This cross-software runs on the multiuser DEC VAX computers and the HP-9000 series computers. Alternatively, a completely decentralized system could be designed using personal computers such as an IBM PC. The personal computer can be used as a software-development station, and for downloading software to a laboratory experiment station containing a bus-based system; or the personal computer can be enhanced with interface boards itself and be used as the real-time system. The essential requirements are that the development system support software development in a high-level language and that the microcomputer be readily interfaced to transducers and actuators in the lab apparatus.

The book by David Auslander and Paul Sagues entitled *Microprocessors for Measurement and Control* (Osborne—McGraw-Hill, 1981) also contains many laboratory exercises well suited for a course based on our book.

ACKNOWLEDGMENTS

We would like to thank the many engineering students who used drafts of this book over the past three years and who gave us valuable criticism, found errors, and carried us along with their interest in the technology. Jim Dukarm of RSI Robotic Systems International contributed very useful comments during the drafting of the book. There were many helpful discussions with Eric Jackson of International Submarine Engineering.

To Réal Frenette who worked out the solutions to the exercises at the end of each chapter, we would like to express our gratitude.

We are also indebted to the reviewers of this book whose many constructive suggestions we have attempted to incorporate into the text.

Finally, we would like to thank our families and friends for their constant encouragement and support.

Peter D. Lawrence
Konrad Mauch

PRELIMINARIES

ONE

REAL-TIME COMPUTER SYSTEMS

The design of a real-time computer system involves the consideration of more than the computer itself. It is necessary to consider the overall task that the system is to perform. In this chapter we will illustrate a method for describing the operation of a real-time system (the ASM diagram) and will introduce the components involved in a complete real-time system. Although each of the topics will be covered in greater detail later in the book, we will use this chapter to provide the reader with some appreciation of the terminology involved and the structure of a real-time computer system.

1.1 WHAT IS A REAL-TIME SYSTEM?

Whenever a computer system is required to acquire data, emit data, or interact with its environment at precise times, the system is said to be a *real-time computer system*. All computer systems are in some sense real-time systems. For control of processes or mechanisms, a real-time system is a system whose temporal performance (response time, data-acquisition period, etc.) is of critical importance to the industrial systems to which it is connected. For example, an automobile computer that generates ignition timing must produce a spark at a certain precise instant, even if there are other tasks that the computer must carry out.

As the number of critically time-dependent tasks in a computer system grows, careful thought must be given to the computer programs (*software*) so that all real-time constraints are satisfied. If, for example, an overpressure signal from a boiler becomes active, the computer must take appropriate action within a specified period of time (the *response time*) to avert possible disastrous results.

On the other hand, in a large time-shared system such as a central university campus computer, no such high-priority response time can be fairly granted to one user's task. In a time-shared computer system, one user's program is allowed to run for a short time, called a "time-slice," and then another user's program runs, and so forth. When all users have had a time-slice of the computer resources, the cycle repeats and the first user's program is continued for another time-slice. Thus as the number of users increases, the total time required to service all users increases and each user will see an increase in the time required to complete the task (the system response time). Such a system is unsuited to real-time applications.

Although the most obvious feature of a real-time computer system is its close temporal coupling with an industrial process or system, another characteristic is the need for bug-free and robust software. Since many microcomputer-based real-time systems are unattended and operating without human supervision, it is desirable that if a failure occurs, it does not produce a catastrophic result. The failure should be detected and a graceful recovery should be executed. For example, no commuter would be pleased to find that a driverless commuter train had just overrun the end of the line.

Real-time systems can be implemented by traditional *analog* electronics, dedicated *digital* hardware, or a computer-based system. An analog signal is infinitely variable or continuous, as are most physical phenomena such as light levels, sound levels, or battery voltages. In a digital system, continuous values are represented to a given accuracy by quantized, or digital, values. For example, the number 3.1415 is one possible digital representation of the constant π. *Binary numbers* are used to represent digital quantities (see App. A for a discussion of number systems), since the binary states 1 and 0 can be conveniently represented by digital electronic devices (see App. B).

1.1.1 Analog Systems

Fully analog electronic systems are also real-time systems. An audio amplifier is an example of an analog system. Analog systems are typically made up of discrete transistors, analog integrated circuits (examples given in Sec. 11.4), and passive components such as resistors and capacitors, all soldered onto one or more *printed circuit boards*. The printed circuit boards have a printed pattern of conductive paths that form the connections between the parts on the board.

One of the important real-time specifications for undistorted waveform amplification is that the time delay of the signal between the input and the output of the amplifier be relatively constant at all audible frequencies. This is necessary in many instrumentation applications where the waveforms are to be analyzed. It is also necessary to some extent in home audio systems.

In audio systems, the audible frequency range extends to about 20,000 Hz. Inexpensive amplifiers have long been available which not only provide voltage and power amplification but also *filter* the signals. Signal filtering provides selective gain or loss in certain frequency bands (e.g., bass or treble controls).

Frequency-band filtering of audio signals can also be achieved digitally by computer, although most general microcomputers would be unable to carry out alone the computations required for this task fast enough to be able to replace an analog system. That is not to say that a microcomputer in conjunction with other high-speed computational elements would be unable to achieve the task, but it does point out that analog systems are serious contenders for industrial applications. An analog system is often used when

- High-frequency signals must be processed
- Simple functions such as signal amplification, signal comparison, or signal filtering is required

However, a custom-designed circuit is usually required, and logic functions (see App. B) are more difficult to implement than with other approaches.

1.1.2 Dedicated Digital Systems

Where rapid decision making is involved, but analog signals need not be processed, a dedicated digital system can be designed. A dedicated digital system is one whose function is determined by logic elements such as *combinational logic gates, flip-flops and registers* (see App. B for a discussion of the basic logic elements), and wired interconnections between elements (i.e., the function is not software programmable). Most of the circuitry necessary for the control of a machine such as a home washing machine is available in a small number of digital integrated circuits. Digital logic devices are most easily employed when their input information is available from binary *sensors* such as on/off switches, and their outputs turn on and off lights, or *actuators* such as solenoids. A sensor is a device which converts a *physical signal* such as the mechanical movement of a switch to an electric signal such as a voltage change across the switch. An *actuator* is a device which converts an electrical signal to a mechanical movement.

Example: a binary counter A counter that counts the pulses on a clock line is a simple form of a dedicated digital system. Let's look at a 2-bit binary counter that counts up whenever a clock pulse occurs on a clock line. The 2-bit binary representation of the numbers 0 to 3 are shown in Fig. 1.1 (00, 01, 10, and 11, respectively). The 2-bit binary number is held in two flip-flops Q_a and Q_b. A sequence of adjacent clocked flip-flops which holds a binary *bit string* (sequence of bits) is called a *register*.

The counter counts from 0 to 3 and then overflows back to 0 again when counting up. The contents of the register is called the *state*. Thus the counter has only four distinct states. In each state the flip-flops are stable. The next state in a counter is determined by the present state only, and the system changes to the next state at the time of the clock pulse.

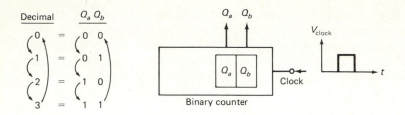

Figure 1.1 A 2-bit binary up counter.

Algorithmic state machine diagram The function of the counter can be described in what is known as an *algorithmic state machine diagram* (*ASM diagram*) (see Fig. 1.2).

In an ASM diagram each circle represents a state. The *state name*, or *label*, is written near each circle. The outputs delivered by the system to the outside world are written inside the circle. The arrows between the circles represent the allowed transitions betweeen states.

The direct implementation of an ASM diagram in dedicated digital hardware, as shown in Fig. 1.3, is called a *finite-state machine*.

In the general finite-state machine shown in Fig. 1.3, the next state is determined at the time of the clock by the present state and any system inputs. Since there are no system inputs into the binary counter, the next state is a function of only the present state. The system outputs in a general finite-state

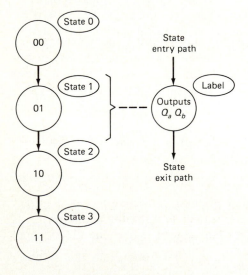

Figure 1.2 Algorithmic state machine (ASM) diagram for the binary counter.

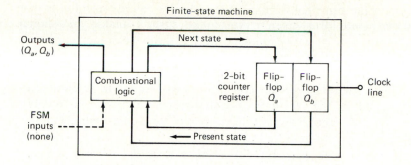

Figure 1.3 A 2-bit counter represented as a finite-state machine.

machine are determined by the present state and the system inputs. In the case of the binary counter, the outputs are exactly equal to the present state.

A dedicated digital implementation is used when one or more of the following conditions apply:

- Highspeed operation is needed
- The application is simple, requiring low circuit complexity
- Function or circuit modifications are not anticipated
- Only simple binary inputs and outputs are required

A custom-designed circuit is usually required for a dedicated digital implementation of a real-time system.

1.1.3 Computer-Based Systems

Many computer families, such as the Digital Equipment Corporation's PDP-11, have decreased in size and cost, through minicomputer to microcomputer. In this evolution the capabilities of the machines have, in most cases, increased. Also, with the decrease in cost particularly, the range and number of applications have increased, so that many process and machine control applications now use microcomputers.

The ease with which a microcomputer can be applied to a particular task has been greatly enhanced by

1. The wide commercial availability of compatible subsystem boards which a system designer can use to assemble a complete microcomputer system that is capable of sensing analog and digital signals and driving actuators
2. The availability of high-level computer languages (such as the C language) that can be used to rapidly create high-speed programs, require modest amounts of computer memory, and provide convenient access to input and output devices

In fact, one of the principal benefits of a computer-based controller is that it can be programmed for a given task. Unlike custom analog and digital hardware, programs can be easily modified during the development phase as circumstances require. A very high level of algorithmic complexity is also possible using a software-based system. In general, however, sophistication is achieved at the expense of speed. The computer must access a large number of memory locations during the execution of each high-level language instruction, and these accesses take time. However, in many real-time-control applications, very high speed may not be required.

Thus when a computer-based real-time system is used,

- A real-time system can be assembled from commercially available subsystem components.
- The operation of the system can be conveniently expressed and modified in software.
- Complex functions and parameter changes can be achieved while the system is operating.
- Operation is slower than an equivalent analog system or dedicated digital system.
- The system is easily compatible with both analog and digital inputs and outputs.

Although the analog systems and hard-wired dedicated digital systems still retain a niche, the number of computer applications is increasing. This is owing to a number of factors, including speed improvements, new hardware architectures for special applications (e.g., digital signal processing), cost reductions, and ease of manufacturing and testing. Another important factor is the growing familiarity of persons in industry with microcomputer technology. Suitable applications range from very simple binary decision controllers, as in the example given in Sec. 1.2, to complex multicomputer systems. For instance, an application such as a robot controller (see Chap. 12) may require a hierarchy of microprocessors in which a single microprocessor is used in the control of each joint in the arm, and a supervisory microprocessor is used to coordinate the actions of all joints.

1.2 AN ASM DESCRIPTION OF A REAL-TIME SYSTEM

To introduce some of the techniques used in designing a real-time system we will look at the design of a very simple real-time system. The controller in this system could be implemented either with dedicated digital circuits or with a microcomputer. We will assume that a microcomputer is used.

In atmospheric monitoring, scientists wish to gather data on the chemical contents of dry-air precipitate and the dissolved chemicals in rain. The example to be described is a machine that allows these two types of data to be gathered separately. The *function* of the system will be described with an ASM diagram. Such diagrams can be used as the basis for a computer program to control the system (or as a guide to the design of the finite-state-machine hardware if a dedicated digital circuit is used).

The machine is to be battery-operated and is to run unattended in remote sites. The machine will uncover its rain collector (and cover its dry-precipitate collector) when it rains and do the reverse when it stops raining (see Fig. 1.4).

A reversible dc motor is used to drive a pair of covers from one side to the other. Limit switches stop the motor at the extreme in each direction.

Rainfall is detected by a conductivity sensor consisting of two metal plates close together that are heated slightly. When rain falls on the plates, one plate becomes shorted to the other, and the output voltage of the conductivity sensing circuit changes from a low voltage (off) to a high voltage (on). The sensor output is monitored by the control system, which drives the motor in the left direction to cover the dry collector. When the rain stops, the heater (which is on all the time) dries the plates, causing them to stop conducting, and the motor is driven to the right.

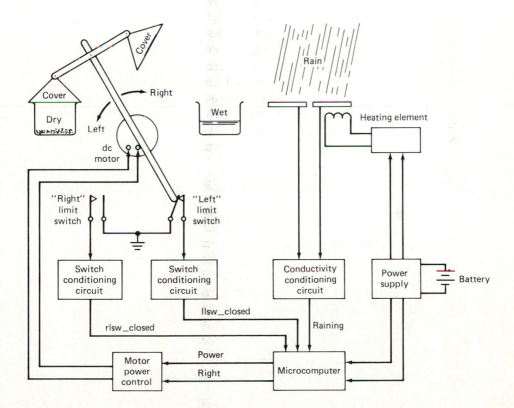

Figure 1.4 Air chemical deposition sampler.

An ASM description of the sampler   In order to simplify the description of the controller function, we will define the notation used for the signal lines in Fig. 1.4. First let's look at the input lines to the controller. The controller senses the conductivity, and the state of both switches.

Note that the signal name was chosen to convey to the reader both the name of the sensor or actuator involved and the voltage level when the line is in the true state. For example, the llsw_closed input line will produce a high voltage on the line when the switch is closed.

Note that the ASM diagram in Fig. 1.5 for the air chemical deposition sampler has five normal states; START, MOVING_LEFT, WET, MOV-ING_RIGHT, and DRY. While inside each state, the output signals (motor direction and motor power) are held constant. The output signals generated in each state are shown inside the state circles.

The ASM diagram also shows the input conditions that must be true for the controller to move from one state to another. The input conditions are listed on the arrows (the *state transition paths*) between states.

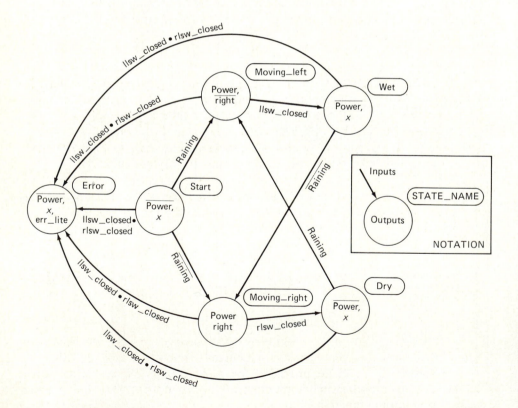

Figure 1.5 ASM diagram of air chemical sampler.

Sensor	Input signal name
Conductivity sensor	raining (true when raining)
Left-limit switch	llsw_closed (true when switch closed)
Right-limit switch	rlsw_closed (true when switch closed)

We will also assign names to the controller output signals.

Control action	Output signal name
Motor power	power (true when power on)
Rotation direction	right (true when the covers move toward the wet-precipitate collector and false when the covers move toward the dry-precipitate collector, i.e., to the left)

START   When powered up, the air chemical sampler begins in the START state. Since no power is applied in this state, the direction output ("right") can be either true or false. This is shown as an × in Fig. 1.5, which indicates a "*don't-care*" situation. The next state is determined solely by the "raining" input logic variable. The digital controller moves to the next state at a clock time.

MOVING_LEFT   If it is raining at power up, the input "raining" will be true and regardless of the limit-switch conditions, the covers will be driven left. In order to move left, the output signal "power" must be on (true) and the direction variable "right" = false (shown by a bar over a variable).
    If the left-limit switch happened to be closed on power up, the MOVING_LEFT state is passed through so rapidly that the motor does not have time to move before the WET state is entered and the motor shuts off.

WET   When the dry collector is covered and the left-limit switch is closed, the controller enters the WET state. In the WET state the motor power is shut off and the direction output is arbitrary (×). The controller remains in the WET state until the conductivity sensor output becomes false.

MOVING_RIGHT   When it stops raining, the controller enters the MOVING_RIGHT state and the covers rotate to the right until the right-limit switch closes.

DRY   When rlsw_closed becomes true, the state changes to DRY and the motor is shut off. This state is maintained until it begins to rain again, and so forth.
    It is advisable to make provision for error conditions. One such error condition occurs if both switches are closed at the same time. This should never occur during normal function of the machine and the desirable action is to have the controller enter the error state shown in Fig. 1.5 from whatever state it was in when the error occurred. The motor power should be off in this state.

Constructing an ASM diagram   The following conventions should be adhered to in the construction of an ASM diagram:

1. Each stable state in the system is represented as a circle.
2. The name of the state (or its number) is written right outside the circle.
3. Output conditions during the state are written inside the circle. Ideally the state of all output variables (on or off) should be listed in the circle, but this may not be practical when there are a lot of output lines. In such a case, a separate list of output variables for each state can be drawn up.
4. Input conditions that select the next state are written on arrows between states. Ideally, the next state for all possible input condition combinations (2^N input combinations, where N is the total number of inputs) should be represented on the ASM diagram. This also may not be practical, but the designer must consider all possible combinations.

Recall that in the example of the binary counter in Sec. 1.1, there were no options for the next state, so no input conditions were shown on the arrows of Fig. 1.3. The counter moves unconditionally into the next state when the clock pulse occurs.

The ASM diagram, shown in Fig. 1.5, describes a machine that consists of binary inputs, binary outputs, and states that can be binary-numbered. The controller can be implemented directly in digital hardware (combinational logic and a register) as a finite-state machine or in software in a microcomputer. A complete example of a software implementation will be given in Chap. 14.

1.3 COMPONENTS OF A REAL-TIME SYSTEM

The system just described is quite simple. However, its basic structure is typical of a real-time computer system. In this section we will introduce the structural components that constitute most real-time systems and we will follow the transformations that a signal undergoes as it progresses in Fig. 1.6 from sensor to processor to actuator.

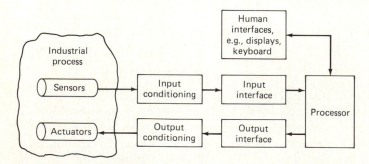

Figure 1.6 Typical real-time computer system.

In general, input signals to a real-time computer system are provided by *sensors* (Sec. 1.3.1), which require *signal conditioning* (Sec. 1.3.2). The signals must usually be converted to computer-compatible codes in an *input interface* (Sec. 1.3.3). The computer, acting on instructions contained in its program, processes (Sec. 1.3.4) the input data and generates output data which requires conversion to suitable control signals in an *output interface* (Sec. 1.3.5). The control signals can be further processed by *output conditioning* circuits (Sec. 1.3.6) before driving *actuators* (Sec. 1.3.7).

1.3.1 Sensor Characteristics

For most physical variables a wide variety of sensor technologies exist, each producing its own characteristic signal as a function of the physical variable it is sensing. In the example shown in Fig. 1.7, the depth of liquid in a tank can be measured using a float to turn the shaft of a potentiometer, thereby generating a continuously variable (*analog*) voltage lying between the two values representing tank empty and full. The depth of the liquid is proportional to the potentiometer voltage V_a.

A second method also employs a float to rotate a shaft. In this case, the shaft is fitted with an *optical shaft encoder* (see Fig. 1.8). An optical encoder contains a disk with many fine radial lines. A light source transmits a beam of light through the disk to light detectors on the other side. As each line on the disk passes through the beam of light, the light detector effectively receives a light pulse which is converted to a voltage pulse on the output signal line by the electronics in the sensor. Thus each output voltage pulse can be related to an increment in shaft rotation and hence to an incremental change in water level in the tank. The sensor shown in Fig. 1.8 uses two side-by-side light detectors arranged so that the electronics will produce a voltage pulse on one line for

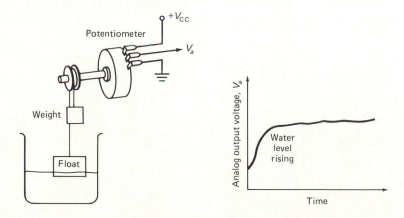

Figure 1.7 Liquid-level sensor using a potentiometer which produces an analog output signal.

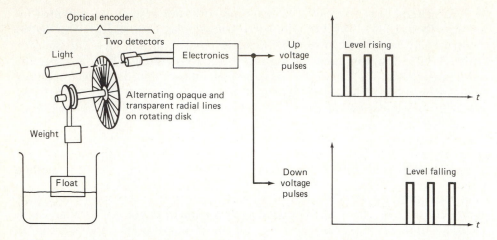

Figure 1.8 Liquid-level sensor using an optical encoder with a two-line digital output.

each increment in rotation caused by a rise in liquid level and a voltage pulse on a second line for each increment in rotation caused by a fall in the liquid level. Thus the liquid level in the tank is proportional to the number of voltage-rise pulses since the tank was filled minus the number of voltage-fall pulses.

A third method (Fig. 1.9) employs an ultrasound transducer mounted in the top of the tank. The transducer periodically sends out a pulse of sound that reflects off the liquid in the tank and returns to the transducer at a time delay

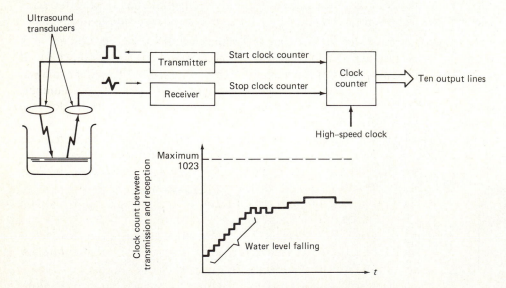

Figure 1.9 Liquid-level sensor using an ultrasound "range finder" to give a 10-line digital output.

determined by the depth of liquid in the tank. When the sound pulse is transmitted, a digital counter starts counting pulses from a high-speed clock; and when the reflected pulse is received, the counter is stopped. The accumulated count is proportional to the distance between the water surface and the transducer. As the liquid level falls, the total count between transmission and reception will increase. Ten output lines carry the binary-coded clock count (interpreted in the figure as a decimal number) to the computer for interpretation.

No single sensor system meets the requirements for all applications. Many factors must be considered in the selection of a sensor, including accuracy, precision, linearity, and temperature stability specifications. Each method described above produces a different type of output signal and uses a different number of output lines. Each signal requires its own special processing considerations, which is the subject of the next section.

1.3.2 Signal Conditioning

Analog signals, such as those generated by the potentiometer in Fig. 1.7, often must have one or more of the following treatments prior to being sampled by the computer:

- Voltage *amplification* to match the full-scale voltage change of the sensor output to the full-scale voltage range of the input interface. For example, the sensor may produce a voltage from $0\,V$ to $1\,mV$ full scale. The input interface may linearly convert any voltage in the range of 0 to $1.023\,V$ to a digital number between 0 and 1023 (10-bit representation; see App. A). Thus in order to use the full digital scale, the input signal must be amplified by a factor of 1023.
- Voltage *level shifting* to align the minimum sensor output voltage with the minimum voltage that can be converted by the input interface. If a sensor produces a voltage between -0.5 and $0.5\,V$, and if the input interface converts a voltage between 0 and $1\,V$ to a digital number, then $0.5\,V$ must be added to the sensor voltage to match the interface requirement.
- *Frequency-range limiting*, or *filtering*, to reduce any unwanted high-frequency signal and noise components in the signal (for example, any 60-Hz power-line electric noise). Filtering can also be used to block a dc voltage component via high-pass filtering or "ac coupling," as it is sometimes called when a simple capacitor is used in series with the input signal.
- *Signal-mode conversions*. A voltage level can be converted to a constant-amplitude pulse train such that either the *pulse rate* or *pulse duration* is proportional to the voltage level (see Fig. 1.10). Conversion to pulses is useful when communicating a signal between two points over a pair of lines in which there is a great deal of noise in the

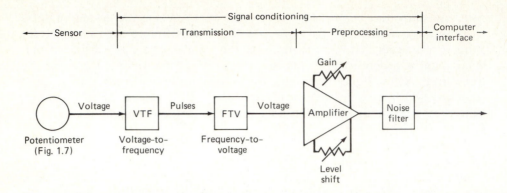

Figure 1.10 Pulse-rate modulation and demodulation of an analog signal.

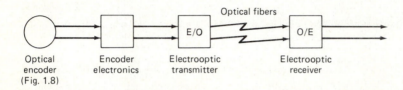

Figure 1.11 Optical transmission of binary pulses.

frequency band of the original signal, or in a system that does not pass direct current, such as in a transformer-coupled circuit. Converters that carry out this transformation are called voltage-to-frequency converters. The pulse train can also be converted back to the original voltage by a frequency-to-voltage converter.

Pulse trains can be conducted either electrically or optically (using optical fiber technology). In the case of optical transmission, the electric signal is converted to a light by an electrooptical device such as a laser. The light is transmitted along a fine glass fiber to the receiving end, where it is converted back to an electric signal by an optoelectrical device such as a phototransistor. Optical transmission is virtually immune to electromagnetic sources of noise. It could be used in the transmission link between the pulse converters in Fig. 1.10; or, as shown in Fig. 1.11, the pulses generated by the optical encoder (see Fig. 1.8) can be transmitted by optical fibers, since they are already in pulse form.

- *Electrical isolation.* The two sensing methods in Figs. 1.8 and 1.9 produce digital signals. It is frequently necessary to electrically isolate signal generators from computer equipment. This is done to reduce the effects of power-line voltage differences and noise that can be generated when there is a common electric circuit connecting two physically separated systems. Isolation can be accomplished by

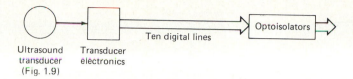

Figure 1.12 Optical isolation of digital lines.

> optoisolators for digital signals (see Fig. 1.12). These devices convert an electric signal to light, transmit the light over a short distance, and then reconvert the light back to an electric signal. Isolation can also be done for analog signals. Isolation is discussed in detail in Chap. 11.

1.3.3 Computer Input

The conditioned input signals are read into the computer by the *computer input interfaces*.

Analog input interface   Signals from the external world enter the computer through what is known as an *interface*. We have been dealing with two kinds of signals—analog and digital. In order to represent an analog variable such as temperature in a computer, we can record its value only at specific instants and to a given number of significant digits. Analog signals must be passed through an interface having an *analog-to-digital (A/D) converter*. The conversion process illustrated in Fig. 1.13 involves *sampling* at one instant of time the continuous analog input voltage, storing the voltage on a capacitor, and before the next sample is taken, converting the stored value to a binary number that lies between the minimum binary number and maximum binary number that the analog-to-digital converter can represent.

For the example shown, the A/D converter has a range of 10 binary bits (see App. A) or can represent $2^{10} = 1024$ different values of input voltage in the voltage range specified (0 to 10 V) for the A/D converter. This gives a resolution of about 10 mV (10 V/1024) in the conversion process, or an accuracy of about 0.1 percent of the full-scale voltage. For each sample converted, 10 binary bits of data are transferred to the processor before the next sample is taken.

The sampling rate and the rate at which the samples are converted (usually the same as the sampling rate) can be controlled by the processor. Rapidly varying signals must be sampled at a higher frequency. A/D converters have maximum sampling-rate specifications that range from a few hundred samples per second (hertz) to several hundred megahertz.

In addition to *resolution* (often specified in terms of a number of bits), *input voltage range* and maximum *conversion rate*, the number of analog input *channels*, or signal lines, per A/D converter is an important design decision. In

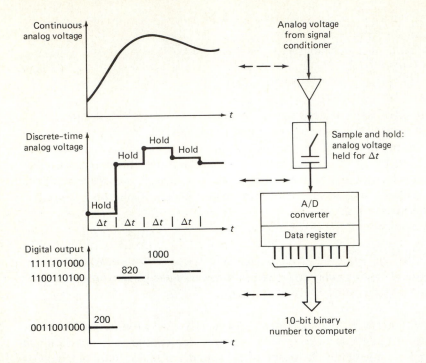

Figure 1.13 Analog-to-digital conversion.

a multichannel A/D, the processor can select which analog channel to convert next. When several analog input channels are to be converted, the A/D converter is shared, or *multiplexed*, by sequentially sampling and converting the voltage on each of the channels. In the case of a 16-channel A/D converter with a maximum conversion rate of 16 Hz, the maximum sampling rate per channel when all 16 channels are sampled in sequence is thus reduced to 1 Hz.

The A/D interface contains a number of *registers* (see Fig. 1.14) that can

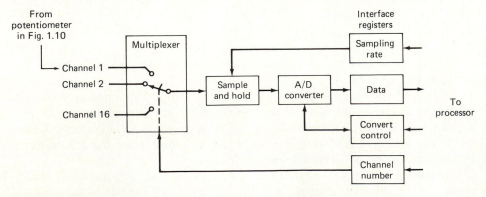

Figure 1.14 Analog-to-digital interface.

be addressed by the processor. The interface registers contain the binary results of the latest conversion and a binary code representing the status of the converter. (Is it through converting?) A code for the desired channel number in a multichannel A/D converter is also stored in a register. A register may also contain a code for the desired sampling rate.

Digital input interface   A digital input interface is usually not as complex as an A/D interface since the information is already in binary form. A register in the interface can load the data from all the input lines connected to it upon command from the processor. The register and connecting lines are usually called a *port*. The number of lines that can be connected to each port can vary between 4 and 16, with each line carrying either a logical 1 or 0 at any time. Many digital interface cards contain more than one port.

Interface connections for the signals produced in Figs. 1.11 and 1.12 are shown in Fig. 1.15. The digital interface has two ports with eight binary bits per port. Thus the ultrasound sensor, which produces 10 bits in this example, requires two ports to accommodate all its connections. The unused portion of port 1 is used for the optical encoder's two lines.

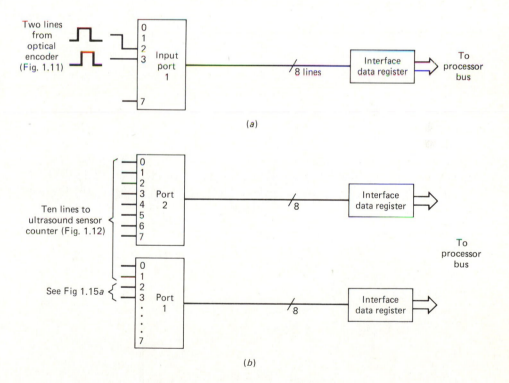

Figure 1.15 Digital input interface. (*a*) Digital interface connections for two lines from optical encoder. (*b*) Digital interface connections for ten lines from ultrasound counter.

1.3.4 The Processor

The portion of the real-time computer system that acts upon information received from the input interfaces we will call the *processor*. For the system shown in Fig. 1.16, one *printed circuit card* contains a *microprocessor* and support circuits; this card is usually called the *processor card*.

The microprocessor is an *integrated circuit* (see Chap. 8) consisting of thousands of transistors organized into all the digital functions needed to carry out the job of a *central processing unit* (*CPU*) of a computer. These functions are

- Instruction fetching from memory
- Data fetching from memory or input interfaces
- Instruction execution
- Data storage in memory or output interfaces
- Response to external requests for attention

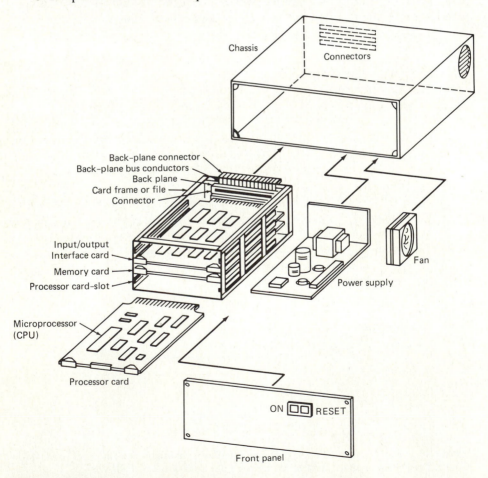

Figure 1.16 Structure of a microcomputer system.

A second card contains the system memory. The memory is a sequentially numbered (*addressed*) list of *memory locations* containing instructions and data. In some systems, the memory is located on the same card (or even inside the microprocessor itself). The memory is usually considered to be part of the "processor."

The third card shown contains the input and output interfaces to the processor.

All three cards slide into a *card frame*. A connector mounted on a *backplane* at the bottom of the frame makes electric contact with the signals from the card's edge and holds the card in place. Communication between boards takes place over a set of parallel wires (called a *bus*) that connects the corresponding pins on each backplane connector together.

The card frame and backplane along with a *power supply* that provides several dc voltages are mounted in a *chassis* with electric connectors for ac power and input/output signals. A *front panel* is also mounted on the chassis. On/off switches and other control switches can be mounted on this panel.

The complete unit is referred to as a *microcomputer*.

1.3.5 Computer Output

Control commands from the CPU are delivered to the external world through an output interface. As in the input interface, the outputs can be either analog or digital.

Analog output interface In order to produce an analog output, the CPU selects a data register in the interface and writes the data to be converted into an analog signal into the data register. The means by which this selection is carried out will be described in Sec. 2.5.

The data register is connected to a *digital-to-analog converter* which produces an analog voltage equivalent to the digital representation (see Fig. 1.17). For example, a 10-V 10-bit D/A converter will generate a voltage that will lie in the range from 0 to 10 V. There are 1024 different possible levels in the 10-V range and thus the conversion will have a resolution of about 10 mV,

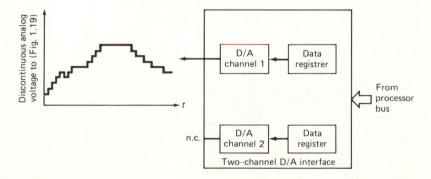

Figure 1.17 Digital-to-analog interface for control of the motor in Fig. 1.19.

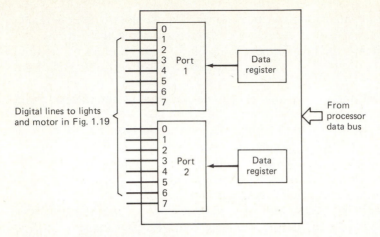

Figure 1.18 Digital output interface.

as in the A/D converter already discussed. When this conversion process is repeated rapidly, the output signal can create an arbitrary "analog" signal shape. When viewed closely, however, the signal is seen to be made up of small voltage steps due to the limitations of the finite digital resolution (10 bits) and the finite-time conversion intervals.

Digital output interface Digital outputs are also made available via an output interface. The CPU selects a register in the interface to which it writes the output data. This output register then holds the data while the CPU does other tasks. The data in the register is available to the output lines until new data is written into the register.

Several digital output ports are shown on the same interface card in Fig. 1.18.

1.3.6 Output Conditioning and Power Control

Returning to the example of the water tank, we will provide a pump operated by an ac motor to pump water out of the tank at a fixed flow rate and a pump operated by a dc motor which pumps water into the tank to maintain the level constant. The computer controls both pumps. Figure 1.19 shows the connections to the output interfaces.

The ac pump is controlled by an optically isolated on/off switch. The dc motor's speed, and hence the flow rate into the tank, is more finely controlled by an analog signal from the D/A converter. The analog signal drives a pulse-width modulator. The pulse-width modulator produces a constant-frequency pulse train in which the width of each pulse is proportional to the analog output signal. The pulses are then passed through optoisolators to the motor power electronics. The power electronic circuits apply voltage to the motor when a pulse is present and remove the voltage when no pulse is

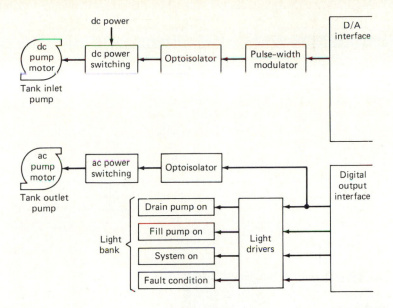

Figure 1.19 Output conditioning and actuators.

present. The average voltage on the motor terminals is proportional to the width of the pulses and thus to the analog signal from the computer. Details of the various blocks of Fig. 1.19 will be discussed in Chap. 12.

In addition to the power controllers, there are some simple lamp drivers shown connected to the digital output interface. The lights are status indicators for the human operators and are very common in the control environment.

1.3.7 Actuators

An actuator is any device that produces a motion. Electric actuators include solenoids, dc motors, stepping motors, torque motors, and ac motors. There is also a wide range of hydraulic and pneumatic actuators. Since we are dealing with computer control, nonelectric actuators must have an electric drive to be computer-controlled.

The example in Fig. 1.19 shows two pumps driven by electric actuators. A wider range of actuators will be discussed in Chap. 12.

1.4 EXAMPLE OF A REAL-TIME COMPUTER SYSTEM

Figure 1.20 shows the overall system design of a microcomputer-based water-level monitoring system. The system was designed to be battery-operated for unattended measuring and recording in remote exposed locations (see Figs. 1.21 and 1.22). The principal application of the instrument is for measuring the

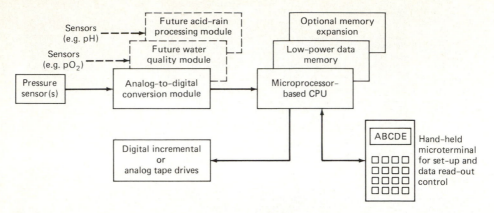

Figure 1.20 Example of a modular data-recording instrument.

Figure 1.21 Outdoor application of data-recording instrument.

water level in rivers and streams. A sufficiently low cost instrument can find applications in dense instrumentation networks for gathering spatially distributed environmental data.

The instrument's sensor is a pressure sensor that provides a voltage between 0 and 5 V to the electronics, over a pressure range of 0 to 3 psi differential (referenced to air pressure). The sensor, located on the bottom of a stream, provides an output voltage that is proportional to water depth via a waterproof submersible cable and connector. An analog-to-digital converter interface receives the signal and converts it to a binary number upon command from the CPU.

In order to conserve power, the CPU is "awakened" only for short periods of time (programmable) to coordinate the sampling, computation, and data-

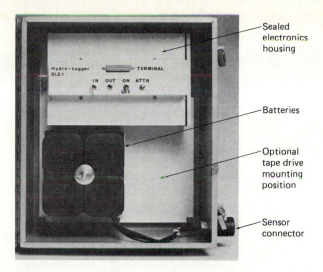

Figure 1.22 Packaging of the data recorder for outdoor use.

storage process. At these times, it orders a sample to be taken, converts the binary quantity from the A/D converter to engineering units, and stores the data in electronic memory (or on an incremental digital tape cassette unit in some applications). After one acquisition cycle, the CPU is again placed in a standby mode.

The system design is modular. The CPU, A/D interface, data memory, and tape cassette interface are each implemented on a card. New cards can be added to the system for both input and output operations. Memory can be expanded, or the basic microprocessor can be changed to a different model. The advantages of this type of modularity and the methods by which designs can be carried out by using standard bus structures and commercially available cards will be one of the major themes of this book.

SUMMARY

- A real-time computer system is one in which the response time is critical to the process or machine interaction with the computer.
- Structure. A computer-based real-time system can be assembled from the following subsystem components which are commercially available:

  Sensors and actuators
  Input and output conditioners
  Input and output interfaces
  Processors

- Function. The function of many simple real-time systems can be described with an ASM diagram.

EXERCISES

1.1 Modify the ASM diagram shown in Fig. 1.5 to produce a signal that turns on the heater whenever it rains and off when it stops raining.

1.2 For each state of the ASM diagram shown in Fig. 1.5, show the most direct next state for all possible combinations of input variables. As an example, if switch llsw_closed is true when the machine is powered up and it is raining, then the next state should be WET. This would be especially important if the state clock were very slow, then the approach taken in Fig. 1.5 would allow the motor to drive the arm farther into the switch. If there were no other switch that interrupted power to the motor, the machine could be damaged.

1.3 Are there any other possible error conditions that might arise in the operation of the air-sampler system other than the one mentioned in this chapter? If so, describe them.

1.4 Draw the ASM diagram for a washing machine. Assume the finite-state controller has the following outputs:

1. A control line called MOTOR which is on when the motor is on and off otherwise.
2. A control line called AGITATE which is on when the motor agitates the tank and off when the motor spins the tank. The MOTOR line must be on for either agitate or spin to occur.
3. A control line called HOT which opens the hot-water valve when it is on and closes the hot-water valve when it is off.
4. A control line called COLD which opens the cold-water valve when it is on and closes the valve when it is off.
5. A control line called PUMP which starts to pump out the tank when it is on and stops pumping when it is off.
6. A control line called TIMEON which resets a timer to zero and starts it timing when the line comes on. The line must be turned off before the preset time expires. Assume that the timer is set to some suitable value for the task.

Assume that the controller has the following inputs:

1. A line called START from the push-button switch that starts the washing machine. The line is on momentarily when the switch is pressed; otherwise it is off. Assume that the line is on long enough to be "read" by the controller (i.e., to cause a change in state).
2. A line called WASHHOT that is turned on *before* pressing START by the user of the washing machine for a hot wash or off for a cold wash.
3. A line called TIMEOFF from the timer that turns on when the preset time expires. Normally this line would be off.
4. A line called HIGH from a level sensor at the top of the tank. The line is on when the water in the tank covers the sensor and the line is off otherwise.
5. A line called LOW from a level sensor in the bottom of the tank. The line is on when the water covers the LOW sensor and off otherwise.

The ASM diagram should cause the machine to do the following:

1. Wait for the START button to be pressed.
2. Fill the wash tub with hot or cold water depending on the setting of the HOTWASH switch. Filling should continue until the high level is sensed.
3. Turn on the AGITATE function for one timer cycle (by turning on the timer and waiting for the TIMEOFF line to turn on).
4. Empty the tank until the low level is sensed.
5. Fill the tank with cold water for the rinse cycle using the HIGH sensor.
6. Repeat 3.
7. Repeat 4.

8. Turn on the spin function for one timer delay time.

9. Go back to step 1 and wait for another start instruction.

1.5 For the sensing circuit of Fig. 1.9, what clock frequency would have to be used to achieve a full-scale depth reading of 1 m. Assume the velocity of sound is 332 m/s.

1.6 An A/D converter is required to convert a voltage that lies in the range of ±1 V to a digital number with at least 0.01 percent accuracy. How many bits would the digital number have to contain?

1.7 In the ASM diagram of a finite-state machine with five binary input lines, what is the maximum number of of state transition arrows that could be leaving a state and entering some other state?

1.8 A temperature sensor outputs a dc analog voltage of 10 mV at 0°C and 100 mV at 100°C. An analog-to-digital converter interface will also be used. The A/D converter produces an 8-bit binary output of 00000000 for 0-V input and a binary output of 11111111 for an input of 10 V. What must be done to the analog signal to allow the A/D converter to convert temperatures between 0 and 100°C?

1.9 A biomedical researcher wants to measure the ability of a person to track a randomly moving dot on an xy display. The subject will use a joystick with two potentiometers in the base—one to control the dot's motion in the x direction and one to control the y motion. Each joystick outputs a voltage in the range from 0 to 10 V. The two voltages will be monitored by a microcomputer system.

The display will have two dots: one dot will move around under the control of the computer in a smooth but relatively random motion and the other dot will be controlled by the joystick. The control signal for each dot enters the display through a separate input channel. The voltage required to deflect either dot from the center of the screen is +10 V to reach the top of the screen and −10 V to reach the bottom of the screen. Similarly for the horizontal motion, +10 V moves the dot to the right edge and −10 V moves it to the left edge.

The motion of the joystick will be sampled by the computer and the error between the computer's dot and the subject's dot will be calculated and displayed on a video display terminal to the researcher. The researcher will also enter on the terminal several parameters at the beginning of each experiment that affect the motion pattern of the computer's dot.

Using the blocks of Fig. 1.6 as a guide, identify the components and their function that would constitute each block.

1.10 You have been asked to design a gasoline-operated heater for a green-house. At the entrance to the green-house are two switches:

1. A FAN_SWITCH which can be turned on by a gardener to circulate either cooling air or air heated by the gasoline combustion unit

2. A HEAT_SWITCH which the gardener can turn on whenever heat is required

The above switches are sampled by a computer which controls an on/off electromechanical GAS_VALVE and a BLOWER_MOTOR that is either on or off.

There are several safety provisions for the heater control. In order for the GAS_VALVE to turn on, the FAN_SWITCH and hence the BLOWER_MOTOR must be on to extract the heat from the combustion unit. Also, if the FAN_SWITCH is turned off while the GAS_VALVE is on, the controller will cool off the unit:

1. Shut off the GAS_VALVE.

2. Trigger a 10 min TIMER_INPUT to on.

3. Keep the BLOWER_MOTOR on until the TIME_UP signal from the timer turns on.

An OVERTEMPERATURE_SWITCH will also cause the cool-off steps just described.

Draw an ASM diagram describing the operation of the controller. Assume that at power up the BLOWER_MOTOR and GAS_VALVE are off and the SWITCHES can be in any state.

BIBLIOGRAPHY

MICROPROCESSORS

Osborne, A. *An Introduction to Microcomputers*, vol. 1: *Basic Concepts*, McGraw-Hill, New York, 1980.
(This is a primer on microprocessors for the novice. It describes computer assembly language, and input/output basics.)
Titus et al., C. A.: *16-Bit Microprocessors*, Howard W. Sams Co., Indianopolis, IN 1981.
(In this book the D.E.C. LSI-11, T.I. 9900, Motorola MC68000, and National NS16000 micro-processors are compared architecturally and with software benchmarks.)

Analog circuit design

Horowitz, P. and W. Hill: *The Art of Electronics*, Cambridge University Press, New York, 1978.
(This book gives a simplified approach to analog circuits that provides some design tools and discusses common analog circuits and their application without a heavy analytical content.)

Dedicated digital circuits

Fletcher, W. I.: *An Engineering Approach to Digital Design*, Prentice-Hall, Englewood Cliffs, NJ, 1980.
(This book provides a comprehensive treatment of component-level digital design—the design of digital circuits using integrated circuits.)

Algorithmic state machine diagrams

Clare, C. R.: *Designing Logic Systems Using State Machine Diagrams*, McGraw-Hill, New York, 1973.
(Using principally a single design example, this book proposes a notation and goes through the design of all types of finite-state machines up to what have become to be known as micropro-grammed controllers used in the control sections of many computers and microprocessors.)

Transducers and signal conditioning

Sheingold (ed.), D.H.: *Transducer Interfacing Handbook*: *A Guide to Analog Signal Conditioning*, Analog Devices, Inc., Norwood, MA, 1980.
(Transduction methods for temperature, pressure, force, flow, and liquid level are described in this book. In addition, appropriate signal conditioning methods for each variable are also treated.)

Analog-to-digital conversion

Sheingold (ed.), D. H.: *Analog-Digital Conversion Notes*, Analog Devices, Inc., Norwood, MA, 1977.
(The basic methods of A/D conversion are covered in this book. The structure of the converters and their nonidealities are described. The specification and application of A/Ds are discussed with a heavy emphasis on Analog Devices, Inc., products.)

TWO

COMPUTER STRUCTURE AND FUNCTION

This chapter introduces the internal architecture of a computer and describes how instructions are stored and interpreted. The rate at which the instructions are executed is important in real-time systems, so this chapter explains how the instruction execution cycle is broken down into its various components and how the execution time can be computed. Readers who are already familiar with this material can proceed to Chap. 3.

2.1 THE BASIC ELEMENTS OF A COMPUTER

At the most basic level, a computer simply executes binary-coded instructions stored in memory. These instructions act upon binary-coded data to produce binary-coded results. For a general-purpose programmable computer, four necessary elements are the *memory*, *central processing unit* (*CPU*, or simply *processor*), an external processor *bus*, as shown in Fig. 2.1, and an *input/output* system.

The memory stores instructions and data.

The CPU reads and interprets the instructions, reads the data required by each instruction, executes the action required by the instruction, and stores the results back in memory. One of the actions that is required of the CPU is to read data from or write data to an external device. This is carried out using the input/output system.

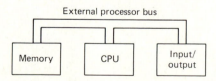

Figure 2.1 Basic elements of a computer.

The external processor bus is a set of electric conductors that carries data, addresses, and control information between the other computer elements.

In the following sections we will examine these building blocks as they appear in a simple hypothetical computer. We will also be able to see some ways to improve the obvious shortcomings of the *instruction set* of this simple computer. The instruction set is the set of allowable machine operations that the computer will recognize and execute.

2.2 THE MEMORY

The memory of a computer consists of a set of sequentially numbered *locations*. Each location is a *register* (see App. B) in which binary information can be stored. The "number" of a location is called its *address*. The lowest address is 0. Each memory location contains the same number of binary bits, a number that is defined by the manufacturer of the microprocessor. The manufacturer also defines a *word length* for the processor that is an integral number of locations long. In each word the bits can represent either *data* or *instructions*. For the Intel 8086/8 and Motorola MC68000 microprocessors, a word is 16 bits long, but each memory location has only 8 bits and thus two 8-bit locations must be accessed to obtain each data word.

Figure 2.2 shows the memory contents for a simple program in an example

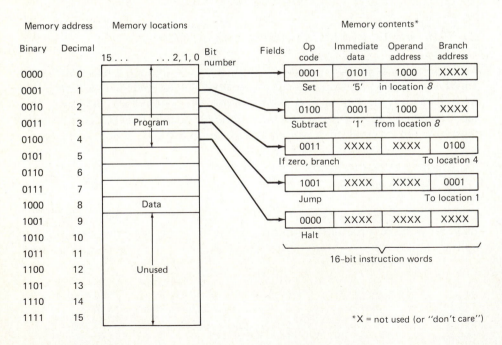

Figure 2.2 Arrangement of program and data in memory.

computer. This computer has only 16 memory locations in total. The word length and the location length are both 16 bits long.

In order to use the contents of memory, the processor must *fetch* the contents of the right location. To carry out a fetch, the processor places (*enables*) the binary-coded address of the desired location onto the address lines of the external processor bus. The memory then allows the contents of the addressed memory location to be read by the processor. The process of fetching the contents of a memory location does not alter the contents of that location.

2.2.1 Instructions in Memory

Instructions stored in memory are fetched by the CPU and unless program branches occur, they are executed in the sequence they appear in memory. Because each memory location can contain a pattern of only ones and zeros, the bits must be organized into meaningful patterns that the CPU can recognize as valid instructions (the *machine language*). An instruction written as a binary pattern is called a *machine-language instruction*. One way to achieve meaningful patterns is to divide up the bits into *fields*, with each field containing a code for a different type of information.

The computer memory in Fig. 2.2 contains a short sequence of instructions which constitutes a simple machine-language program. The program first deposits a count of 5 in location 8 and then repeatedly subtracts one until the contents of location 8 are 0, whereupon the processor halts (see Table 2.1).

Each instruction in our simple computer can be divided up into four fields of 4 bits each. Each instruction can contain the following:

Table 2.1

Memory address	Meaning of the contents of each memory location
Location 0	"*set the data*, in location 8, to '5'." The immediate data (5) is deposited into location 8.
Location 1	"*subtract* '1' from the data in location 8 *and put the result back in the same location*." The contents of location 8 are brought to the CPU. The immediate data (1) is subtracted from it and the result is deposed back into location 8. The CPU stores, whether the result is positive, zero, or negative, in a *status register*
Location 2	"*if the result is zero, get the next instruction from* location 4." The CPU checks the status of the last instruction. If it isn't 0, the instruction in the next location is fetched. If it is 0, then the instruction in location 4 is fetched.
Location 3	"*go to* location 1 to get the next instruction."
Location 4	"*halt*"

Operation code. Each instruction has a unique code called an *operation code*, or *opcode*. In the examples given in Table 2.1 the operation is indicated. In Fig. 2.2 a 4-bit code located on the left side of every instruction word is assigned to the operation. The 4-bit operation-code is located in what is called the *operation-code field*. The CPU will carry out the correct operation whenever it encounters this code in the first 4 bits of an instruction word. Since only 16 unique codes can be represented in 4 bits, this machine can have only a maximum of 16 instructions.

Operand address. The operand address is the memory address from which data can be fetched. The address of the data is located in a 4-bit *operand address field*. In this simplified machine there are only 16 memory locations holding either instructions or data and thus 4 bits in the operand address field is sufficient. In the example instructions given in Table 2.1, the data used by instructions is in locations at or above location 8. The data is usually separated from the instructions in memory.

Immediate operands. One exception to the separation of instructions and data occurs when an instruction itself contains a field for special constants that are used by the instruction. Constants contained in the instruction are often called *immediate operands* because they are located in memory along with the instruction and are fetched from memory at the same time as the instruction itself. In Fig. 2.2 the immediate operand field could represent one of 16 values in its 4 bits.

Branch address. The execution of an *unconditional branch* instruction results in the subsequent execution of the instruction located at the address stored in the branch address field of the unconditional branch instruction. A *conditional branch* instruction will have the same result if the condition tested is true. In the third sample instruction, if the sum is not zero, the next instruction in sequence will be executed. If the sum is zero, a branch in the instruction sequence takes place to the branch address location (location 4).

In a real instruction set there are many more instructions. There is also a much larger number of memory locations in which to store instructions and data.

Improvements The instruction format just described allows only a 4-bit address to appear in any instruction field. In order to increase the number of memory locations, the number of address bits must be increased. Thus the address fields and hence the instructions must be longer than 16 bits if we use the same approach.

There are a number of ways to increase the addressing range of the microprocessor without increasing the instruction length. These involve the use of variable field assignments, multiword instructions, multiple address modes, and variable instruction lengths. The design of a microprocessor involves making difficult choices among the various options. Some of these considerations are discussed here.

1. Variable instruction fields The number of bits in an address field is a measure of the maximum size of program that will be able to execute in the memory at one time and the amount of data that we will be able to store in memory. In the previous example, the use of 4 bits in the address field means that only 16 memory locations can be addressed. For the Zilog Z80 micro-processor, some instructions have an address field that is 16 bits wide; thus the processor can address $2^{16} = 65,536$ memory locations. The allocated address field of the Motorola MC68000 microprocessor is 32 bits wide, giving $2^{32} = 4,294,967,296$ locations (4 Gbytes) potentially. However, only 24 address bits are actually available for connection to physical memory, giving 16,777,216 locations (16 Mbytes).

By inspecting the instructions in the example, it can be seen that one way to improve the address range from only 16 locations (without increasing the word length) is to allow different instructions to have different field assignments. For example, each instruction has at least 4 unused bits ($\times\times\times\times$). These can be associated with the address (either branch or operand address) required by the particular instruction to allow an 8-bit address, or a range of 256 locations. The cost is that the instruction fields are no longer the same for each instruction, which makes the instructions slightly more complex to interpret.

2. Multiword instructions Another way to increase the size of the address field for the example processor is to devote the next memory location after the one holding the operation code to an address (branch or operand address) associated with the present instruction. This provides a 16-bit address, or an *address space* of 65,536 locations, and also frees up some bits in the first instruction word for other purposes such as increasing the operation-code size (and therefore the number of instructions). But if one memory location is used for the operation code and a second memory location for either an operand address or branch address, at least two memory locations must be read before an instruction can be carried out, and each memory access takes time. Programs will run more slowly.

3. Multiple addressing modes The need to reduce the number of memory reads where possible, to speed up execution time for a program, can lead to the use of several *addressing modes* for each instruction. If, for example, we wish to branch to a memory location that is nearby, we need only a few address bits to express the memory address difference (called an *offset*) between the present instruction address and the address of the desired instruction. This is a form of *relative addressing* (see Chap. 7). The form of addressing used earlier is called either *direct addressing* or *absolute addressing* (when the address in the instruction is the actual address in the memory). Absolute addressing would then be used whenever a desired memory location was out of range of the maximum address difference that can be expressed in the number of bits allocated to the relative addressing field. Of course, when several addressing modes are possible, bits in the instruction are required to indicate to the CPU which mode is being used in the current instruction.

4. Variable instruction length   The use of multiple addressing modes means that instructions will have a variable length. In the example just discussed, each memory location was 2 bytes long and the word length was also 2 bytes long.

In the Intel 8085 microprocessor architecture, each memory location is a single byte. An 8085 "word" is also a single byte, but the address field is 2 bytes long. The instruction length for the Intel 8085 can be 1 to 3 bytes long, as shown in Fig. 2.3.

In some microprocessors, such as the Motorola MC68000, each memory location contains a single byte, the word length is 2 bytes, and the instruction length (in bytes) depends on the particular instruction. The minimum instruction length for the Motorola MC68000 is 2 bytes and the maximum is 10 bytes. In contrast, the Intel 8086/8 has a minimum instruction length of 1 byte and a maximum of 6 bytes.

The example instructions given here so far do not give a complete view of a typical instruction set. More will be covered in the chapter on assembly language (Chap. 7). Not only are there a greater number of CPU operations but instructions have a broader range of addressing modes and can have different source and destination addresses, as in the instruction:

> *add the contents* of register 2 to the data in memory at location 9 and put the result in register 2.

The number of bits in the operation code field is an indication of the richness of a CPU's instruction set. The more bits, the larger the potential number of instructions in the instruction set. The benefit can be difficult to assess, however, because several instructions on one machine may be needed to replace a single "powerful" instruction on another.

2.2.2 Data in Memory

Data is information that is represented in memory as a code. An instruction fetches the contents of the memory locations it addresses and operates on the contents, assuming that the contents are represented in the way the

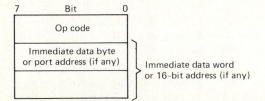

Figure 2.3 Intel 8085 instruction format.

instruction expects. For example, one instruction might add 8-bit binary operands from memory, while another instruction in the same instruction set might add 16-bit operands. For efficient use of the memory space and processing time, most computers provide the capability of manipulating data of different lengths and representations in memory. The various different representations recognized by the processor are called its *data types*. An elementary data type is distinguished by its length and the CPU operations to be carried out on it.

Let's assume that our example processor defines only three data types: binary-coded decimal (BCD) 4-bit digits, bytes, and 16-bit words. The representation of these types is shown in Fig. 2.4.

In a 16-bit word we could have

The BCD digits 4567	= 0100 0101 0110 0111
	| 4 | 5 | 6 | 7 |
Two 8-bit binary integers 45	= 0000 0100 0000 0101
	| 4 | 5 |
The 16-bit binary integer 45	= 0000 0000 0010 1101
	| 45 |

The bits in a location are usually thought of as being numbered from the right end (the least-significant bit is bit 0 in Fig. 2.4).

As another example, the data types recognized by the MC68000 microprocessor are:

- Bit
- Binary-coded decimal digit (4-bit *nibble*)
- Byte (8 bits)
- Word (2 bytes)
- Long word (4 bytes)

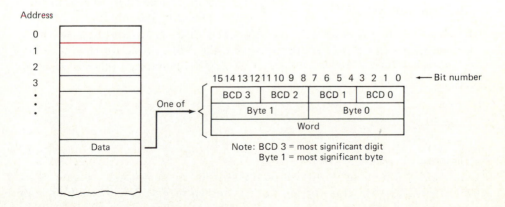

Figure 2.4 Three data types in example processor.

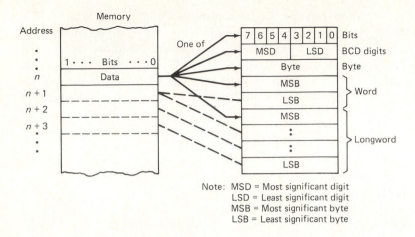

Note: MSD = Most significant digit
LSD = Least significant digit
MSB = Most significant byte
LSB = Least significant byte

Figure 2.5 Data types supported by the MC68000 family.

The representation of each of the data types in memory is important because the processor assumes a particular organization of the data. Different processors have different organizations for the same data (see Chap. 7). It is up to the programmer to ensure that the right data type is in the expected place. For the MC68000, the memory is addressed by byte; that is, there are 8 bits in each memory location. Figure 2.5 shows the assumed representation of the various data types.

Other processors provide instructions that manipulate other data types. For example, the LSI-11/23 provides a floating-point instruction-set option that includes single-precision floating-point data types (32 bits) and double-precision floating-point data types (64 bits).

There are many other possible data types that are recognized by other CPUs. Examples of other data types are arrays and character strings.

Character data is also usually represented in 8 bits. Each computer terminal key and key combination (such as shift and control functions) on a standard terminal keyboard has a 7-bit code defined by the American Standard Code for Information Interchange (ASCII). This code is stored in the 7 least-significant bits of a byte. For instance, the ASCII code for the letter "a" stored in 8 bits is 01100001. A table of ASCII character codes is given in App. A.

2.3 THE CPU

The CPU's job is to fetch instructions from memory and execute these instructions. The structure of the CPU is shown in Fig. 2.6. It has four main components:

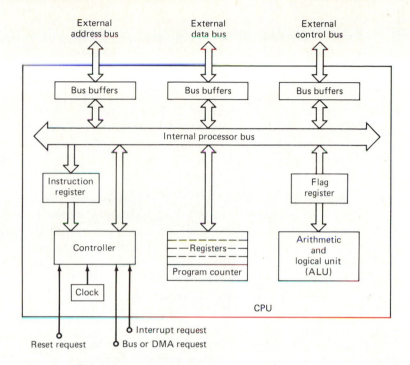

Figure 2.6 Central processing unit (CPU).

- An *arithmetic and logical unit (ALU)*
- A set of *registers*
- An *internal processor bus*
- A *controller*

These and other components of the CPU and their participation in the instruction cycle are described in the following sections.

2.3.1 Arithmetic and Logical Unit (ALU)

The ALU provides a wide variety of arithmetic operations, including add, subtract, multiply, and divide. It can also perform boolean logic operations such as AND, OR, and COMPLEMENT on binary data. Other operations, such as word compares, are also available. The majority of computer tasks involve the ALU, but a great amount of data movement is required in order to make use of the ALU instructions.

2.3.2 Registers

A set of registers inside the CPU is used to store information. Although some of these registers, such as the *instruction register* described in the next subsection, exist, knowledge of their contents is not important to the programmer. Other registers can be addressed by the programmer and can be used for temporary data storage, for example. These registers are included in what is known as the *programmer's model of the CPU*.

Instruction register When an instruction is fetched, it is copied into the *instruction register* (Fig. 2.6), where it is *decoded*. Decoding means that the operation code is examined and used to determine the steps of the execution sequence. This will be discussed in Sec. 2.3.4.

Programmer's model of the CPU The collection of registers that can be examined or modified by a programmer is called the *programmer's model of the CPU*. The only registers that can be manipulated by the instruction set, or are visibly affected by hardware inputs or the results of operations upon data, are the registers represented in the model.

For example, the Intel 8085 has the model shown in Fig. 2.7. The *accumulator register* is used for data and receives the results of ALU activity. The register pairs 00, 01, and 10 can be either data registers or address registers. The *program counter* contains the address of the next instruction to be fetched. When set, the *interrupt enable bit* permits the instruction sequence to be interrupted by external devices requesting service. Results of certain instructions affect the *flag register* bits (such as positive, negative, or zero results). The registers just mentioned are further described in the following subsections. The *stack pointer* can be used to store the address of a block of memory data called a *stack*. Data in the stack can be manipulated with special stack instructions. This will be discussed in Chap. 7.

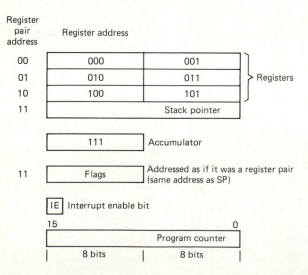

Figure 2.7 Programmer's model of the Intel 8085 CPU.

Zero	Positive	Negative	Overflow	• •	• •	• •

Status (or flag) register: each bit set to 1 if condition shown in box occurs

Figure 2.8 Intel 8085 flag word.

Flag register The execution sequence is determined not only by the instruction but also by the results of the previous instructions. For example, if an addition is carried out in the ALU, data on the result of the addition (whether the result is positive, negative, or zero, for example) is stored in what is known as a *flag register*, *status register*, or *condition register*, as illustrated in Fig. 2.8. If the next instruction is a conditional branch instruction, the flag word is tested in that instruction to determine if a branch is required.

For example, in a "branch if zero" instruction, if the flag bit for zero was set by the previous instruction, the branch address in the branch address field would be chosen as the source of the next instruction; otherwise (if the zero bit was not set), the next sequential instruction would be taken. Thus the controller can use the flag register to determine which instruction will be executed next.

Program counter The address of the next instruction is located in a register called the *program counter*. At the completion of any instruction that does not involve an instruction branch, the program counter is incremented by 1 to get the next instruction if the instructions are only one memory location long (as in the example processor instructions) or by more than 1 to get past any remaining locations used by the present instruction (see Fig. 2.9).

In the event of a program branch or subroutine call, the program counter will be loaded with the address specified in the branch instruction (if the branch was taken) or the address specified in the subroutine call instruction.

Data registers When an instruction uses the registers to store data, the reference to the register in the instruction is called *register addressing*. There

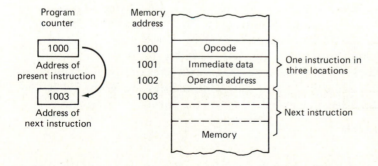

Figure 2.9 Incrementing the program counter.

are several reasons for making use of the internal registers to store data. First, they can be stored and retrieved in a subsequent calculation. Since there are only a few registers, only a few bits are needed in the instruction code (e.g., for eight registers, 3 bits) to refer to the register address to be used. Thus an instruction can be *shorter*.

The execution of an instruction can also be *faster* for two reasons. If fewer memory locations are used in each instruction code, the instruction fetch time will be reduced. Also there are decreased delays associated with using the internal CPU bus as opposed to the external bus to memory. There are no other devices connected to the internal bus and thus the controller can put (*gate*) data onto the bus whenever data are needed.

Address registers   The internal registers can also be used for the storage of *addresses* of data in memory. In such a case, the instruction word contains a register number (i.e., a register address). In the register is contained the address of memory data to be used in the instruction. This form of addressing is called *register indirect addressing*. The contents of the register are said to *point to* the data in memory.

The use of register indirect addressing in the instruction word decreases the number of address bits required by an instruction, but there is now a data fetch from memory (sometimes called a *memory cycle*) required in order to get the actual data, and this takes time.

One advantage of the use of address registers in memory addressing is that the address in a register can be easily modified during the instruction processing cycle. For example, if data is being accessed from an array of data stored in sequential locations in memory, one form of the instruction can be used to automatically increment the data address in the register to point to the next location after each datum is fetched from memory.

2.3.3 Internal Processor Bus

The internal processor bus moves data between internal registers. A *bus* is a set of closely grouped electric conductors that transfers data, address, and control information between functional blocks of the CPU. Data from a *source* register can be passed to a *destination register* when both are *enabled* onto (connected to) the bus. A data bus carries words of width D bits (D parallel conductors), an address bus carries addresses of width A bits, and a control bus carries C control lines. Among other tasks, control lines are used for enabling registers onto and off the bus lines. The control lines are governed by the controller. The value of A, D, and C depends on the particular processor.

2.3.4 Controller

The controller provides the proper sequence of control signals for each instruction in a *program cycle* to be fetched from memory. We use the term

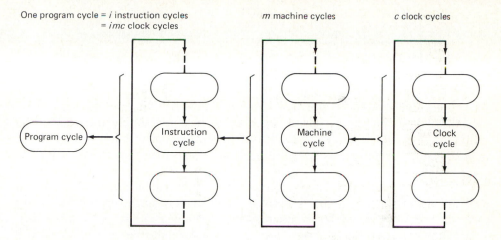

One program cycle = *i* instruction cycles = *imc* clock cycles   *m* machine cycles   *c* clock cycles

Figure 2.10 A program cycle and its components.

"program cycle" here to emphasize the fact that many real-time programs are cyclic and repeat at a rate determined by the sum of the execution times of each instruction in the program cycle. It is very useful to be able to determine the execution time of a program cycle. To do this, we will look at the component parts of the program cycle.

As can be seen in Fig. 2.10, a total program cycle comprises *i* *instruction cycles*. An instruction cycle consists of the total time required by the controller to *fetch* the instruction itself and then to execute it using the facilities of the CPU and any data required from memory or I/O (see Fig. 2.11). Depending on the result of the execution (e.g., branch conditions), a new instruction address will be chosen and the next instruction will be executed.

Each instruction cycle can be divided up into its component *machine cycles* (*m* machine cycles in Fig. 2.10). An expanded instruction cycle is shown in Fig. 2.12. The "read" machine cycle obtains the necessary operands required by the

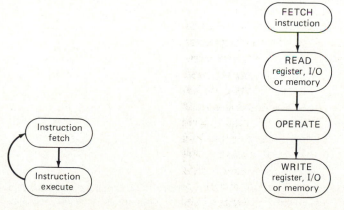

Figure 2.11 An instruction cycle.

Figure 2.12 Expanded instruction cycle.

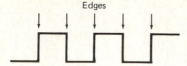

Edges

Figure 2.13 A CPU clock waveform.

operation specified in the instruction operation code. "Operate" carries out the specified operation (e.g., divide, add, etc.). In some simple processors, the operate phase does not add significantly to the total instruction time. Finally, the "write" machine cycle stores the results in registers or memory, or into an I/O address.

Each machine cycle comprises a number of *clock cycles*. The timing of all modern computer systems is derived from a single crystal-controlled *clock*. The clock generates a square wave, as shown in Fig. 2.13. The clock feeds the controller, as shown in Fig. 2.6, and the controller uses the edges of the waveform to sequence the various events in the *machine cycle*. Each machine cycle involves a characteristic number of clock cycles (c cycles in Fig. 2.10).

The phases of the machine cycle for an "opcode fetch" machine cycle are shown in Fig. 2.14. The same information in the form of a *timing diagram* is shown in Fig. 2.15. Most manufacturers provide a timing diagram for each machine cycle. The timing diagram is required for designing microprocessor interfaces to memory and peripheral devices.

In order to fetch an instruction, the address in the program counter is placed on the address lines of the external bus (AB) at the onset of clock cycle C_1. Simultaneously, using a code on the control lines of the bus (CB), the CPU informs all devices attached to the bus that an "opcode fetch" machine cycle is being executed by the CPU. The memory allows the memory address to select the memory location containing the instruction.

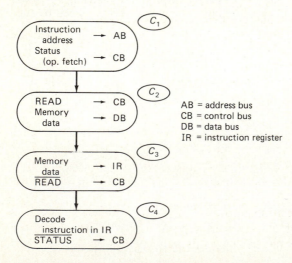

AB = address bus
CB = control bus
DB = data bus
IR = instruction register

Figure 2.14 Simplified ASM diagram for "operation-code fetch" machine cycle.

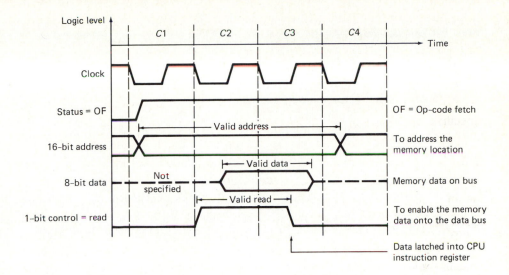

Figure 2.15 Logic levels on the data, address, and control buses for each clock cycle during an "operation-code fetch" machine cycle.

At C_2 the controller places a "read" command onto the control bus which allows the memory data to be placed onto the data bus (called "gating"; see App. B for a description of bus operations). The contents of the memory location are unchanged in the process.

The controller then gates the data into the instruction register and removes the read command from the control bus in C_3.

At C_4, the controller removes the address from the address bus and begins to decode the operation-code portion of the instruction to see what steps are required for execution. The decoding operation may take several more clock cycles at the end of which the "opcode fetch" status code will be removed from the control lines terminating the operation-code fetch machine cycle.

In Fig. 2.15, whenever a signal is "high," a logic 1 voltage will be seen on the microprocessor pins corresponding to the labeled signals. The actual voltage value of a logic 1 or 0 voltage will be defined by the manufacturer in the data sheets describing the CPU. The address and data lines are shown both high and low when valid, since some will be high and some low.

Instruction timing The amount of time taken for the execution of each *instruction cycle* depends on the instruction. A "multiply" instruction may take 100 μs, where an "add" instruction may take only 1 μs. The exact timing is often important both for hardware considerations and for the application software of real-time systems.

For software design, the timing information is usually summarized in a table that lists the number of clock cycles for each instruction. Since the number of clock cycles can vary depending on the source of the data to be used

in an instruction (i.e., the addressing mode), the instruction times are sometimes broken out into

- A basic time consisting of fetch and execute times
- A source time to read data operands
- A destination time to write results.

Examples   For the Digital Equipment LSI 11/23 CPU, the MOV (data move)
instruction has a basic time of 1.72 μs. With register indirect addressing, the
time to get a source word is 1.12 μs and the time to store a word in the
destination address is 1.84 μs. The total time is 4.68 μs for the instruction.

For the Intel 8085A microprocessor, an ADD instruction that adds the
contents of a register to the contents of a directly addressed memory location
takes seven clock cycles. For a 1-MHz clock frequency, the instruction requires
7 μs.

External attention requests   It is often necessary to stop the normal instruction
processing sequence. Several ways to do this by external signals have been
shown in Fig. 2.6.

One type of external attention request is the *reset request*. The reset line is
activated when the computer is powered up in order to put the machine into a
well-defined starting state. Once a system is processing instructions, it would be
unusual to cause a reset. In the case of an unrecoverable error, a computer
system may be required to reset itself. This would have the effect of initializing
all important registers in the system and starting instruction execution from a
standard memory location—usually location 0.

An input that is more commonly activated during the normal course of
events is the *interrupt request*. Many input/output devices use an interrupt
request line to the CPU to indicate a need for service. An interrupt request
signal from an external device can cause the CPU to immediately execute a
service subroutine which carries out the necessary actions. After completing
the service subroutine, the processor will continue the task from which it was
originally interrupted.

For example, a video terminal interface can be provided with a capability
to cause an interrupt to the CPU for every character that is struck on the
keyboard. In this way the CPU can go about its other tasks until it receives the
interrupt. It then processes the character and returns to its other tasks. A more
complete description of this subject will be given in Chap. 7.

The third type of input is the *bus request*, or *direct memory access* (*DMA*)
request. It is possible, for example, to have a terminal interface that stores up
all the characters in a line of text until it receives a "carriage return." Then the
interface requests the use of the system bus, at which time the complete line of
data is transferred to memory as fast as possible. This is achieved in the
interface by generating the memory addresses and depositing the data directly
into the memory using the data bus. Then the bus is released to the processor

again. In this way the processor simply becomes inactive until the transfer is completed. This process is much faster than that of having the CPU execute the necessary instructions to read the data and transfer it to memory.

2.4 BUSES

The *bus* is the most important communication system in a computer system. Rather than having all devices interconnected in a central exchange as in a telephone system, computer systems have the same communications wires pass by each device. Under control of the CPU, a data source device and a data destination device are "enabled" onto (equivalent to being connected to) the bus wires for a short transmission. It is possible to have other methods of communicating data between two devices; but when there are a large number of bus conductors (e.g., 100) and there is close physical proximity of the devices, and when there has been careful consideration of the amount of bus "traffic," bus systems are efficient.

External processor bus   The internal processor bus described in Sec. 2.3 is connected to the *external processor bus* by a set of *bus buffers* located on the microprocessor integrated circuit (see Fig. 2.6). The purpose of the micro-processor's buffers is to provide sufficient current to drive a modest number of inputs of other integrated circuits. These integrated circuits are located on a single *printed circuit board* (called the *microcomputer board*) and provide basic computer functions. A printed circuit board is usually in the range of 100 to 1000 cm². The *integrated circuit chips* are soldered onto metallic conductive traces bonded to the board. The traces form the electric interconnections between the chips (i.e., the wires).

System bus   The microcomputer board can communicate with other boards by connecting its bus to an external *system bus* through a connector. Since there may be other boards in the system, buffers are again required.

The timing of the events taking place on the various buses is controlled by the CPU (see Fig. 2.15), except during a DMA operation, as described in the last section.

2.5 COMPUTER INPUT AND OUTPUT

So far we have talked about several different types of registers whose contents can be manipulated by the machine instruction set and so far all the registers have been in either the CPU or memory of the microprocessor (considering memory locations as equivalent to a register).

Another set of registers external to the CPU is associated with what is known as the *input/output* (*I/O*) *system*. Like the memory system, the I/O system is connected to the external processor bus using control, address, and data buses through an *I/O interface* described in Chap. 9. It is necessary to know how to address the I/O registers in an interface using the machine instruction set, since all data to and from the computer must pass through the I/O system. There are basically two ways that are used to address I/O registers.

In the first method, called *I/O-mapped input/output*, the operation code itself has special I/O instructions that address a numbered register in the interface called an I/O *port*. For example, the instruction

> *input byte* from port address 255 to CPU register 3.

The operation code for "input byte" would cause the CPU controller to set an "I/O control bit" on the external control bus to 1 at the time it places the address 255 on the external address bus. This will tell the memory system to ignore the address and will tell the I/O system to place the contents of port 255 onto the data bus after which the CPU reads (*strobes*) the data into CPU register 3.

Thus the I/O registers have an *address space* that is outside the memory address space in I/O mapping. For the Intel 8085 microprocessor, the address space contains 256 ports (8-bit address); and for the Intel 8086, it is 65,536 ports (16-bit address).

The second method of addressing I/O registers gives the I/O ports addresses that lie within the memory address range of the CPU. This is called *memory-mapped I/O*. Of course there must not be any memory locations at the same address as I/O locations. In microprocessors such as the MC68000 that has 16 Mbytes of memory, it may not be a problem to lose a few memory locations to I/O registers. In a processor with 64KB addresses, it could be a real sacrifice to lose program or data storage space.

One of the benefits of the memory-mapped approach is that the full range of memory addressing modes is available to the addressing of I/O registers.

SUMMARY

- The memory of a computer contains both data and instructions.
- Each data referencing instruction contains an operation code that specifies the operation to be used to transform data in specified source addresses to data in a specified destination address. The instruction also contains bits that identify what type of source and destination are being addressed in the instruction (CPU register, memory location, or I/O register) and how the instruction address bits are to be interpreted (direct memory address, immediate address, or register indirect address). The operation code of the instruction also implicitly or explicitly defines the expected data

type—BCD, byte, word, longword, floating point, character, character string, etc.) and the position of the data in a location or word (i.e., right- or left-justified).

- Other instructions can control the *program flow* by altering the program counter (branch, conditional branch, and subroutine calls).
- The computer's processor, or CPU, is a finite-state machine that fetches instructions from memory and executes them. The execution of these instructions requires a fixed number of clock cycles that is specified by the manufacturer in a set of *data sheets*. The number of clock cycles can be used to accurately determine the execution time for a complete instruction or program cycle.

EXERCISES

2.1 Sketch the structure of a typical CPU and describe the function of each of the component parts.

2.2 Memory addresses of data sources or destinations may appear as binary codes in an instruction. What other types of addresses can appear in the instructions?

2.3 The number of instructions can be increased by adding operation-code bits to the instruction, and the size of the memory can be increased by adding bits to the address fields of the instruction. For a microprocessor that reads one 16-bit word at a time, what limits the practical instruction length?

2.4 (*a*) Show the right-justified 16-bit binary integer equivalent to the decimal numbers 6, 27, 359, and 4095.

(*b*) Using the table of ASCII characters in App. A, show the decimal numbers in part *a* as ASCII character codes.

(*c*) Write the hexadecimal equivalent of each of the numbers.

2.5 The Texas Instruments TI9900 16-bit microprocessor has its registers in memory rather than in the CPU itself. This has some advantages in some processor operations known as "context switching." What disadvantages can you see to storing having the registers in memory.

2.6 For the five-line program shown in Fig. 2.2, make a list of all the machine (bus) cycles required for each instruction where the possible cycles are:

Operation-code fetch and execute (OF): fetches one word
Memory read (MR): reads memory data
Memory write (MW): writes memory data

(*Note*: Assume that the operation-code fetch cycle fetches the whole instruction word.)
If each machine cycle is composed of four clock cycles and the clock is 1 MHz, what will be the total execute time for the *complete* program.

2.7 If twos complement representation is used for the operands in Fig. 2.2, rewrite the memory contents to increment immediate data of -5 to 0 using the same operation codes.

2.8 When a memory system is powered up, the contents of memory are unpredictable. Using the operations given in Fig. 2.2, can you write the instructions to clear location 8?

2.9 Can you list several reasons why the inclusion of a large number of registers in the CPU integrated circuit is an advantage to the user?

2.10 In real-time systems, the ability of the computer to respond to an interrupt (or "request-for-service") in an appropriate way necessitates that a higher priority be placed on events that require immediate service. Briefly describe what you would do in the following situations if you put yourself in the place of the computer. Assume that it takes you some time to carry out the required actions whenever an interrupt occurs.

(*a*) A low-priority interrupt occurs.

(*b*) A low-priority and high-priority interrupt occur at the same time.

(*c*) A low-priority interrupt occurs, but while it is being serviced, a high-priority interrupt occurs.

(*d*) A high-priority interrupt occurs, but while it is being serviced, an equally high-priority interrupt occurs.

(*e*) Two equally high-priority interrupts occur at the same instant.

BIBLIOGRAPHY

The best source of information about individual microprocessors is the manufacturer's data books. These are available through the local representative of the manufacturer. The data books contain at the minimum a description of each of the machine instructions, the programmer's model of the CPU, the data types assumed by the instruction set, an external bus timing diagram for each of the machine cycles and external request cycles (interrupt, DMA, etc.), the electrical description of the connections to the processor, and available chip packaging materials and shape.

For the Motorola, Intel, and Digital 16-bit microprocessors referred to in this chapter, the following data books are available:

Motorola MC68000 16-bit Microprocessor User's Manual, 3d ed., Prentice-Hall, Englewood Cliffs, NJ, 1982.
Intel iAPX 86, 88, 186, 188 User's Manual, vols. 1 and 2, Intel Corp., Santa Clara, CA, 1983.
DIGITAL Microcomputer Processor Handbook, Digital Equipment Corp., Maynard, MA, 1980.

There are similar data books available for the 8-bit processors such as:

Intel Corp. *Intel MCS-85 Users Manual*, Santa Clara, CA, 1978.

Another source of information of a comparative nature for the various processors are the books containing an overview of several manufacturer's products. One such overview for the 16-bit processors is,

Titus, C. A., and J. A. Titus, A. Baldwin, W. N. Hubin, and L. Scanlon: *16-bit Microprocessors*, Howard W. Sams & Co., Inc., Indianapolis, IN, 1981.

There are a number of textbooks of a more general nature that describe computer structures:

Hamacher, V. C., Z. G. Vranesic, and S. G. Zaky: *Computer Organization*, McGraw-Hill, New York, 1978.
Siewiorek, D. P., G. Bell, and A. Newell: *Computer Structures*: *Principles and Examples*, McGraw-Hill, New York, 1982.

An insight into computer architecture can be derived from examining the inner workings of existing computers which can be *microprogrammed* by the user. This means that users can create their own instructions and add them to the instruction set of the computer. To do this, they must have a knowledge of all the internal registers of the CPU. Also, users must be able to manipulate the way in which each instruction is interpreted. In a microprogrammed CPU each step in the machine cycle is determined by one microinstruction. Each microinstruction is contained in a *microprogram memory* and there is a primitive *microcontroller* that executes each microinstruction.

It is really a small processor within a large processor. The advantage is that very complex machine instructions ("macroinstructions") can be made up from many microinstructions. Also when an error is made, only a microprogram need be revised, not some dedicated hardware. The LSI-11 and MC68000 microprocessors were designed and constructed using this approach.

An example of a microprogrammed machine is described in:

Hewlett-Packard Company: *Microprogramming the 21MX Computers: Operating and Reference Manual*, Cupertino, CA, 1974.

THREE

PRELUDE TO THE DESIGN PROCESS

The first objective of this chapter is to describe the factors that must be taken into account in specifying a suitable microcomputer system for an industrial environment. The industrial setting will considerably constrain design options. (The design process will be discussed in detail in Chap. 4.) The second objective is to stress the importance of selecting an appropriate technology for the implementation of a design in both software (Chaps. 5 to 7) and hardware (Chaps. 8 to 12). The final objective is to outline the choices for tools used in the development of software and hardware. This is a very important decision and should be carefully thought through at the time of project planning. The development environment is discussed in more detail in Chap. 15.

3.1 THE SYSTEM COMPONENTS

Before the execution of a detailed system design, a general block diagram of the major components of the real-time system should be drawn showing the major physical blocks in the system. Each block should have a label. The interconnecting signals between the blocks can then be drawn and labeled with a descriptive name. The names of the blocks and signals can then be referred to in the design specifications, hardware diagrams, and microcomputer software.

Example 1: water-level recorder As an example of a simple system structure, the water-depth-monitoring data logger introduced in Chap. 1 is shown in Fig. 3.1 In the figure each line has a meaningful name. This allows a list to be compiled of all the signals entering and leaving the microcomputer system. Later in the design process, the selection of a mode for the signals (analog or digital) will determine the types of processor interfaces required. The principal components are

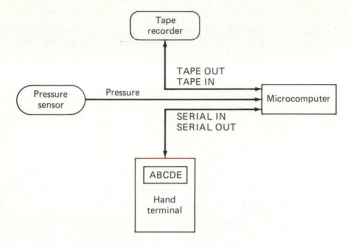

Figure 3.1 Components of a water-level recorder.

- Microcomputer
- Pressure sensor connected to the microcomputer
- Detachable hand terminal to set up the system parameters and take on-site readings
- Detachable tape recorder for use when the operator comes on-site to read out data recorded in the memory of the instrument

Example 2: subsea telerobot system   As a second example, the system components of a subsea telerobot are shown in Fig. 3.2. A telerobot is a combination
of a master-slave manipulator and an industrial robot.

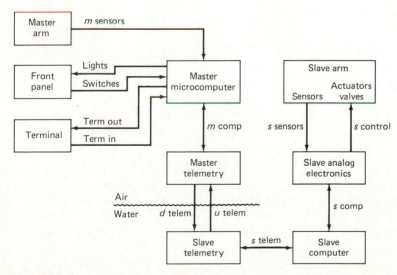

Figure 3.2 Components of a subsea telerobot.

In its subsea applications, a human operator in a surface ship controls the movement of the slave manipulator on the ocean floor (with visual feedback from a set of remote TV cameras) by moving the master control shown in Fig. 3.3*a*. The slave manipulator is shown in Fig. 3.3*b*. The slave manipulator is mounted on a submersible that is connected to the surface ship by up to 2400 m of umbilical cable. The slave is capable of carrying out a number of pre-programmed (robotic) movements as well as human-controlled tasks.

The principal components of the system are:

- A master hand control that has sensors on each axis of movement
- A microcomputer that receives the master's command signals, digitally formats these commands, and transmits the signals as a serial pulse train to the electronics canister in the submersible
- Another microcomputer in the submersible receives the digital pulses which are converted to analog signals that drive the hydraulic control servos for each axis of movement of the slave arm
- A front panel that provides power up and power down and emergency halt buttons and status lights
- A terminal with keyboard and display for interactive communication with the processors, calibration, and status, and for maintenance

Once the system components are envisaged (these may change during the design), the specifications for the system and its components should be determined.

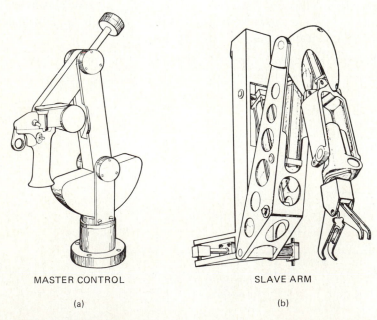

MASTER CONTROL

(a)

SLAVE ARM

(b)

Figure 3.3 Mechanical components of a subsea telerobot system. (*a*) Master hand control. (*b*) Slave manipulator arm. (*Courtesy RSI Robotic Systems International, Sidney, B.C.*)

3.2 THE DESIGN SPECIFICATION

One of the major tasks facing the engineer before the design process can get underway is to ensure that complete design specifications are available. These will be obtained in a number of ways depending on the circumstances.

Sometimes the requirements to meet a specific need in the designer's own plant will be passed down from a previous study to an in-house engineer.

Alternatively, the engineer might be a consultant who has been asked by a client to design the water-depth data-logging system (see Sec. 1.4). In this case the requirements may be very specific in some areas and rather vague in others. The client may state very specifically that the measurement must be at least 0.1 percent accurate, the battery life at least 3 months, and the depth-sampling interval selectable from 1 min to 24 h. On the other hand, the client may not know enough initially about what is feasible with a microprocessor system to be able to efficiently express a set of requirements; and with discussion, the engineer may find that the requirements are shifting as the client sees the possibilities.

A client may wish to constrain the design by specifying certain design components. For example, the client may require water depth to be measured with a specific transducer—for good reasons (compatibility with an existing network of units) or possibly bad reasons ("There's a nice little disposable pressure transducer I heard of that would be good to use here for measuring water depth."). In this case the designer must work with the client to clarify and document a reasonable and complete set of requirements. What constitutes a complete set of requirements is the subject of the next several subsections.

3.2.1 Functional Specifications

Functional specifications answer the question, "What does the system do?" There are three levels of functional specifications:

1. Essential capabilities
2. Additional capabilities
3. Future capabilities

The description of the essential capabilities should provide a concise overview of the system. For the example of the data logger, the essential capabilities are to

- Sample water pressures from a sensor every minute for 3 months
- Store the samples in a solid-state memory
- Read out the data into a portable tape recorder upon command from a portable terminal

The additional functional capabilities elaborate on the essential functions, completing the functional specifications for the immediate system. For the data logger, some of the additional functions are to

- Enter program parameters using a portable terminal
- Power down the electronics between samples
- Convert the samples to engineering units
- Be able to read the water level while the person setting up the unit is on-site
- Perform error checking for valid data in all operations

When a system is designed for production, provisions are usually made so that a number of future enhancements can be added to the existing system without costly redesign. There are several kinds of reasons for making these provisions:

Marketing strategy. It may be more important to offer the customer an inexpensive system rather than a high-performance one.

Technological. The system components are either not available yet or are too expensive.

Sales. The possibility of upgrading the system in the field to a higher-performance system will have sales advantages.

For the data logger, future enhancements include

- Increasing the maximum number of samples that the unit can store
- Expanding the programmability of the unit
- Expanding the number and type of environmental variables that the unit can acquire

3.2.2 Performance

Performance specifications answer the question, "How well does it do it?" The types of performance measures for a data logger, for example, include accuracy, reading *drift* (changes with time), temperature effects, linearity, and maximum sampling rate.

3.2.3 Characteristics and Constraints

The characteristics of a device and the specification constants are related. To use a trivial example, we can specify that the cabinet of a piece of industrial monitoring equipment must be white (a characteristic). On the other hand, we could say that the cabinet must be any color but black (a constraint). For the purposes of brevity, the remainder of this discussion will use the term "characteristics" to refer to both characteristics and constraints.

Physical characteristics The physical characteristics of a device need careful attention. If the device is to be portable, weight, weight distribution, size, and proportion are important. If the device is to be wall-mounted, flanges must be provided. Although these seem like unimportant items, in a competitive market, they may determine whether your instrument or the competition's will be successful. Other forms of installation include rack mounting and tabletop installation.

Environmental characteristics Environmental hazards for the system are many. The most likely problems in the industrial environment are the following:

Temperature. Some applications may demand a tolerance for wide temperature fluctuations. These can cause problems not only with selecting electronic components but also with packaging.

Moisture. Moisture can range from humidity through spray and rain to total immersion.

Corrosive liquids and gases

Electrical noise. The system may be subject to nearby high electric and magnetic fields (such as in an aluminum smelter where high currents are used in the process, or in the vicinity of electric motors). Even small hand-held transceivers used by security personnel are potential sources of problems when measuring very small electric signals (e.g., bioelectric signals). High voltages used in electrosurgical equipment can damage sensitive input amplifiers in biomedical equipment. Similarly, lightning can be a real hazard to field-mounted electric instruments, so that protection may be required for the inputs.

Vibration. In vehicular electronics, vibration can lead to failures in many parts of the system, so that a packaging system must be designed with vibration in mind in these applications. For any application, the vibration involved in shipping the device to the point of installation must be considered when designing the shipping container.

Sunlight. Even exposure to sunlight must be considered when selecting displays, for instance. Light-emitting diodes cannot be easily seen in bright sunlight, whereas liquid crystal displays are visible under bright light.

Power. Power consumption (especially in battery-operated units), heat dissipation, noise immunity from disturbances in the power line, and power-line noise generation by the equipment to be designed are all factors to keep in mind with respect to the design of the power supply particularly.

Human factors Human factors are an important part of electronic equipment that is used interactively by a human being. The most well known examples are associated with terminal monitors and keyboards. The monitors should tilt to a comfortable angle and the keys should not be too small and should have tactile feedback when depressed. Good human factors are most important when there are many manufacturers competing for the same product sales (e.g., the small computer market) or when a design may meet resistance from a set of users that are firmly established in a different, perhaps older, technology.

Maintenance and reliability   It is unfortunately true that even good equipment breaks down. Some idea of the necessary reliability of an instrument can have an influence on the design. The reliability is often quoted in terms of the *mean time between failure*. Another relevant specification is the *mean time to repair*. The mean time to repair can be reduced by design provisions that allow failures to be isolated to a particular board.

3.3 THE DEVELOPMENT ENVIRONMENT

The approach taken to the development will be affected by three constraints:

1. The development time allocated to the project
2. The personnel (number and experience) available for the various tasks
3. The projected costs of the tasks (labor), parts of the system (materials), and services (e.g., computer time)

The resources available to the designer for the selection or use of each component of the microcomputer system are also an important aspect of the predesign process that can act as a significant constraint upon the designer. Thus we will consider the components and resources for

- *Hardware* (equipment) development
- *Software* (programs) development

3.4 HARDWARE DEVELOPMENT

There are two factors related to the choice of hardware to be used in development: the *semiconductor technology* and the *level of integration*. The semiconductor technology is the fabrication technique and the organization of the transistors used to implement the idealized logic gates and flip-flops in the integrated circuits.

Semiconductor technology   The various technologies differ in their properties. These will be important particularly if a design requires very low power, very high speed, high electric noise immunity, or immunity from the effects of radiation. For example, CMOS (complementary metal oxide semiconductor) integrated circuits are used when low power and high noise immunity are required. (The semiconductor technologies will be discussed in Chap. 8 in more detail.) The types of equipment used for the development of very high speed circuits are different, and more expensive, than that required for lower-speed systems.

Level of integration   In addition to the semiconductor technology factor, there is *level of integration* to consider. This will be discussed in more detail later (Chap. 8), but for now the various alternatives are listed below in order of increasing amount of labor given to design.

1. Purchase of a complete turnkey system to solve the design problem.
2. Purchase of a microcomputer with single vendor-supplied hardware. Software must be designed.
3. *Board-level design*, where the designer chooses boards from a variety of vendors and performs the *system integration* (i.e., packages the system using a system bus, card cage, power supply, cabinet, and software).
4. *Component-level design*, where the designer must design and fabricate the boards using discrete integrated circuits and other components in addition to the tasks at the board level.
5. *Semicustom and custom-integrated circuit design*. In this case, even the integrated circuits themselves are designed in addition to the tasks described above. There is a distinction between semicustom design and custom design because in semicustom design, only an interconnection pattern between transistors on a semiconductor wafer must be specified, whereas in the case of full-custom integrated-circuit design, even the transistors themselves must be designed.

3.4.1 Board-Level Hardware

This book will be primarily concerned with design at the board level, because such a variety of boards has become available that in many cases the advantages to be gained by designing one's own boards are not significant enough to justify the time and costs involved when only a small number of the same systems are being made. For this situation the hardware is obtained from an outside vendor and the software is developed in-house to the desired specifications. The resources in terms of equipment and personnel to perform this type of development are less than are required for any of the lower levels of integration.

A board-level approach makes sense in both of the following cases:

- In manufacturing industrial real-time systems of moderate complexity, the cost of the microcomputer hardware components can be a relatively small fraction of the cost of the complete system. This is especially true in systems with a large mechanical component such as the subsea manipulator system. Here the cost of the materials, machining, and hydraulic components exceeds the microcomputer hardware and software costs.

- In systems where there is a large software component, and where only a few systems of one type are built, the design time and software costs dominate the hardware costs.

There may also be other reasons for using a board-level approach, such as the need to get a prototype product completed quickly or the need to install a new controller on an existing industrial process.

When larger numbers of similar systems are manufactured, and price competition arises, the board-level design approach may not be appropriate. The water-level monitor design, for example, used a combined approach—a standard bus structure which allowed the use of commercially available cards plus several specially designed boards. An estimate of the acceptable cost for the various components of the system must be made to be able to determine the feasibility of one approach or another.

The following subsections will outline the various components required to assemble a microprocessor-based real-time system using the board-level approach.

Cabinet, card frame, and backplane As described in Chap. 1 and illustrated in Fig. 3.4, a system designed from commercially available boards would be housed in a cabinet that meets the desired environmental and physical specifications. The cabinet contains a *card frame* into which the various circuit boards can slide. The card frame supports a *backplane* on which multipin connectors are mounted in parallel rows.

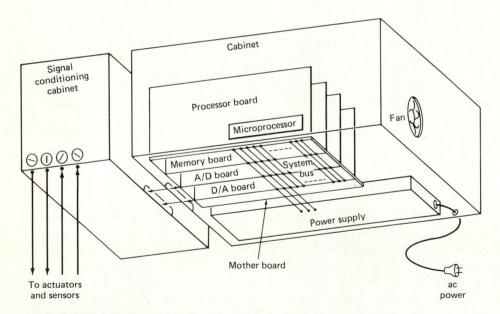

Figure 3.4 Hardware components of a microcomputer system.

System bus   Each connector pin is *bused* to the corresponding pin on all other connectors by means of the printed circuit wiring on the backplane board. The bus distributes signals and power to all boards. Thus the microprocessor board can talk to all the other boards on the bus.

The various commercially available bus structures differ with regard to

- Physical dimensions and arrangement
- Functional definition of the pins
- Electrical properties of the signals

Some of the available bus structures are discussed in detail in Chap. 13.

A consideration in the selection of a particular bus may be the availability within the designer's company of an operating microprocessor system using the same bus structure. This resource will facilitate the testing of boards and the development of new systems.

Microprocessor board   One of the key components in the design will be the selection of the microprocessor card. This will have an effect on both the hardware and software performance of the system. Although many factors may affect the selection of a microprocessor (see Chap. 8), an important factor in practice is the familiarity of in-house staff with a particular microprocessor or processor board.

Processor auxiliaries   Several types of cards are closely associated with the processor function. They are shown as processor auxiliaries in Fig. 3.5. They include program memory and data memory that may not be present on the main CPU board. For applications requiring high precision or high-speed mathematics functions, *math boards* are available with many of the functions that would be available on a calculator. Since the timing of events takes on a special significance in a real-time system, *counter-timer boards* are available with time-of-day registers, presettable timers that interrupt the processor or cause a signal on an external event line when the set time elapses, and counters that accumulate the time from either an internal or external set time. Finally, some systems have a need for external *interrupt arbitration* capability. These boards allow only the highest *priority* external device to interrupt the CPU at any given time. The priority of each device can be selected.

Interfaces   The third class of boards shown in Fig. 3.5 are the interface boards for the real-time signals and *peripheral devices*. The term "peripheral device" includes any device that is locally attached to a microcomputer and supports its function. The boards can be grouped into four classes:

- External *mass memory* system interfaces called *controllers*. These include controllers for magnetic disk and tape storage subsystems with capacities ranging from several hundred thousand bytes per *drive* to

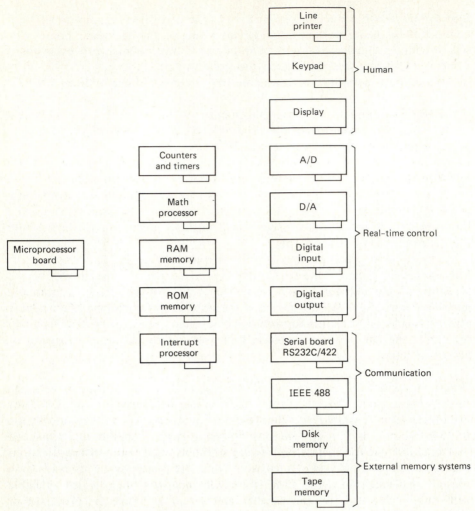

Figure 3.5 Types of system bus cards.

several hundred megabytes. These interfaces are often of a form called *direct memory access*, meaning that they are capable themselves of addressing data memory and transferring blocks of data into data memory for use by the processor.

- Communication interfaces. These support remote devices that are connected to the CPU. They have special provisions to handle data streams that are sent serially over only a few wires. Two of the most common standardized protocols are called RS232C and IEEE-488.
- Interfaces to human-operated devices. These devices include terminals, printers, and displays and are relatively low speed compared with the mass-storage data rates.
- Real-time control interfaces. These analog and digital interfaces have been introduced in Chap. 1.

3.5 SYSTEM SOFTWARE

Although the system hardware components just described will be acquired from commercial sources and the availability and in-house experience will be the main resource considerations, the software will have a large in-house component.

Particularly in the design of software the concept of *tradeoffs* occurs. The various software components will be described by outlining the tradeoffs that the designer must make.

To some extent there is a tradeoff between hardware and software that the board-level designer should be aware of. This is indicated conceptually below on a line diagram. The designer must choose the proportion of hardware and software in the system:

100 percent hardware   100 percent software

In practice there is a limited amount of choice at the board level. An example would be whether to use a math processor board or program a math package in software. The decision would depend on costs, computation speed required, development time saved by the purchase of the board, and other factors.

3.5.1 Operating Systems and Application Programs

An *operating system* is a master control program that coordinates all computer software tasks. It usually provides a set of resources that the application programs can make use of, such as a *file structure* (an orderly arrangement and index of information in memory and mass storage) and *interrupt handling routines* for some devices. Interrupt handling routines are described in Chap. 7. Many standard operating systems are available commercially. These tend to run only on specific processors. For example, the operating system called CP/M runs on the Intel 8080, 8085, and 8088 and the Zilog Z80 CPUs but is not available for the Motorola 6809 CPU. The converse is true for the OS-9 operating system.

An *application program* is written by the designer to meet the needs of a particular task. An example would be a program for closed-loop control of an industrial process using proportional, integral, and derivative control. This program would be invoked either by other application programs or by the operating system.

Another application program is a program that formats data for a file. In this case, however, this capability might be made a part of the operating system. Hence a tradeoff can exist here too.

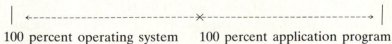

100 percent operating system   100 percent application program

It is possible for a system to have no operating system at all. In this case, the application programs must control all microcomputer activities. Many small dedicated applications do not require file support, terminal handling, disk I/O, and other tasks that are supported by many operating systems and their associated *device handlers*. A device handler is a routine that communicates with specific I/O devices and passes data to the operating system, ideally in a device-independent format.

Because of the complexity of modern operating systems, the time invested in learning the system can become a dominant factor in the selection. Thus if your staff is familiar with Intel's RMX operating system, it may not be efficient to move to another operating system even though there may be benefits in so doing. Because RMX runs on Intel products, this decision may override all hardware considerations as well. Again, the in-house resources may have the final word in the design process.

3.5.2 High-Level Language and Assembly Language

The options for a programming language component can be broadly grouped into two categories: *assembly language* and *high-level language*.

Assembly language is one step above machine language, introduced in Chap. 2. In the assembly language for the machine, the manufacturer assigns a name called a *mnemonic* to each binary operation code because the name is easier to remember. ADD, for example, is easier to remember than 11010. Of course, the computer must end up with 11010 in its program memory if we expect it to add, so a program called an *assembler* translates the ADDs to 11010s wherever they occur in a program. In addition to the mnemonics, the addressing modes are specified symbolically in the assembly language. This will be discussed in Chap. 7.

The definition of the mnemonics can be done by anyone who is willing to write an assembler. For example, the assembly language for the Motorola MC68000 defined by Motorola differs slightly from the assembly language defined by an independent software supplier, Whitesmiths, who have written their own assembler for the MC68000.

High-level language (e.g., FORTRAN) is a step above assembly language. In a high-level language, the commands more closely resemble the words and structure of natural language. The program written in high-level language is usually more concise than the equivalent assembly-language program which has one machine-language instruction for each assembly-language instruction. A translator, called a *compiler*, must be written to produce either assembly language or machine language from the high-level language before the program can be used. Several high-level languages will be reviewed in Chap. 5 and the principal language used in this book will be the language called *C* because of its advantages in real-time applications.

There is also a tradeoff in the choice of high-level or assembly language for a given task:

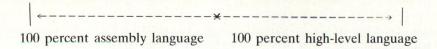

100 percent assembly language   100 percent high-level language

A time delay, for example, might have to be written in assembly language if the delay is to be very short (by writing a "do-nothing" loop) but could be written in high-level language for a longer delay because there will generally be more instructions in the translated assembly-language program than in the handwritten assembly-language program. For that matter, a delay can often be accomplished using some operating system facilities or, in hardware, invoking the tradeoffs mentioned before.

As in the case of operating systems, the resources for high-level language development in-house may determine the language chosen and the amount of assembly and high-level code written.

3.5.3 Development Systems

A software development system is a computer with programs that support the development of a wide range of different microprocessor applications.

The main use of a software development system is to provide facilities to prepare the control program for a real-time system. A significant set of hardware and software resources may be required to do this preparation. For example, a program written in a high-level language such as FORTRAN requires an editing or word-processing program to write and modify the original program. A compiler is required to translate the program into assembly language. Then an assembler and possibly other programs such as a linker and loader (these will be described in Chap. 5) may be required to produce the actual machine instructions (*machine code*).

The hardware facilities required to support the software development just described usually consist of a computer system with some sort of disk storage (floppy disk or hard disk). This *may* be a necessary part of the final real-time system, but it is often not required or even desirable.

The usual situation is that the software is developed on a larger *host* computer or a commercially available *microprocessor development system* and *downloaded* (sent through a digital communication link) to the microcomputer system under development (the *target* system). However, the host can be a personal computer. Also, the target system itself can be augmented with the necessary disk resources which can later be removed when the system is delivered.

Development systems can have other uses such as project management, documentation, and support for hardware development. It is possible, for example, to remove the microprocessor from a processor board being tested and replace it with a connection to the development system. The development system has a special board that has the same kind of microprocessor on it as the one removed from the system under test, only *this* microprocessor is under the full control of the development system. Programs resident in the memory of the development system can now be run to exercise all aspects of the system under development. This process is called *hardware emulation*.

Various forms of development systems will be covered in Chap. 15, but it should be evident that the existence of an in-house development system for a particular microprocessor family can generate a certain amount of inertia which resists the change to a new microprocessor.

SUMMARY

In preparation for the design, the following steps should be taken:

1. Draw a system block diagram showing the major physical components and label their interconnections with a meaningful name.
2. List the functional specifications (essential, additional, and future).
3. List the performance specifications.
4. List the device characteristics and constraints (including physical, environmental, and human factors and maintenance and reliability).

Consider the constraints of the development environment including:

1. Time
2. Personnel
3. Costs
4. Hardware-development resources
5. Software-development resources

This book concentrates on system designs for which a board-level design approach is appropriate.

EXERCISES

3.1 The water-level recorder shown in Fig. 3.1 is intended to be mounted out of doors in an exposed site. It is connected by a cable to a submerged pressure transducer at the bottom of a stream, lake, or well. Since the cable and connector are expensive and any connector is potentially unreliable, the pressure transducer is mounted inside the electronics package with a pressure orifice to pass the ambient pressure to the transducer diaphragm. Then the whole package can be lowered into the body of water to the desired monitoring depth. What would be the advantages and disadvantages of this packaging concept?

3.2 You are working in a consulting enginering office that has never developed any microprocessor-based systems but has developed a large amount of personal computer-based engineering software on various consulting engineering jobs in the past. You have just been asked to develop a microprocessor-based flow-control system for a gas pipeline. Since the system is to replace existing analog controls, the analog sensors and analog valve controls are already present and the specifications for the job are pretty well defined. Since only several devices will be built, you have decided that a board-level system is appropriate.

What steps are required to complete the project? Start with a general outline and once that is complete, start to fill in the details in point form.

3.3 Your company, A-Z Microdesigns, has been asked by a client to design a weather station for a remote monitoring site. The site consists of a small heated building with electric power and a telephone connection. A weather station stands several hundred meters outside the building and has the following devices attached to it:

1. An anemometer (wind velocity sensor) that outputs one voltage pulse for each revolution of the rotating part
2. An air-temperature sensor that outputs an analog voltage proportional to temperature
3. A wind direction sensor that outputs a voltage proportional to the incident angle of the wind from true north
4. A rain-gauge sensor that outputs one voltage pulse each time a very small tilting bucket empties its collected rain

A microcomputer system is desired to sit in the building and continuously monitor each variable. It must compute various functions such as average rainfall for the hour, day, and month, average wind velocity, peak wind velocity, and other functions. Periodically, a central office computer will communicate with the system over the telephone lines, at which time the system will transfer all the calculated statistical information to the remote host computer.

(*a*) Follow the steps outlined in the summary of Chap. 3 where sufficient information is available above.

(*b*) Make a list of questions that you must ask the client to complete the specifications.

(*c*) Assuming a board-level design, list as many components of the system as you can, including the board functions (e.g., analog input board, or A/D)

3.4 A reservoir is located in a mountainous region at some height above a power-generating station. The water leaves the reservoir either through the turbines generating electricity or is deliberately run off through the spillway. The utility operating the station has awarded a contract to your consulting company to develop the controls for the spillway gate on the reservoir dam. You have been selected to manage the design task.

The utility has several requirements:

1. The computer-based controls will be located in the power-generating station.
2. The spillway gate on the dam will be controlled to maintain the reservoir water depth constant in normal circumstances. The water level varies due to the combined effects of *input* from rainfall and streams and *output* of water into the turbines and through evaporation from the surface of the reservoir.
3. The normal control will sense the *water depth* in the reservoir and the *gate-opening size* at the spillway entrance. The gate will be driven by a motor through a set of gears.
4. Water depth, gate-opening size, and "gate motor on" are to be displayed to personnel in the control room.
5. A manual override of the control program should allow the operator to open and close the gates by observing the gate-opening readout and holding either an open or close button down until the desired size is achieved.
6. Water depth over a certain maximum should sound an audible alarm and flash the water depth display.
7. The utility personnel wish to have the ability to modify your software and add some of their own programs later.

The tasks: Outline in general terms first the various stages of system design that you would have to go through. Keep in mind what we have recommended in terms of design tools and practice.

For each of the stages above, itemize in more detail the design choices that you would have to make for each stage.

3.5 Repeat the tasks of Prob. 3.4 for the greenhouse control problem in Chap. 1 (Prob. 1.10).

BIBLIOGRAPHY

Microcomputer Data Books

Databooks are available from the manufacturers of microcomputer and interface boards for particular bus protocols. Bus protocols will be discussed in Chap. 15, but at this point a look at the functions provided by the various boards is interesting. Some of the manufacturers are Pro-Log, Mostek. Intel, Motorola, Zilog, Intersil, Analog Devices, and Digital Equipment Corp.

IEEE Standards for Buses

Gustavson, D. B.: "Computer Buses—A Tutorial", *IEEE Micro*, vol. 4, no. 4, August 1984, pp. 7–22.
(This is a good overview of some of the issues involved in computer bus design and some of the major bus standard developments.)
IEEE S696-1983, *Interface Devices*, IEEE.
(This is the standard that emanated from the S-100 bus.)
IEEE S796-1983, *Microcomputer System Bus*, IEEE.
(This bus came from Intel's Multibus specification.)
IEEE P961/D2, IEEE.
(This is a proposed bus standard based on Pro-Log STD bus.)
IEEE P1014, *Versatile Backplane Bus*, IEEE.
[This is the proposed standard based on the VME bus. (See W. Fischer, IEEE P1014, "A Standard for the High-Performance VME bus," *IEEE Micro*, vol. 5, no. 1, 1985.)]

Some Important Corporate Buses

Intel's Multibus II.
IBM's PC buses.
Digital Equipment Corp.'s Q-bus.

HIGH-LEVEL DESIGN

DESIGN OF REAL-TIME SYSTEMS

The design of a real-time computer system is a complex task. It involves the design of both hardware and software components and often requires the coordinated efforts of several people to complete. It is imperative that the design be approached as an engineering project with the normal engineering requirements for specifications, documentation, design reviews, and acceptance testing. Many of the scheduling and project management techniques used in other engineering disciplines are applicable to real-time systems engineering.

In this chapter we examine the development cycle for a real-time system and present design techniques used in different stages of the development cycle. The design techniques presented in this chapter deal with the design at the system level. Later chapters deal with the detailed design of the software and hardware components of the system.

4.1 THE SYSTEM-DEVELOPMENT CYCLE

Figure 4.1 shows the development cycle for a real-time computer system. The development process proceeds in a top-down fashion, with the design progressing from the general to the specific. The design starts with the development of specifications for the entire system. Then a very general preliminary design that shows the major functions to be performed and the basic control sequence to be followed is produced. The specification and preliminary design are verified by the end user to ensure that the system being designed does what is wanted. The preliminary design is also used to prepare cost estimates for the project and to identify potential problem areas in the design.

After the specifications and preliminary design have been validated, the design is continued by partitioning the system into successively more detailed *modules*. A module is a largely self-contained portion of the system which performs a specific function or subfunction. At some point in this partitioning process it will be possible to assign some modules to be implemented in

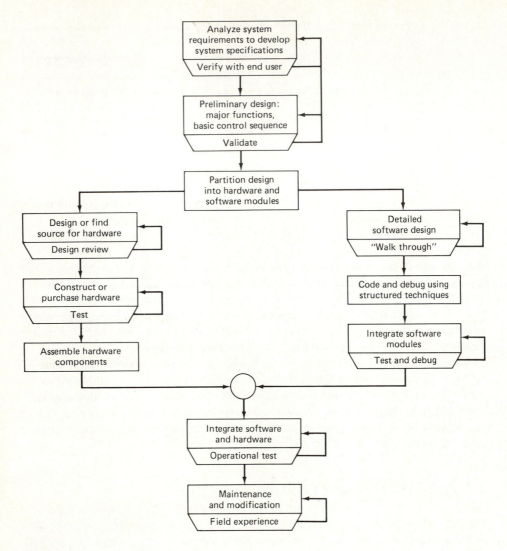

Figure 4.1 System-development cycle.

software while others are assigned to hardware. The design process then splits up into parallel paths of hardware and software development. In each path, detailed design of each module is performed and subjected to a design review (often called a program "walk through" for software designs). Following approval of the module design, construction of the module can proceed. Note that in the software path, this is the first point where any actual writing of program statements occurs. Starting to write program statements prior to this point is somewhat akin to starting to build a bridge before the plans are drawn up.

The individual hardware and software modules are tested to the greatest extent possible as individual modules and are then gradually integrated into the complete system. This system is subjected to preliminary testing by the developers and is then given to the end user for operational or acceptance testing. Once the system is accepted and operational, further changes will inevitably be made as field experience uncovers errors or indicates the need for additional functions. In many real-time systems, the maintenance and modification activity represents the largest fraction of system life-cycle costs.

4.2 ANALYSIS OF SYSTEM REQUIREMENTS

As we emphasized in Chap. 3, proper analysis of system requirements is very important to the success of a real-time computer system. If the developers have misunderstood what the end user wants, they may produce a system that works but is not useful. The goal of the analysis is a written document, called a *system-specification*, or *system-requirements*, document that specifies in considerable detail what the system must do but not how to do it. This distinction between the end result and the means used to achieve it is important. The designers of the real-time system must know at the beginning of the design process what functions the system must perform, but they shouldn't be unduly constrained in how they implement the required function.

The document is written so that the customer can understand it and can participate in the verification process. If the customer has understood and agreed to the specifications, the developer can have more confidence that there will be no last-minute changes. A requirements document signed by the customer can often be used as a basis for the acceptance tests that the system must pass.

4.2.1 Requirements Document

In this chapter we will use as an example the development of a controller for a battery charger for large lead-acid batteries. The controller controls the charging current so that it follows a current-vs.-time profile which recharges the battery quickly but doesn't overcharge or undercharge the battery. A proper charge current profile extends the life of the battery significantly.

Figure 4.2 shows the first page of a requirements document for this controller. The section shown covers the overall operation of the system. Succeeding sections, which are not shown in the figure, expand on each of the operations in considerable detail. A complete document includes specifications on the following:

1. Functions to be performed by the system
2. Techniques or formulas required to perform the functions in cases where the designers do not have the required specialized knowledge
3. The interface between the real-time computer system and the outside world

Acme Battery Charger
System Requirements Document

1.) General Operation:

The charger is equipped with a control panel which allows the following data or commands to be entered:

    a) Type of battery to be charged (Section 2.1)
    b) Present time (Section 2.2)
    c) Time to start charging (Section 2.2)
    d) Start (Section 2.3)   In subsequent pages of the document
    f) Stop (Section 2.3)

The control panel also has the following displays:

    a) 4 digit clock display (Section 2.2)
    b) Charging indicator lamp (Section 2.4)
    c) Battery Charged indicator lamp (Section 2.4)
    d) Malfunction indicator lamp (Section 2.4)

When the Start command is entered, the charger waits for the commanded time to start charging. If the clock is not set, then the charger begins charging immediately. The charger lights the Charging indicator lamp when it begins charging.

The charger begins charging at a high constant current determined by the battery type (Section 2.1). The charger continues charging at this rate until the battery voltage reaches the gassing voltage which is calculated according to the procedure given in Section 3.1. The charger then begins to control the charging current so that the battery voltage is held at the gassing voltage (constant voltage mode). This continues until the charging current has dropped to a level determined by the battery type (Section 2.1). The charger now continues to charge at this low current level for a period of three hours. After this period, the charger lights the Battery Charged indicator lamp. extinguishes the Charging lamp, and ceases charging.

The Stop command causes charging to cease at any time. During the charging cycle, the charger checks for battery temperature and voltage conditions that are out of limits (Section 5.1). If a limit is exceeded, the charger ceases charging and lights the Malfunction indicator lamp.

Figure 4.2 First page of a requirements document for a microcomputer-controlled battery charger.

4. The performance to be achieved in terms of speed of response and accuracy or some other objective measure of performance
5. Cost constraints divided into cost goals for development, production, and maintenance

Relating this list to the requirements document for the battery-charger controller, we find that the functions to be performed are described in a general way on the first page of the document (Fig. 4.2). More detailed descriptions of the functions would come in later pages. The charging of lead-acid batteries is a specialized area of knowledge. Therefore, information, such as the formula to calculate the battery-charging voltage at which hydrogen gas is evolved (the gassing voltage), is included in the requirements document.

There are three interfaces between the battery-charger controller and the outside world. There is the interface to the control panel and the human operator; an interface to the power circuit in the battery charger which allows the controller to set the charging current and turn the charger on and off; and finally, an interface to the battery which allows the controller to measure the battery's charging current, voltage, and temperature. Specifications must be developed for all three of these interfaces. For example, in the case of the interface to the battery, the expected ranges of the charging current, battery voltage, and battery temperature must be specified. In addition, the accuracy with which these parameters are to be measured must be specified.

Performance specifications must also be developed for other aspects of the controller. For instance, the resolution of the internal timer (i.e., to the second or to the minute) must be specified. Response times must also be determined. For example, the maximum time allowed between measurements of the battery parameters must be specified. As another example, the allowable time to detect a malfunction and to respond must be determined.

4.2.2 Response-Time Specification

In real-time systems, the response time of the system is often of critical importance. All critical response times must be listed in the specification. In addition, any actions that must be performed in a precisely timed sequence must be listed. However, it is also important to avoid overspecification of time-critical responses since this adds to the complexity and cost of the design. For example, in the case of the battery-charger controller, it is apparent that the battery temperature cannot change very quickly, since the battery is large and heavy. Therefore, a requirement that the battery temperature be measured five times a second would needlessly complicate the design. A 5-s interval between temperature measurements would probably be perfectly adequate and would give the designers an easier job.

4.2.3 Specification of the Human Interface

The specification of the human interface between the user and the real-time system is also of importance. It consists of hardware elements, such as switches, keyboards, and displays, as well as more abstract elements, such as the layout of a control panel or the format of a display.

In many cases the users of a real-time computer system are not computer experts and don't have the time or inclination to become experts. The human interface must be designed so that these users can make full use of the system. One common technique is to design the human interface so that it resembles a system the user is already familiar with. For example, in a computerized process-control application, the displays of process variables would be set up to resemble the meters and strip chart recorders used in older process-control systems. One way to get an idea of the acceptability of a proposed user interface is to write a user's manual for the system during the specification stage of the design. If potential users can understand the manual and are happy with the operating procedures described in it, then the user interface is probably acceptable.

The production of a system-requirements document is an iterative process. An initial draft of the document is written and used as the basis for the first attempt at a preliminary system design. The result of the preliminary design may show that some cost or performance goals are impractical or that some specifications are unclear. A revised specifications document is then written to deal with these problems and the system design is continued.

4.3 PRELIMINARY SYSTEM DESIGN

The goal of the preliminary system design is to begin to specify the nature of the control flow, data elements, and functions in the real-time system.

4.3.1 Block Diagram

In Chap. 3 we recommended that the designer draw a simple block diagram of the system as one of the first steps in the design process. Now that the system

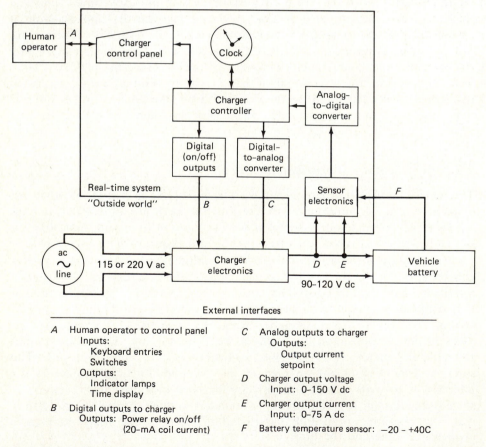

External interfaces

A	Human operator to control panel	C	Analog outputs to charger
	Inputs:		Outputs:
	Keyboard entries		Output current
	Switches		setpoint
	Outputs:		
	Indicator lamps	D	Charger output voltage
	Time display		Input: 0–150 V dc
B	Digital outputs to charger	E	Charger output current
	Outputs: Power relay on/off		Input: 0–75 A dc
	(20-mA coil current)	F	Battery temperature sensor: −20 – +40C

Figure 4.3 Initial block diagram of the battery-charger system showing the interfaces between the system to be designed and the outside world.

requirements have been formally established in a specifications document, the block diagram should be updated to incorporate any changes or additions. The block diagram should then be divided into two sections:

- The outside world, which represents all portions of the system which are already designed or determined in some way
- The system to be designed, which represents the portions of the system which are the subjects of the design effort

The interfaces between the two portions of the system are now drawn onto the block diagram and the basic nature of each interface is noted on the diagram. Figure 4.3 shows such a block diagram for our battery-charger example. The block diagram acts as a supplement to the specification document in the preparation of the preliminary design.

In order to carry out the preliminary design we need *design tools* to allow us to represent the control flow, functions, and data in the system. One design tool that is commonly used is the flowchart. Unfortunately, for systems of realistic complexity, flowcharts tend to obscure the required information, since they combine information on control flow, functions, and data in one diagram. An alternative approach is to use different design tools for the three different elements of the design.

4.3.2 Representation of Control Flow

The control flow can be presented quite clearly with an algorithmic state machine (ASM) diagram of the type discussed in Chap. 1. The diagram consists of circles, each of which represents a *state* and lines indicating *transitions* among states. We will loosen the rigorous definition of a state given in Chap. 1 and instead define it as an operating mode in which the tasks being performed do not change. The outputs from the real-time system during each state are listed within the circle corresponding to the state. The lines indicating transitions are annotated with the condition that causes the change of state. The condition could be the value of an input or program variable or simply the completion of the tasks within the state.

Figure 4.4 is an ASM diagram which shows the control flow for the battery-charger controller. This diagram can be derived almost directly from the description of the charger's operation given in Fig. 4.2. When the power is applied to the system, the first state entered is the INITIALIZE state. This state is used to ensure that all the internal variables in the controller have a defined initial value and that the outputs all have a defined value.

Once initialization is complete, a transition is made to the IDLE state. In this state the controller simply monitors the control panel. The operator sets the present time and the time to start charging and then presses the START button. When the controller senses that the START button has been pressed, it makes a transition to the WAIT state.

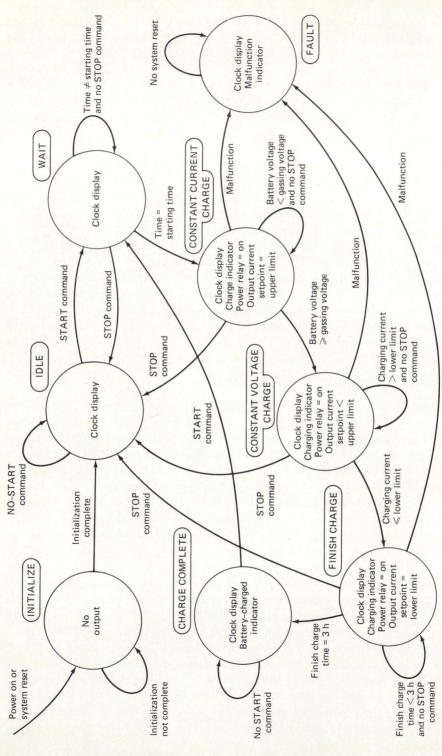

Figure 4.4 ASM diagram for battery-charger controller.

In the WAIT state the controller monitors the internal clock. When the time on the internal clock equals the preset starting time, the controller makes a transition to the CONSTANT CURRENT CHARGE state. If the starting time has not been preset, the controller will immediately make the transition to the CONSTANT CURRENT CHARGE state. Alternatively, if a STOP button is pressed, the controller sequences back to the IDLE state.

The CONSTANT CURRENT CHARGE state handles the operating mode where the charger is delivering the maximum current to the battery. The

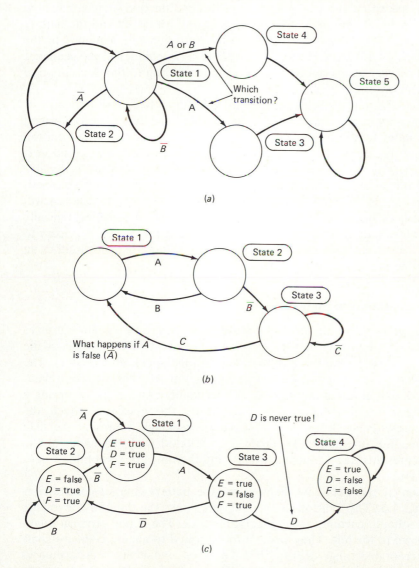

Figure 4.5 Errors in control flow. (*a*) Inconsistent or ambiguous ASM diagram. (*b*) Incomplete ASM diagram. (*c*) ASM with unreachable state.

output-current set-point, which controls the charger output current, is set to its upper limit. The power relay which connects ac power to the battery charger is turned on so that charging can commence. The charging indicator lamp on the control panel is turned on to inform the operator that charging is in progress. The controller stays in this state until one of three conditions occurs. If the STOP button is pressed, the controller sequences back to the IDLE state. If a malfunction is detected, such as the battery overheating or the battery voltage going outside acceptable limits, the controller sequences to the MALFUNC-TION state, where charging is stopped and the malfunction indicator light is turned on. Finally, if the previous conditions haven't occurred and the battery voltage has risen to the gassing voltage, the controller will sequence to the CONSTANT VOLTAGE CHARGE STATE. The remainder of the sequence through the charger states can be understood by referring to the description of charger operation in Fig. 4.2.

The ASM diagram explicitly shows the conditions that cause a change of control while suppressing much of the detail about the functions performed within each state. It is also possible to identify errors in control flow from the ASM diagram. Figure 4.5*a* is *inconsistent* because more than one transition is possible from state 1 when condition *A* is true. Figure 4.5*b* is *incomplete* because the transition which occurs when variable *A* is false is not specified. Finally, Fig. 4.5*c* is an ASM diagram with an *unreachable* state. State 4 will never be reached, since *D* is always false in state 3. Real-time systems typically have many states and many possible transitions among them. It is very easy to design a system with all the above errors if an ASM diagram is not used as an aid in detecting and eliminating them.

4.3.3 Representation of Data Flow

The ASM diagram shows the flow of control in the real-time system but gives little information about the flow of data from the input to the output of the system. A *data-flow diagram* is used to represent this aspect of the system. The diagram consists of rectangles which represent operations that transform data. Arrows are drawn entering and leaving the rectangles. The arrows entering a rectangle are the input data items required by the operation which the rectangle represents, while the arrows leaving a rectangle represent the data resulting from the operation. An arrow is annotated with the name of the variable it represents, the variable's format, and the range of acceptable values for the variable.

A portion of the data-flow diagram for the battery-charger controller is shown in Fig. 4.6. This portion of the diagram shows the data flow involving the battery voltage, current, and temperature data. The actual physical quantities are shown at the left. They are first transformed to analog voltage signals having an acceptable range of 0 to 10 V. Then the signals are transformed to binary integers by an analog-to-digital (A/D) conversion. The data format is specified as a binary integer with an allowable range of 0 to 1023 (i.e., a 10-bit

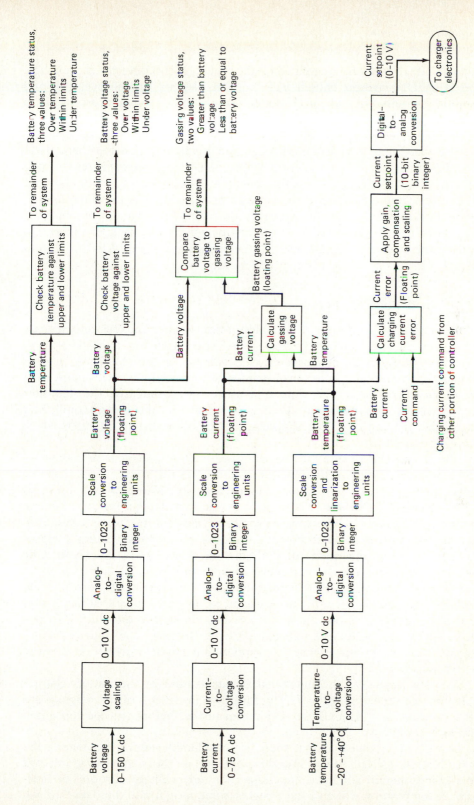

Figure 4.6 Portion of data-flow diagram for battery-charger controller.

number). This format is determined by the type of A/D converter chosen. The type of A/D converter chosen will, in turn, depend on the accuracy and resolution required in the measurement of the battery parameters. (These issues will be discussed in more detail in Chap. 9.)

The binary integers representing the battery voltage, current, and temperature are then transformed into floating-point numbers which represent these quantities in engineering units. For example, the battery current is represented in amperes and the battery temperature in degrees Celsius. These variables are distributed to a number of operations which produce some of the data required for the operation of the battery-charger controller. For example, the battery-current variable is subtracted from the charging-current command variable to create a current error variable which is eventually transformed to the current set-point output sent to the charger electronics.

4.3.4 Functional Decomposition

Up to now we have suppressed any information on the functions that actually do the work in the real-time system. The ASM diagram lets us define a set of states which the system sequences through but contains no real detail on what is done within each state other than defining what outputs are active and what conditions will cause a transition to another state. Similarly, the data-flow diagram allows us to define the data elements in the system and their relationship to each other but suppresses information on the functions which transform the data.

Therefore, at the same time as the ASM diagram and data-flow diagram are being drawn, a list of the functions required in the system should also be drawn up. This list can be developed in a top-down fashion through a process of *functional decomposition*. The process starts by defining a few very general functions that carry out the major operations described in the system specifications. Each function is then broken down into a set of more detailed subtasks that carry out the operations of the function. These subtasks can then be broken down into even more detailed functions in turn.

Obviously this process of functional decomposition must stop at some point. This should be when the functions are sufficiently simple that there is no point in breaking them down any further. It is difficult to give hard and fast rules about what constitutes a sufficiently simple function. One guideline is that functions are simple enough when their operation can be described in detail in one written page.

The first stages of the functional decomposition of the battery-charger controller are shown in Fig. 4.7. The functions on level 3 of the diagram are still quite general and should be divided up further before the decomposition process is complete.

It must be emphasized that there is considerable interaction among the three different design tools. The development of the data-flow graph will provide information about the functions required, while the process of functional decomposition will give the designer some insight into the data that will

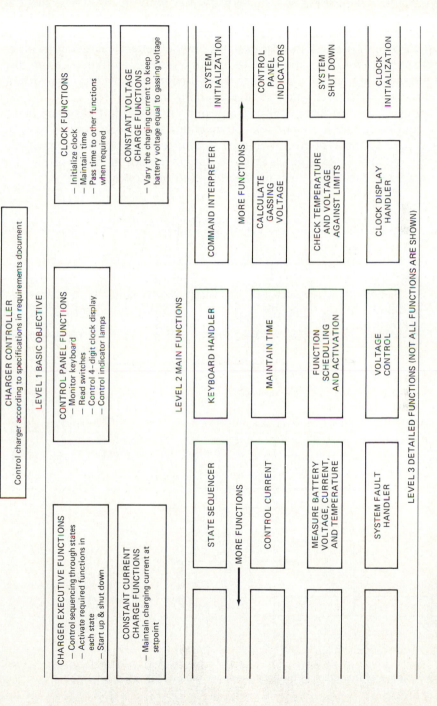

Figure 4.7 Functional decomposition of design into functions.

be required by the functions. Similarly, the designer can use the ASM diagram to help define the executive functions in the system. The designer can also determine which functions must be active in each state in the ASM diagram.

As functional decomposition continues, the number of functions does not necessarily keep on increasing, because many of the functions can be used more than once in the system. For example, in the battery-charger controller, the functions to calculate the battery gassing voltage and to measure the battery temperature, voltage, and current are used in both the constant-current charging mode and the constant-voltage charging mode.

A principal goal of functional decomposition is to divide the system up so that it consists as much as possible of functions which are used in several different parts of the system. A real-time system consisting of a few functions used over and over again is much easier to design than a system consisting of a large number of functions, each of which is used only once.

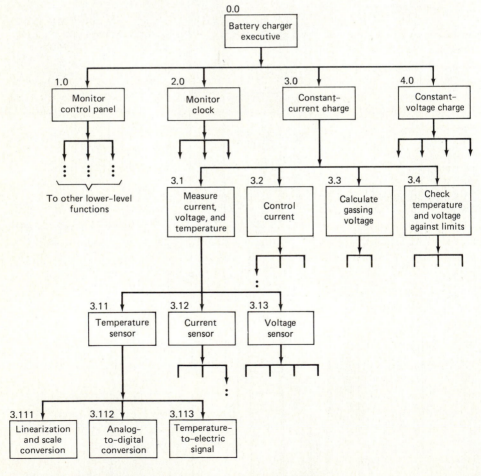

Figure 4.8 Portion of tree diagram for battery-charger controller.

4.3.5 Relationships among Functions

The simple decomposition of functions into successively lower-level functions does not do much to illustrate the relationships among functions. The functions should be organized in a *hierarchical* fashion through the use of a *tree* or *hierarchy* diagram. A partial tree diagram for the battery-charger functions is shown in Fig. 4.8. The root of the tree, at the top of the diagram, represents the highest-level functions in the system—the functions that control the execution of all the other functions. In the battery-charger controller these functions are aptly called the executive functions. They sequence the charger through the states on the ASM diagram and activate the functions required in each state. The executive functions directly control the four functions immediately below them in the tree diagram. These functions in turn make use of functions lower down in the hierarchy. In the diagram, only the functions controlled by the constant-current charge function are shown to avoid excessive clutter.

The basic organization of the system is now becoming clear. The high-level functions (closer to the root of the tree diagram) are basically sequencers which invoke lower-level functions and pass information from one function to another. They represent the control structure of the design. The lower-level functions do all the actual work. For instance, the conversion of the battery temperature to an electric signal, then to a digital quantity, and finally to a number in a format understandable by the computer is all done at the very lowest level of the diagram.

4.4 DIVISION INTO MODULES

As we have mentioned, one of the primary goals of the hierarchical decomposition of the system is to break the system down in such a fashion that the system tasks can be carried out by a minimum number of relatively simple functions. The goal is a *modular* design consisting of a number of independent function blocks, called *modules*.

4.4.1 Definition of a Module

A module is a function block that comes close to meeting the following ideal goals:

1. The module "hides" information required to implement the module functions from the outside. This simplifies the use of the module, since no knowledge of the module's internal operation is required. Also, the module can be modified without affecting the rest of the system as long as the specified "public" interface to the outside remains unchanged.

2. The connections that pass data between the module and the rest of the system can be minimized and rigorously defined. These connections and the specification of the data passed constitute the public interface to the module that users of the module see. Ideally, only data is passed; the passing of "flags," or "switches," that modify the flow of control should be avoided.
3. The contents of the module are oriented toward one specific type of operation.
4. The function performed by the module is simple enough so that it can be described on one written page. In the case of a software module, the program to carry out the module functions is no more than about 50 or 60 lines long. This keeps the module short enough so that it can be easily understood in detail by a single designer.

To illustrate these concepts, consider the functions required to measure battery temperature in the battery-charger controller. Figure 4.9 shows how these have been divided up into software (program) and hardware (electronic circuit) modules. The design shown has a few frills, such as the ability to measure temperature in different scales, which aren't required in the battery-charger controller but would be useful in a more general-purpose data-acquisition system.

If some other part of the program wants a temperature measurement, it requests the data from module 1 (temperature sensor) and passes the desired temperature scale (i.e., Celsius, Kelvin, or Fahrenheit) to the module. The module returns a floating-point number equal to the temperature in the desired scale. Module 1 makes use of lower-level modules (modules 2 to 4) to get the raw temperature sensor data which it converts to the correctly scaled floating-point value.

Note that the remainder of the program need not know anything about how the temperature value is actually obtained. For example, information about the temperature transducer, such as whether its output is linear or nonlinear, is completely hidden from the remainder of the program. Therefore, if the temperature transducer is changed, only module 1 and module 4 (the transducer itself) must be changed. The rest of the program is unaffected. Similarly, module 1 can get the raw temperature data from module 2 (the A/D converter interface module) without knowing any of the hidden information about the A/D converter and its interface. All module 1 must do is pass the A/D channel number of the temperature transducer to module 2.

Consider a nonmodular design where each part of the program that requires temperature data goes directly to the A/D converter to get the raw temperature data and does its own conversion to the actual temperature. If any change occurred in the type of temperature transducer, the A/D channel to which it is connected, or the A/D converter used, we would then have to find every point in the program where a temperature measurement is performed and make the required changes. In addition, if the development project involves more than one person, every person who plans to make a temperature

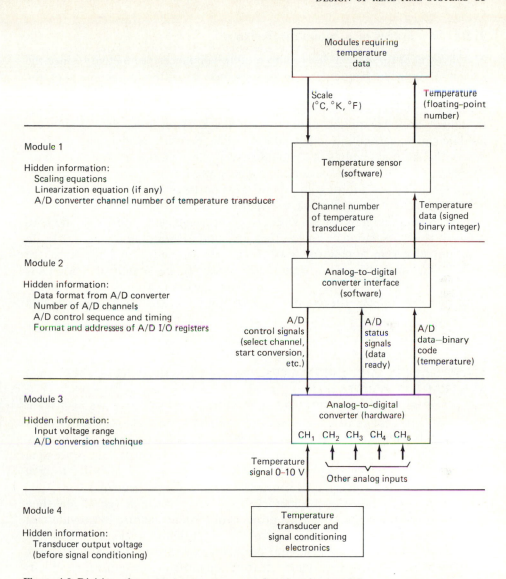

Figure 4.9 Division of temperature-measurement functions into modules.

measurement in his or her portion of the design must be aware of all the details which are listed as hidden information in Fig. 4.9. In contrast, if the modular approach is used, only the person designing the modules containing the hidden information need be aware of these details.

The design tools used in the preliminary systems design provide much useful information for the design of modules. The data-flow diagrams show what data must be passed to a module and what data the module produces. The tree diagram shows how modules are interconnected and which modules make use of other modules to accomplish their task.

4.4.2 Advantages of a Modular Design

The division of the system design into modules enhances productivity, since the design of individual modules can be assigned to different members of the design team who can work independently. The interface to each module is carefully specified at an early stage in the design and placed in a design document available to the entire design team. After this, the design of a module can be carried out independently, since no other module has access to any of the internal features of a module except through the specified interface. Productivity is enhanced because designers need not spend as much time exchanging detailed information about their portions of the design to ensure that different portions will work together.

A modular design is usually easier to get working, since individual modules can be tested in isolation as long as they are supplied with the required data. In addition, errors in a modular system are usually easier to find, since the module in which the error occurs can normally be identified reasonably quickly. As we have already seen, a modular design is also easier to modify, since a change usually affects only one module rather than having effects that propagate throughout the system.

4.5 PRELIMINARY ESTIMATES OF DEVELOPMENT COST AND SYSTEM PERFORMANCE

In an industrial environment it is normally necessary to make at least a rough estimate of the cost and performance of the real-time system at an early stage in the design process. Estimates of the memory capacity and processing throughput of the computer system are often required also in order to give some indication of what type of computer system will be needed.

If the organization has considerable experience in the design of real-time systems, it can draw upon previous projects of a similar nature to provide cost and performance guidelines. This approach is of little use to organizations that are undertaking a real-time computer project for the first time or are designing a totally new type of system. Unfortunately, quantitative methods for estimating software development costs at an early stage in the design are not well-developed.

4.5.1 Required Development Time

Studies have shown that the cost of software development is linearly related to the length of the program that is the end product of the software-development project. It is therefore possible to determine a unit cost for software on the basis of cost per program statement and estimate total cost from an estimate of program length. For real-time programming projects the cost per program statement has been estimated as $20 to $30. Therefore, a software-development project that resulted in a 1000-line program would cost from $20,000 to $30,000. Obviously these figures are approximate and will vary with labor rates and inflation.

Estimates of the length of time required for a software-development project can also be made if the length of the program is known. Studies show that software productivity is about 10 program lines per worker-day in real-time system-design projects. This figure may appear rather low, but software productivity has actually been considerably lower in some real-time systems projects.

4.5.2 Estimating Program Length

Given that the software development cost can be estimated directly from the length of the program produced, the next problem is to estimate the length of the program at an early stage in the design. Computer scientists have used statistical techniques and probabilistic arguments to estimate the lengths of programs from the number of *operators* and *operands* used in the program. Operators are the elements of programming languages that act on variables or control program flow. Familiar examples are +, −, *, >, =, GO TO, and IF-THEN-ELSE. Operands are the elements acted on by the operators. These include program variables, constants, and subroutine names.

Shooman [1] has used the results of these studies to create a formula which gives a rough estimate of the length of a program from preliminary design information. The procedure is as follows:

1. Estimate the total number of distinct operators in the programming language being used. The estimate can be made by studying a manual or textbook on the programming language. For most programming languages the number of distinct operators is about 30 to 40. Assembly languages which reflect the machine language of a particular computer will have 50 to 60 distinct operators.
2. Estimate the number of distinct operands that will be required by the design. This is the sum of all the input variables, output variables, internal variables, constants, and function names. This estimate can be made using the results of the preliminary design.
3. Sum the estimates of steps 1 and 2 to get a value k which is equal to the estimated number of operators in the programming language plus the estimated number of distinct operands in the program. Insert the value for k into the following formula:

$$N = k(0.5772 + \ln k)$$

   The result N is an estimate of the total number of operands and operators in the program.
4. Each line in a program has an average of 3 to 5 operators and operands:

Estimated operand count for battery-charger controller

Inputs:
1. Control panel
    Battery type command
    Present time command
    Time to start charging command
    Start command
    Stop command

2. A/D Converter
    Battery temperature
    Battery voltage
    Charger current

3. Clock
    Clock pulse
                                                    9

Outputs:
1. Control panel
    Clock display
    Charging indicator lamp
    Battery-charged indicator lamp
    Malfunction indicator lamp

2. Clock
    Initialize clock
    Start clock

3. D/A converter
    Charger current setpoint

4. A/D converter
    Initialize
    Choose input channel
    Start conversion
                                                    10

Internal variables:
Present time
Time to start charging
Gassing voltage
Time at which constant current
charging began
Battery type
Battery temperature
Battery voltage
Charger current
Charger current command
Charger current error
Clock display·status
Charging indicator status
Battery charged indicator status
Malfunction indicator status
Temporary variables — 20
                                                    34

Constants:
Battery types — 3
Constant-current setpoints — 3
Low-current setpoints — 3
Gassing voltage equation constants — 6
Temperature conversion constants — 2
Voltage conversion constant
Current conversion constant
Battery temperature limits — 2
Battery voltage limits — 2
                                                    23

Modules:
From Fig. 4.8 we might estimate a total
of 25 distinct modules   25
                                                    ____
Total: 101

Figure 4.10 Estimated operand count for battery-charger controller.

88

$$A = Y \qquad \text{Two operands and one operator}$$

$$A = Y + B \qquad \text{Three operands and two operators}$$

The number of lines in the program can therefore be calculated by dividing N by (say) 3.

Figure 4.10 shows the procedure for estimating the number of operands in the battery-charger controller design we have been developing. Using the requirements document, the ASM chart, tree diagram, and the data-flow diagram, a list of inputs, outputs, internal variables, constants, and modules is prepared. The sum of these elements is 101. In the final design we are likely to have more than this number of operands, since it is impossible to identify them all during a preliminary design. Therefore we will increase our estimate by 50 percent to 150 operands. This should produce a reasonably conservative estimate.

Assuming the use of a standard programming language, we estimate that the number of operators is about 40. Thus our value for k is 190. Applying the formula for the total number of operators and operands,

$$N = 190(0.5772 + \ln 190) = 1106.6$$

we estimate that the program has approximately 1100 operators and operands. Next we estimate the number of program lines:

$$\frac{1100}{3} = 367 \text{ lines}$$

It should be stressed that this estimate is very approximate. We can safely say that the program is likely to be more than 200 lines in length and likely to be less than 500 lines in length. Applying a cost figure of $30 per program line to our calculated program length, we arrive at an estimate of $11,000 for software development. To estimate the software-development time, we divide the number of lines in the program by the 10-line-per-day productivity figure. Our 367-line program-development project should require about 35 to 40 worker-days. The cost and development time for the hardware portion of the system must now be calculated to get a total for system-development cost and time.

4.5.3 Estimating Memory Requirements

This technique can also be used to make an initial rough estimate of memory requirements. Each line of a high-level language program typically requires 5 to 20 bytes of memory once it is translated into machine language. Therefore our 367-line program will require 7200 bytes of memory if we make a worst-case estimate. The program will be stored in read-only memory (ROM). The amount of read/write (RAM) memory required can be estimated by examining the estimated number of internal variables. Each integer variable will normally require 2 bytes of storage (4 bytes in some systems). Floating-point variables will require 4 or 8 bytes. Storage requirements for arrays and strings can be calculated by multiplying the number of elements by the number of bytes required to store an individual element. It is extremely important to make generous estimates of the amount of memory required. If a program must be squeezed into an inadequate amount of memory, development costs will rise dramatically. The cost of memory is declining so rapidly that there is little excuse for a memory-limited design in most situations.

4.5.4 Estimating Execution Speed

In many real-time systems the response time of the system is of primary importance. Therefore a rough estimate of system speed must often be made early in the design to determine if the design has any chance at all of meeting the response-time requirements. A modern 16-bit microcomputer such as the Motorola 68000 or the Intel 8086 typically takes 5 to 20 μs to execute a simple programming-language statement that uses 16-bit binary integers as operands. Statements that use floating-point (real) numbers as operands slow the computer down considerably (by a factor of 10 to 100) unless the computer is equipped with a hardware floating-point processor. The 8-bit microcomputers are normally one-half to one-tenth as fast as the 16-bit microcomputers. Again, the figures above are very approximate.

If there are time-critical functions in the real-time system, it may be necessary to design and program these functions at an early stage in the system design to determine if the timing specifications can be met. This is a departure from the top-down design philosophy we have been advocating, but it represents a necessary compromise in some cases. If the detailed design of these time-critical functions were left to the later stages of the design process, then the discovery that the timing specification cannot be met could be disastrous. If the discovery is made early in the design process, it is possible to adapt by using a faster computer, employing a better algorithm, or substituting hardware for software.

Time-critical functions can often be identified from the data-flow diagram. For instance, referring back to the data-flow diagram in Fig. 4.6, we find that if the time between the measurement of BATTERY CURRENT and the output of the current set-point signal to the charger electronics is important, then following the flow of the data from the BATTERY CURRENT input to the current set-point output will reveal all the time-critical functions.

4.6 DIVISION BETWEEN SOFTWARE AND HARDWARE

Some aspects of the division between software (computer programs) and hardware (electronic circuits) are obvious. Transducers and A/D converters are hardware items, but calculations are most likely carried out in software. Other functions could be done either in software or hardware. For example, in the battery-charger controller, a temperature sensor with a nonlinear relationship between the temperature and its output may be selected. In that case the sensor can be linearized with analog circuits (hardware) or with software. A process control loop could be done in hardware with software supplying the set-points or the entire control section could be done in software.

4.6.1 Hardware-Software Tradeoffs

Considerations of speed, flexibility, and cost come into the decision when making these *hardware-software tradeoffs*. Normally a software implementation of a module is slower than a hardware implementation but the software implementation allows more flexibility. The relative costs of the two implementations depend on a variety of factors. Hardware costs can be quite low if a specialized integrated circuit is available to carry out the function. Design and production costs can be quite high if hardware must be custom-designed and requires many components. A software module has a high fixed cost for the initial development but has essentially no production costs.

4.6.2 Hardware-Software Tradeoff Example

Consider, for example, the clock which keeps track of time in the battery-charger controller we have been using as a design example. The hardware could be reduced to a timer which emits a pulse at regular intervals. Software would be responsible for counting the pulses and keeping track of the seconds, minutes, hours, and days. At the other extreme, we could install a real-time-clock integrated circuit such as the National Semiconductor MM58167. This circuit counts clock pulses from a crystal oscillator and keeps track of seconds, minutes, hours, and days in internal registers which can be read by a computer. In that case the software design for the clock is limited to functions which initialize the clock chip and read the time from the clock chip.

The choice between the two alternatives is not easy to make. The software-intensive approach will tie up computer time that could be used by other functions and will require a reasonable amount of software development. On the other hand, the clock chip, while relatively cheap (about $10), requires a crystal oscillator which runs at a frequency that is different from the clock frequency of most microcomputers. The chip also requires some care in the design of the interface to the microcomputer. Thus the hardware design costs of adding the clock chip could be fairly high.

If the microcomputer is already overloaded, then the choice should probably go to the clock chip so that system response specifications can be met. Otherwise, the only rational way to decide between the two alternatives is to sum the incremental development and production costs of the software-intensive and hardware-intensive approaches and then choose the cheaper alternative. To make the cost comparison really accurate, the incremental production costs should be calculated as a net present value to account for the fact that future production costs are less important than present development costs.

4.7 SOFTWARE DESIGN

Leaving hardware design for later chapters, we continue with the software-design process. Each module to be implemented by software must be designed to the point where the programming of the module is a relatively straight-forward process.

4.7.1 Structured Flowcharts

A restricted version of the flowchart (a *structured flowchart*) is useful in integrating functions, control flow, and data. The structured flowchart restricts the choice of *control structures* to the ones shown in Figs. 4.11 and 4.12. These structures allow only one entry and one exit for control flow. As such, they tie in neatly with the use of ASM diagrams and modules. Since there is a very clear flow of control in these flowcharts, it is possible to check them against the ASM diagrams developed earlier in the design process. The single exit and entry points to structures encourage the design of modules with well-defined interfaces to other modules. Branches from the middle of one module to the middle of another module are discouraged by the nature of these control structures.

The SEQUENCE construct implies an unconditional flow of control in which all the tasks indicated in the blocks are executed.

The IF-THEN-ELSE structure is a method of selecting between two paths of control flow depending on the results of a test on some condition. When the tested expression is true, the THEN branch is executed. If the tested expression is false, the ELSE branch is executed. It is possible that the ELSE branch will not contain any task to execute.

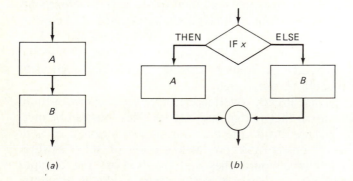

(a)   (b)

Figure 4.11 Structured flowchart control constructs: (*a*) sequence (*b*) if-then-else.

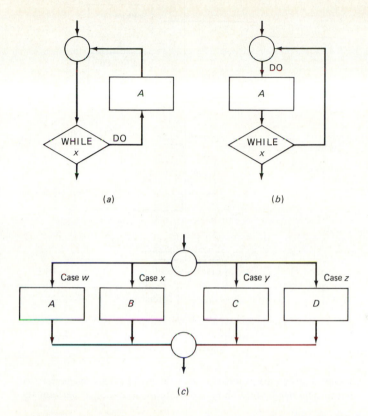

Figure 4.12 Structured flowchart control constructs: (*a*) while-do; (*b*) do-while; (*c*) switch.

The WHILE-DO construct is used to control program looping. An expression is tested and if it is true, the function in the block is performed. Looping continues until the result of the test is false. At that point control passes on to the next construct in the flowchart. The DO-WHILE construct is almost identical except that the function in the block is executed before the expression is tested. Therefore the function will always execute at least once.

The CASE construct is a method of selecting among several paths of control flow depending on the value of a variable. When using the CASE construct it is important to make provision for the possibility that the variable has a value different from any of the enumerated possibilities. Some kind of default must be provided.

4.7.2 Nesting and Stepwise Refinement

Structured flowchart constructs can be nested. It is possible to break down a block in a structured flowchart into a new flowchart which represents the subtasks necessary to perform the function described in the original block. Figure 4.13 illustrates this subdivision process.

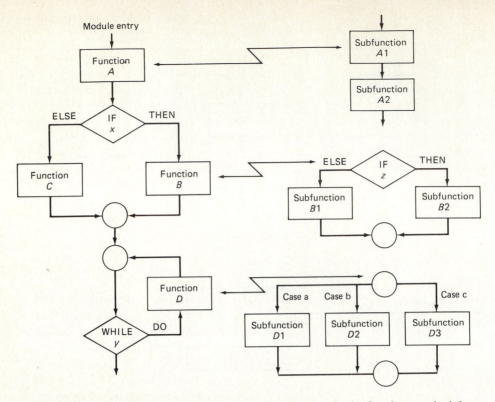

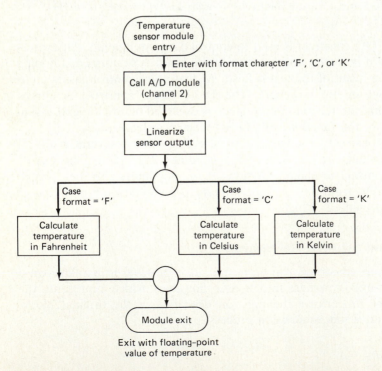

Figure 4.13 Subdivision of functions within a module. Functions in the flowchart on the left are broken down into more detailed flowcharts which show the subtasks necessary to perform the functions.

Figure 4.14 Structured flowchart for the temperature-sensor module.

The design of a module should start with a simple flowchart showing the primary functions to be performed and the overall flow of control. The various function blocks in this master flowchart are then decomposed into structured flowcharts which represent the details of the function blocks. This design by *stepwise refinement* is carried out until the final level of detail is such that the translation of the design to a program is a direct and obvious process.

Figure 4.14 is a structured flowchart for the temperature-sensor module discussed earlier. The flowchart is at about the level of detail required for translation to a program. There are no further control constructs within each block of the flowchart and the functions described in the blocks can be implemented with a few lines of a high-level programming language.

4.8 DESIGN REVIEW

Before a program is written, it is good practice to review the design to verify that it meets its specifications. The designer first reviews the design and works through it by hand with sample data to see if the correct result is obtained. After this a more formal design review is normally performed. This design review is often called a *design walk through* in the software business. As with other engineering design reviews, the designer presents the design to other members of the design team and perhaps to management. The designer goes through the flowchart for the module under review, following the flow of control and explaining what each function does. The reviewers are obviously on the lookout for mistakes in program logic but they also watch for module interfaces that do not meet the previously defined specifications and for designs that are obscurely formulated or poorly documented. If the reviewers cannot understand a design when it is explained to them by the designer, it is unlikely that the design will be understood by the personnel who have to maintain the software in the field.

4.9 TRANSLATING THE DESIGN INTO A PROGRAM

Once the module design has passed its design review the designer begins to write the program. In some organizations the designer gives the design to a programmer or programming technician who does the actual coding. The programming task is fairly simple if the design is clear and the programming language used supports program modules and the control constructs used in structured flowcharts. Programming languages suitable for real-time applications are discussed in Chap. 5.

Organizations that develop a considerable amount of software will have standards for the format and documentation of programs. If such standards are not available, *The Elements of Programming Style* by Kernighan and Plauger [2] is an excellent guide to the art of writing clear and well-documented programs.

4.10 TESTING THE MODULE

After the module is written in a programming language, the designer begins to *debug* and *test* the program. Debugging is the process of finding and correcting program logic and *syntax* errors. Syntax errors are errors in the "grammar" of a programming language. Normally the translation from the programming language to machine language and the first few attempts at execution will reveal the existence of a fair number of these types of errors. After the program has been debugged to the extent that it runs without apparent failures, the testing process begins. Testing examines the behavior of the program by executing it with sample input data. Testing is intended to reveal any remaining logic and syntax errors in the program. However, it is also used to reveal fundamental flaws in the design. For example, a program can be bug-free in the sense that it is a correct implementation of a design, but the design itself may not do what the system specifications require.

4.10.1 Limitations on Testing in Real-Time Systems

The extent to which the designer can debug and test a software module prior to its integration with the rest of the real-time system depends on the nature of the module and on the computer being used to develop the software. If the module directly controls some hardware, it may be difficult to perform realistic tests until both the hardware and software are installed in the real-time system. Modules that interact only with other software modules can be tested prior to system integration. It can sometimes be difficult to perform tests if the computer used to develop the software is different from the *target* computer (the computer on which the software is to execute). However, if the designer uses a high-level programming language that can be executed on either computer, some testing should be possible. These issues will be covered in greater depth when development systems are discussed in Chap. 15.

4.10.2 Planning for Testing

Planning for the testing phase starts at the beginning of the project when the system specifications are being defined. Sets of test data are designed to test the system for compliance with the system specifications. As the design is divided into modules, sets of test data are designed to test each module for adherence to interface standards. Finally, the designer assigned to a particular module designs test data to ascertain whether the module properly performs its function.

Testing can prove that there are no errors in a program only if the testing is *exhaustive*. This means that the set of test data contains all possible combinations of input variables and the correct result is known for all these possible inputs. Exhaustive testing is possible only in cases where there is a limited number of input variables with a very limited range of possible values.

For example, a module that converts 8-bit character codes from one format to another can be exhaustively tested since there is only one input variable (an 8-bit character) which has only 256 possible values. In addition, the expected results can be determined from a code-conversion table. By contrast, a module that has two 16-bit integer variables as its inputs has more than 4 billion possible input data combinations. If the module takes only 100 μs to execute, it will take over 100 h to test all the combinations.

Since exhaustive testing is not normally possible, the designer must choose a set of test data that is small enough to make testing practical yet test the module thoroughly enough that there is some confidence that the module is error-free. The designer must also determine what the program should do when it executes using the test data set. Some guidelines can be given for the choice of test data. Choose test data items that

1. Represent the maximum and minimum values for the input variables
2. Represent "illegal" inputs that the program must detect and handle
3. Result in the output variables taking on their maximum and minimum values
4. Cause all branches in the module's structured flowchart to be executed or cause all states in the module's ASM chart to be reached

It is good practice to let another member of the design team choose some of the test data to ensure that the module designer does not consciously or subconsciously avoid test data that exercises potential weak spots in the design.

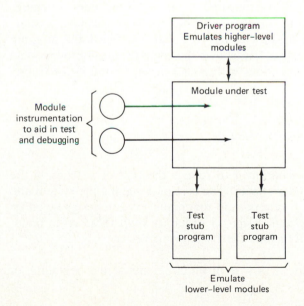

Figure 4.15 Software tools used to test program modules.

4.10.3 Test Software: Drivers and Stubs

The designer normally must design some additional software in order to test the module. A *driver* program (see Fig. 4.15) is written to emulate the higher-level module that calls the module under test. The driver program duplicates the module calling sequence and passes any required data to the module under test. *Test stubs* (see Fig. 4.15) are written to emulate lower-level modules that are used by the module under test. The complexity of these test stubs varies. In some cases, it is a simple program statement to print out a message that the lower-level module has been successfully entered. In other cases, the stub returns some test data to the module under test. In some situations, the test stub simulates a hardware module in considerable detail so that a software module can be tested prior to the integration of software and hardware.

4.10.4 Program Instrumentation

The designer also *instruments* the module under test in order to generate a *trace* of its execution. A *trace* is a listing of the key program statements that are executed and of the values of important internal variables. This listing allows the designer to follow the execution path of the program for a particular combination of test data. Instrumenting the module basically consists of adding program statements to print out messages indicating which branch the program is executing after every branch in the program and adding program statements to print out the values of important variables. These statements are removed once the module testing is completed. The program trace is obviously useful in debugging the program, since the designer can see where the program fails. However, the trace is also used to determine whether the test data set causes the program to execute all the possible branches in the flowchart. If some branches are missed, extra test-data elements are added to cause the program to execute these branches. Figure 4.16 shows the temperature-sensor module we have discussed in previous sections with added instrumentation for testing and debugging.

SUMMARY

Advances in technology have continued to reduce the cost of purchasing the hardware in a real-time computer system. The cost of designing the system and its software has not declined. It is important that managers and designers adopt an organized approach to system design in order to keep these costs under control. The approach should include the basic principles of

1. Performing a detailed analysis of system requirements at the beginning of the design

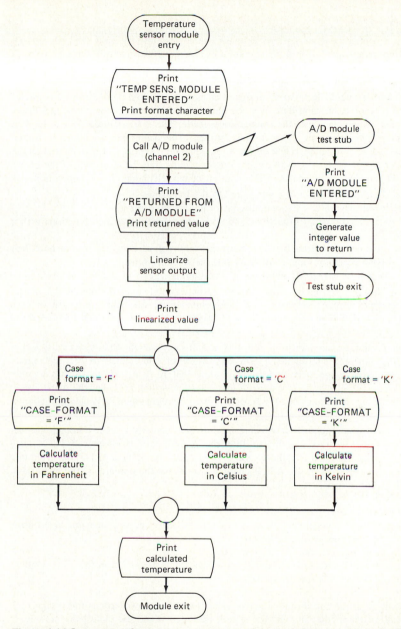

Figure 4.16 Instrumented temperature sensor module used during testing and debugging.

2. Using a top-down design methodology when possible
3. Dividing the design into modules
4. Using structured programming techniques in the design of software modules
5. Using design reviews and testing to verify that a design meets its specifications

EXERCISES

4.1 Determine the requirements for a microcomputer-based elevator-control system. Write a system-specification document for the system and draw a block diagram of the system showing the major elements of the system and the signals flowing between these elements.

4.2 An automatic telephone dialler is to be designed as part of a remote data-acquisition system. When the system is ready to transmit data over the phone lines to a central computer, the dialler dials a preprogrammed number. Once the connection is made, data is transmitted. When the data transfer is complete, the telephone connection is broken. The dialler has an interface to the telephone line that allows it to take the line off the hook and send dial pulses or tones to the phone system. The dialler can detect dial tones, busy and ringing signals, and the answering of the phone by the central computer. Draw a *detailed* ASM diagram of the telephone dialler's operation.

4.3 Continue the design of the elevator controller. Do the following;

(*a*) Draw an ASM diagram for the overall control flow in the system.

(*b*) Break the system down into functions and show the relations between functions with a tree diagram.

(*c*) Make a preliminary division between hardware and software in the system and then estimate the size of the program required to carry out the software functions.

(*d*) Identify the functions that are involved in time-critical aspects of the system.

4.4 Obtain listings of several programs (the longer, the better) and count the number of distinct operators and operands in each program. Use these numbers to calculate the length of each program according to formulas given in this chapter. Compare the results with the actual lengths of the programs.

4.5 Part of a controller for an industrial robot is a software module that converts a set of coordinates in the spherical coordinate system into a set of coordinates in the rectangular coordinate system. Write a specification for the module that is clear enough so that another programmer could use the specification as a user's manual when writing programs that make use of the module. Write the specification using the terminology of a programming language that you are familiar with.

4.6 A data-acquisition system requires a software module that converts a positive integer number into an array of ASCII character codes so that the number can be displayed on a small alphanumeric display incorporated into the system. For example, the integer 132 is converted into a three-element array consisting of the ASCII character codes for the numeral 1, the numeral 3, and the numeral 2. Other specifications are:

The maximum value of the integer is 65536.
Inputs to the module are the integer variable or constant to be converted and the name of the array in which the character codes are to be stored.
The array must be large enough to hold the character codes.
The module places the character codes in the array so that the first element in the array contains the code for the most-significant digit of the integer.

Design the module from the specifications above using structured flowcharts as the design tool. Design a test plan for the module and list the sample data you would use to test the module.

REFERENCES

1. Shooman, M. L.: *Software Engineering*: *Design, Reliability, and Management*, chap. 3, McGraw-Hill, New York, 1983.
2. Kernighan, B. W., and P. J. Plauger: *The Elements of Programming Style*, 2d. ed., McGraw-Hill, New York, 1978.

BIBLIOGRAPHY

System Specification and Requirements Analysis

DeMarco, T.: *Structured Analysis and System Specification*, Yourdon Press, New York, 1978.

Haase, V. H., and G. R. Koch (eds.): "Special Issue on Applications Oriented Specifications," *Computer*, vol. 15, no. 5, May 1982.

Ross, D. T. (ed.): "Special Collection of Requirement Analysis," *IEEE Trans. on Software Engineering*, vol. SE-3, no. 1, January 1977.

Thurber, K. J.: *Tutorial: Computer Design Requirements*, IEEE Computer Society, Los Alamitos, CA, 1980.

System Design

Allworth, S. T.: *Introduction to Real-Time Software Design*, Springer-Verlag, New York, 1981.

Bergland, G. D., and R. D. Gordon (eds.): *Tutorial: Software Design Strategies*, IEEE Computer Society, Los Alamitos, CA, 1979.

Carson, J. H. (ed.): *Tutorial: Design of Microprocessor Systems*, IEEE Computer Society, Los Alamitos, CA, 1979.

Freeman, P., and A. I. Wasserman (eds.): *Tutorial on Software Design Techniques*, IEEE Computer Society, Los Alamitos, CA, 1977.

Jensen, R. W., and C. C. Tonies: *Software Engineering*, Prentice-Hall, Englewood Cliffs, NJ, 1979.

Myers, G. J.: *Composite/Structured Design*, Van Nostrand Reinhold, New York, 1978.

Ogdin, C. A. (ed.): *Tutorial: Microcomputer System Design and Techniques*, IEEE Computer Society, Los Alamitos, CA, 1980.

Estimating Software-Development Costs

Boehm, B. W.: *Software Engineering Economics*, Prentice-Hall, Englewood Cliffs, NJ, 1981.

Brooks, F. P.: *The Mythical Man Month: Essays on Software Engineering*, Addison-Wesley, Reading, MA, 1974.

Christensen, K., et. al.: "A Perspective on Software Science," *IBM Systems Journal*, vol. 20, no. 4, 1981.

Putnam, L. H. (ed.): *Tutorial: Quantitative Management: Software Cost Estimating*, IEEE Computer Society, Los Alamitos, CA, 1977.

Program Design

Aho, A. V., J. E. Hopcroft, and J. D. Ullman: *The Design and Analysis of Computer Algorithms*, Addison-Wesley, Reading, MA, 1977.

Basili, V., and T. Baker (eds.): *Tutorial: Structured Programming*, IEEE Computer Society, Los Alamitos, CA, 1977.

Bentley, J. L.: *Writing Efficient Programs*, Prentice-Hall, Englewood Cliffs, NJ, 1982.

Dahl, O. J., E. W. Dijkstra, and C. A. R. Hoare: *Structured Programming*, Academic, Orlando, FL, 1972.

Metzger, P. W.: *Managing a Programming Project*, 2d ed., Prentice-Hall, Englewood Cliffs, NJ, 1981.

Williams, G.: "Structural Programming and Structural Flow Charts," *Byte*, March 1981, pp. 20–23.

Wirth, N.: *Algorithms + Data Structures = Programs*, Prentice-Hall, Englewood Cliffs, NJ, 1976.

Debugging and Testing

Bruce, R. C.: *Software Debugging for Microcomputers*, Reston, Reston, VA, 1980.
Huang, J.C.: "Instrumenting Programs for Symbolic Trace Generation," *Computer*, vol. 13, no. 12, Dec. 1980, pp. 17–23.
Myers, G. J.: *The Art of Software Testing*, Wiley, New York, 1979.

PROGRAMMING LANGUAGES

Programming languages run in a hierarchy from machine language at the very lowest level to applications-oriented high-level languages at the top. One of the choices that must be made in the design of a real-time computer system is the choice of the appropriate language level. Once the language level is chosen, there may be several different programming languages available at that level. The designer must then choose the language most suitable for the project.

5.1 MACHINE AND ASSEMBLY LANGUAGE

5.1.1 Machine-Language Programming

The use of machine language, as discussed in Chap. 2, involves programming the computer directly with the binary patterns that correspond to its internal representation of program instructions and data. The location of every data element must be specified and the address for every program branch calculated. The following program is written in the machine language of the Intel 8085 microprocessor:

```
00111010
00000000
11100000
01010111
00111010
00000001
11100000
10010010
00110010
00000010
11100000
11000011
00000000
00000001
```

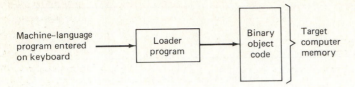

Figure 5.1 Machine-language programming with the aid of a loader program.

The program takes two 8-bit numbers from memory, subtracts one number from the other, stores the result in memory, and then jumps back to the beginning of the program. Obviously this program is difficult to read. It is also difficult to write this program without making an error, and the work of looking up the binary codes for the instructions and calculating the addresses is boring. A machine-language program of this type can be loaded into some computers via front panel switches. The loading process is even more tedious than writing the program. Other computers have loader programs (see Fig. 5.1) that accept data in hexadecimal format entered from a keyboard and place it in specified memory locations.

5.1.2 Assembly-Language Programming

The use of machine language is almost never justified, since *assembler* programs, which allow programs to be written in *assembly language*, are available for most computers. Assembly language has the computer take over most of the mechanical details of creating machine-language programs. Each basic instruction that the computer can execute is given a short name, called a *mnemonic*. Mnemonics are also assigned to the registers in the computer. The programmer can assign names to memory addresses or data that are used in the program. The programmer writes the program using these mnemonics and names and enters it into a text file in the computer using a text editor program. This program is called the *source program*, or the *source code*. The text file containing the source program is then processed by the assembler program to produce the machine-language program, called the *object program*, or *object code*, which can be executed by the computer (see Fig. 5.2).

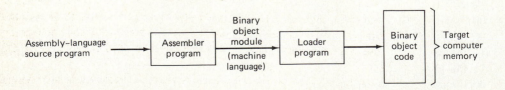

Figure 5.2 The assembly process.

The assembly-language equivalent of the machine-language program in the previous example is written as follows:

```
;Any text following a semicolon on a line (like the text you're
;reading now) is a comment which is
;ignored by the assembler program but is helpful to humans
;
;Program to subtract the byte stored in VAR1 from the byte stored
;in VAR2 and store the result in DIFF
;
;******The following section is used by the programmer to define some names***
;
DATA        EQU    0E000H      ;define a name for the first data address
PROGRAM     EQU    100H        ;define a name for the first program address
;
;****The following section (the code segment) contains the actual program****
;
            ORG    PROGRAM     ;tell the assembler to place the following
                              ;in the program area of memory starting at
                              ;the address represented by PROGRAM
START:      LDA    VAR1        ;get byte in memory location VAR1 and put it
                              ;in the A register
            MOV    D,A         ;move the contents of the A reg to the D reg
            LDA    VAR2        ;now get byte in location VAR2 into A reg
            SUB    D           ;subtract contents of D reg from A reg with
                              ;result remaining in A reg
            STA    DIFF        ;store contents of A reg in memory location DIFF
            JMP    START       ;jump to the memory location START
;
;****The following section (the data segment) is used to assign memory******
;*************locations for program variables************************
            ORG    DATA        ;tell the assembler to place the following
                              ;in the data area of memory starting at
                              ;the address represented by DATA
VAR1:       DB     15H         ;tell the assembler to place the number 15
                              ;(hexadecimal) into a memory location
                              ;and give the address of the location the name VAR1
VAR2:       DB     68H
DIFF:       DS     1           ;tell the assembler to reserve one byte of memory
                              ;for data and name the address of the byte DIFF
;
            END
```

This program is considerably easier to read and write than the machine-language program, particularly after one has gained some familiarity with the assembly language of the Intel 8085 microprocessor.

5.1.3 Format of Assembly-Language Programs

An assembly-language program is made up of lines of text. Each line is either a

- Comment statement which will be ignored by the assembler
- *Directive*, or *pseudoinstruction*, to the assembler program which gives the assembler information required to translate the assembly-language program into machine language.
- Microprocessor instruction which will be translated into machine language

In most assembly languages each line is divided into four areas or *fields*, as follows:

Label field	Mnemonic field	Operand field	Comment field
EXAMPLE:	MOV	A,D	;move reg D to reg A

The label field is optional. It allows the programmer to give a name to an instruction. This name can subsequently be used by another instruction to address the labeled instruction. For example, in the subtraction program, the JMP START statement causes a transfer of program execution to the instruction labeled START. If we didn't have a label facility, we would have to calculate the address of the instruction we wish execution to branch to.

Mnemonic field The mnemonic field contains the mnemonic for a computer operation which will be translated into machine language. For example, if the JMP mnemonic was put in the mnemonic field of an Intel 8085 assembly-language program, the assembler would insert the binary *operation code* for the 8085 "jump to memory location" instruction into the machine-language program (see Fig. 2.2).

Alternatively, the mnemonic field may contain the *pseudo-op* for an assembler directive. A pseudo-op is a name for an operation to be performed by the assembler program. It is called a pseudo-op because it resembles a normal computer instruction mnemonic but is not part of the computer instruction set and is not translated directly into machine language. For instance, in the assembler directive

DB   68H

the pseudo-op DB (define byte) in the mnemonic field directs the assembler to convert the number 68 hexadecimal in the operand field of the directive into an 8-bit binary number and place it into the machine-language program. This pseudo-op allows the assembly-language programmer to store constants in memory locations. A similar pseudo-op, DW, allows the programmer to specify constants larger than FF hexadecimal. These are converted into 16-bit binary numbers and stored in two successive memory locations.

In the example program, LDA, MOV, and SUB are examples of mnemonics for operations performed by the Intel 8085 microprocessor. EQU, DS, and ORG are examples of pseudo-ops for operations to be performed by the assembler program.

Operand field The operand field in an assembly-language instruction contains information needed to complete the computer instruction or assembler directive. For instance, it may specify the memory location where data to be operated on by the instruction is stored. For example, the Intel 8085 SUB instruction subtracts the contents of a register from the *A* register and stores the result in the *A* register. The register to be subtracted must be specified in the operand field. Thus,

SUB   C

will subtract the contents of the *C* register from the *A* register.

Comment field The comment field, which is optional, allows the programmer to add explanatory text to the program. Comments are ignored by the assembler program but are printed out in program listings.

5.1.4 The Assembly Process

An assembler program normally makes two *passes* through the assembly-language source program. During the first pass it does no translation; it simply scans the source program for names defined by the programmer and determines what numerical *value* is to be substituted for each name. Then on the second pass, it translates the assembly-language source program into the machine-language object program.

The assembler program maintains a *location counter* and a *symbol table* to help it translate the assembly-language source program into machine language. The location counter is used to keep track of where the machine-language program will be located in the computer's memory. The symbol table is used to store the names defined by the programmer and their corresponding numerical values. The symbol table is filled on the first pass of the assembler program. When the assembler encounters a name on the second pass, it looks the name up in the symbol table and gets its numeric value to substitute into the machine-language program it is producing.

When the assembler makes its first pass through the example program it begins with an empty symbol table and with the location counter set to 0000. It first encounters the two statements

DATA   EQU   0E000H
PROGRAM EQU   100H

The EQU pseudo-op allows the programmer to assign a name to a numeric constant. During the second pass of the assembler, every occurrence of the name in the source program will be replaced by the value of the numeric constant. During this first pass, the assembler simply puts both the name, given in the label field, and its assigned numeric value, given in the operand field, into the symbol table.

The assembler then encounters the

ORG PROGRAM

assembler directive. This directive commands the assembler to set its location counter to the value of PROGRAM, which is 100 hexadecimal. The ORG pseudo-op is used to set the address of the first memory location (*origin*) of a section of the machine-language program. Therefore, in the example, the machine-language program will be stored in memory locations starting at 100 hexadecimal.

The assembler now proceeds to the

```
START:  LDA VAR1
```

instruction. This is the first Intel 8085 instruction encountered. The label START is stored in the symbol table and is assigned the current value of the location counter (100 hexadecimal). Since the location counter keeps track of the address where an instruction is stored, the label START is assigned the address of the instruction it is associated with. We can see that the rules for assigning values to labels are quite simple:

1. If there is an EQU pseudo-op in the mnemonic field, the value of the label is the numeric constant in the operand field.
2. Otherwise, the value of the label is the current value of the location counter.

The assembler doesn't generate machine-language code for the LDA VAR1 instruction during this first pass. One reason it doesn't is because the name VAR1 isn't in the symbol table yet. This is an example of a *forward reference*, where a name is used in an assembly-language program before it has been assigned a value. Since forward references are common, machine-language code generation occurs on the assembler's second pass through the source program once all the symbolic names in the program have been entered into the symbol table. On the first pass, the assembler simply allocates space (three memory locations) for the LDA instruction by adding 3 to the contents of the location counter. The assembler allocates space for the rest of the executable instructions in the program in the same fashion.

Once the assembler has passed through all the Intel 8085 instructions in the program, it encounters the

```
ORG    DATA
```

assembler directive. This forces the location counter to be set to the value of the symbol DATA (E000 hexadecimal), which is the first address of the memory area where the data used by the program will be stored. If the program writes data to this area, then the area must be composed of read/write (RAM) memory.

The assembler now arrives at two directives of the form

```
label: DB    numeric constant
```

On the first pass, these directives simply cause the assembler to put the label in the symbol table and assign it the current value of the location counter and then increment the location counter. On the second pass, the assembler will place the binary equivalent of the numeric constant into the object code it generates. The net effect is to place a number into a memory location and assign a name to the address of the location so that the number can be accessed by a symbolic name within the assembly-language program.

Next the assembler encounters the

```
DIFF:  DS  1
```

assembler directive. During the first pass, the assembler puts the label DIFF into the symbol table and assigns it the current value of the location counter, which is E002 (hexadecimal). The assembler then adds the contents of the operand field (1 in this case) to the location counter. No action is taken by the assembler on the second pass. The effect of the assembler directive is to reserve one memory location for a data item and assign a name to the location so that the data can be accessed by a symbolic name within the assembly-language program. In the subtraction program, this location is used to store the result of the subtraction. By putting a number greater than 1 into the operand field, larger blocks of memory can be reserved for data storage.

Finally, the assembler reaches the END pseudo-op, which signals the assembler that there are no more lines to process. The symbol table is now complete, so the assembler makes the second pass through the program in which it generates the actual machine-language program.

A listing showing the machine-language program and symbol table generated by the assembler is shown below.

Location counter contents	Object code	Source code		
0000		DATA	EQU	0E00H
0000		PROGRAM	EQU	100H
		;		
0100			ORG	PROGRAM
0100	3A E000	START:	LDA	VAR1
0103	57		MOV	D,A
0104	3A E001		LDA	VAR2
0107	92		SUB	D
0108	32 E002		STA	DIFF
010B	C3 0100		JMP	START
		;		
E000			ORG	DATA
E000	15	VAR1:	DB	15H
E001	68	VAR2:	DB	68H
E002		DIFF:	DS	1
		;		
		END		

```
Symbol table
E000  DATA      E002  DIFF      0100  PROGRAM
0100  START     E000  VAR1      E001  VAR2
```

Examining one line of the listing,

```
0104     3A E001          LDA     VAR2
```

we find that the left-most column contains the contents of the location counter. It specifies the memory address (104 hexadecimal) where the first byte of the LDA VAR2 instruction will be stored. The next two columns (3A E001) contain the machine-language instruction in hexadecimal format. The code for the LDA instruction is 3A and the value of the VAR2 operand is E001, requiring a total of 3 bytes of memory to store the instruction.

Besides producing a listing like the one given here, the assembler program will also create a file containing only the machine-language program and the memory addresses where the program is to be stored. This *object module* can then be loaded into the correct area of memory by a loader program.

5.1.5 Relocating Assemblers

The assembler program we have been discussing is an example of an *absolute* assembler. With an absolute assembler, the programmer must specify where the program and data will be stored by the use of ORG directives. In addition, all names must be defined within the program, either implicitly from context or explicitly by use of the EQU directive.

These requirements are inconvenient when a large program is being developed in a modular fashion by several programmers. All the ORG directives in the different modules must be coordinated so that there will be no overlap of storage locations. If a module refers to a name defined in another module, the module cannot be assembled alone, since the assembler has no way of determining the value of the name defined in the other module. The assembler can look up only the values of names in the symbol table for the module it is currently assembling. Concatenating all the modules into one program prior to assembly to avoid this problem is inconvenient, since a change in one module then requires reassembling the entire program. Also, all the modules must be completed before an assembly can be attempted.

A *relocating assembler* makes the development of large assembly-language programs much easier. With a relocating assembler the generation of the machine-language program is divided into two phases: the assembly phase and the *linking* phase. The programmer divides a program module into a *code segment*, where the computer instructions are written, and a *data segment*, where space is reserved for the data items; but ORG directives are not used to specify where these are located in memory.

The programmer uses a PUBLIC or GLOBAL pseudo-op to specify any names defined in the module that will be referred to by other modules. For example, the directive

```
GLOBAL MOD1
```

indicates that the name MOD1 whose numeric value is defined within the present module will be referred to by other modules in the program. The programmer uses an EXTERNAL pseudo-op to specify names used within the module that are defined in another module. Thus,

EXTERNAL MOD2

indicates that the name MOD2 is used in the present module but has its value defined in another module in the program. A name that is marked GLOBAL in the module where its numeric value is defined will be marked EXTERNAL in all the other modules which refer to that name. During the link phase, when an EXTERNAL name is encountered in a module, the numeric value of the corresponding GLOBAL name is substituted.

The assembly phase generates a *relocatable object module* in which all addresses and label values are relative to the beginning of the code and data segments. Thus the first instruction in the module will have an address of 0000 with respect to the beginning of the code segment, and the first data item in the program will have an address of 0000 with respect to the beginning of the data segment. The object module also contains special markers for symbols which are GLOBAL or EXTERNAL.

The relocatable object module must first be processed by the linker program before the program can be loaded and executed. Figure 5.3 shows the stages in the development of a relocatable assembly-language program.

During the link phase, the programmer specifies which modules are to be included into the final program and the beginning addresses where the program and data are to be stored in memory. The *linker program* concatenates the modules together into one module with one code segment and one data segment. It recalculates the addresses and labels within the module so that the code and data segments start at the addresses specified by the programmer. Then the linker matches EXTERNAL names with GLOBAL names to determine the values of the EXTERNAL names. The linking process is illustrated in Fig. 5.4. The object module produced by the linker program can then be loaded using a *loader program* and executed.

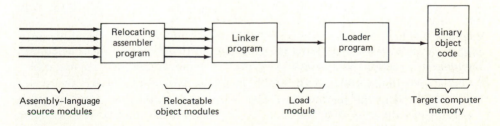

Figure 5.3 Assembly with a relocating assembler.

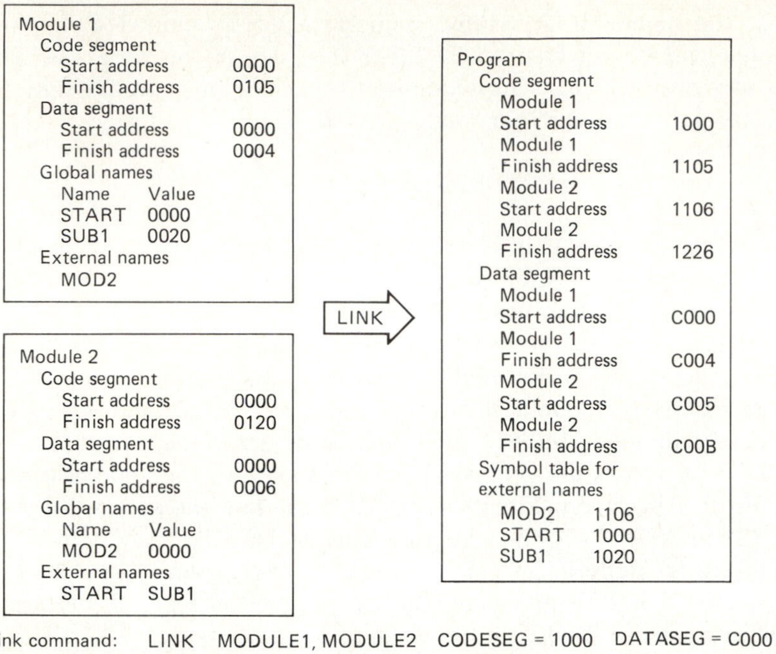

```
Module 1
   Code segment
      Start address     0000
      Finish address    0105
   Data segment
      Start address     0000
      Finish address    0004
   Global names
      Name     Value
      START    0000
      SUB1     0020
   External names
      MOD2
```

```
Program
   Code segment
      Module 1
      Start address      1000
      Module 1
      Finish address     1105
      Module 2
      Start address      1106
      Module 2
      Finish address     1226
   Data segment
      Module 1
      Start address      C000
      Module 1
      Finish address     C004
      Module 2
      Start address      C005
      Module 2
      Finish address     C00B
   Symbol table for
   external names
      MOD2     1106
      START    1000
      SUB1     1020
```

LINK ⟹

```
Module 2
   Code segment
      Start address     0000
      Finish address    0120
   Data segment
      Start address     0000
      Finish address    0006
   Global names
      Name     Value
      MOD2     0000
   External names
      START   SUB1
```

Link command: LINK   MODULE1, MODULE2   CODESEG = 1000   DATASEG = C000

Figure 5.4 Example of a linking operation where two modules are linked to form a program with a code segment starting at memory location 1000 (hex) and a data segment starting at memory location C000.

5.1.6 Native and Cross Assemblers

Assembler programs can be either *native assemblers* or *cross assemblers*. A native-assembler program executes on the same type of computer for which it produces machine-language code. Thus a native assembler for the Intel 8085 microprocessor will run on Intel 8085 based microcomputers and produce machine-language code for the 8085. A cross assembler runs on a different type of computer than the one for which it generates machine-language code. A cross assembler for the Intel 8085 microprocessor might run on a large main-frame computer but produces machine-language code for the Intel 8085.

5.1.7 Macroassemblers

Macroassemblers represent useful upward extensions of basic assembler programs. A macroassembler allows the programmer to define a group of assembly-language instructions as a *macro* by giving it a name. Every time the assembler comes across that name in an assembly-language program, it inserts the group of assembly-language instructions. The programmer can thus emulate more powerful instructions that are not in the computer's instruction set. For instance, the subtraction program we have been using as an example, minus the JMP statement that causes continuous looping, can be defined as a macro as follows:

```
SUBMEM  MACRO     VAR1, VAR2, DIFF

        LDA       VAR1
        MOV       D,A
        LDA       VAR2
        SUB       D
        STA       DIFF

        ENDM
```

In the preceding macro definition, the names VAR1, VAR2, DIFF are called dummy names. When the macro is used in a program, any other name can be substituted for the dummy name. After the macro has been defined in a program, it can be used by entering a program statement such as

```
SUBMEM    DATA1, DATA2, DATA3
```

which will cause the assembler to generate machine-language code for

```
LDA    DATA1
MOV    D,A
LDA    DATA2
SUB    D
STA    DATA3
```

To the programmer it appears that the 8085 has a new instruction, SUBMEM, which subtracts two numbers in memory and stores the result in memory.

5.1.8 Structured Assemblers

A *structured assembler* adds the constructs of structured programming such as IF-THEN-ELSE and WHILE-DO to the basic instruction set of the computer. These constructs are often added through the use of macros. The subtraction example written in a structured assembly language might look something like

```
LOOP:
        A = M(VAR1)
        D = A
        A = M(VAR2)
        A = A - D
        M(DIFF) = A

REPEAT(LOOP)

VAR1:   STORE 15H
VAR2:   STORE 68H
DIFF;   RESERVE 1
```

5.1.9 Limitations of Assembly Language

Even structured assemblers have severe limitations from the viewpoint of ease of programming. The data types supported by assemblers are limited to those recognized by the machine, typically binary numbers of a few different lengths. More complex data types such as real (floating-point) numbers or arrays must be created and managed by the programmer. The programmer must constantly deal with the low-level details of the computer. For example, to perform a simple operation such as subtraction of two variables, we had to move the variables from memory to the computer's registers, shuffle the variables between registers, perform the subtraction, and then move the result from a register back to memory. The details of passing data to functions (subroutines) are also up to the programmer. This can cause problems with consistency when several programmers are working on the same project. Finally, an assembly language inevitably reflects the architecture and instruction set of the computer for which it produces machine language. Assembly language is therefore not *portable*. An assembly-language program written for one type of computer will not run on another type. For example, the subtraction program written in the assembly language for the Motorola 68000 microprocessor looks like this:

```
DATA        EQU         $E000
PROGRAM     EQU         $100

            ORG         PROGRAM

START       MOVE.B      VAR1, DO
            MOVE.B      VAR2, D1
            SUB.B       DO, D1
            MOVE.B      D1, DIFF
            BRA.S       START

            ORG         DATA

VAR1        DW.B        $15
VAR2        DW.B        $68
DIFF        DS.B        1
```

In a longer and more complicated program, the differences between Intel 8085 assembly language and Motorola 68000 assembly language would be even more pronounced, since the Motorola 68000 has many instructions for which the Intel 8085 has no equivalent.

5.2 HIGH-LEVEL LANGUAGES

High-level languages such as C, Pascal, or BASIC are designed to solve some of the problems of programming in assembly language. A high-level language supports a variety of control and data structures and handles the details of "mechanical" operations such as maintaining index variables in loops or passing parameters to functions. As a result, high-level language programs are easier to write and can be written faster than assembly-language programs. For instance, the program to subtract two 8-bit numbers, when written in the C high-level language, looks like this:

```
#define ALWAYS 1
char var1 = OX15, var2 = OX68, diff;
main( )
{
while(ALWAYS)
        diff = var2 – var1;
}
```

A high-level language program will run on different types of computers as long as a translation program is available to convert the high-level language *source code* to machine-executable form. As a result, programs written in a high-level language are said to be more portable than programs written in assembly language. For example, the C program above can run on both Intel 8085 microprocessors and Motorola 68000 microprocessors, since translation programs are available for both machines.

5.2.1 Compiled, Interpreted, and Intermediate Code Languages

A number of methods are available to carry out this translation process. A *compiled* high-level language program is translated into machine or assembly language by a *compiler* program. The end result of the translation process is machine-language *object code* which is stored in the memory of the computer on which the program is to run (the *target* computer) (see Fig. 5.5).

In an *interpreted* high-level language, the high-level-language source code is stored in the target computer's memory. A machine-language *interpreter* program also stored in the target machine's memory *interprets* and executes the stored source code line by line (see Fig. 5.6).

An interpreted program will run more slowly than a compiled program, since the program must be both translated and executed each time it is run. On the other hand, interpreted programming languages are said to be more *interactive* than compiled languages. Portions of interpreted programs can be

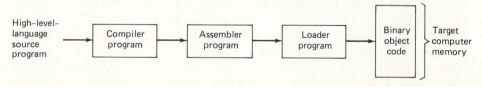

Figure 5.5 The compilation process.

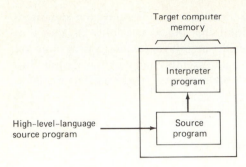

Figure 5.6 Interpreted high-level language.

run for testing and debugging purposes, and the execution of the program can be stopped at any time with the program variables available for examination. The BASIC language is an example of a language that is commonly interpreted (although compiled versions are available), while FORTRAN is an example of a language that is commonly compiled.

In some cases both compilers and interpreters are available for a language. In such a case the programmer can develop a program using an interpreter, which makes debugging easier, and when development is finished, the program can then be compiled to produce object code for the target system. This technique is becoming increasingly popular with users of the C high-level language since both compilers and interpreters are now available for that language.

A combination of compilation and interpretation is used in high-level-language systems based on an *intermediate level of code*. The basic idea of this technique is that a compiler is designed with an imaginary abstract machine as the target computer (see Fig. 5.7). The output of this compiler is the "machine language" for the imaginary machine. This is called *intermediate code*. Interpreters are then written to interpret the intermediate code on the various real target machines (path 1 in Fig. 5.7).

This technique has several claimed advantages for cases where a high-level language is to run on many different machines. The compiler portion of the translator is the most complex part; it must be written only once. The interpreters, which must be written for each different type of target computer, are relatively easy to write, since the intermediate code is fairly close to machine language already. Also, since the interpretation process is fairly simple, programs will run more quickly than is the case where a high-level language is interpreted directly.

The intermediate code can be compiled to machine language for the target computer by a *native code generator* in cases where maximum execution speed is desired (path 2 in Fig. 5.7). It is also possible to design a real machine to directly execute the intermediate code (path 3 in Fig. 5.7). The best-known example of an intermediate language system is the UCSD (University of California at San Diego) p system, which allows Pascal and FORTRAN to run on a variety of microcomputers and minicomputers.

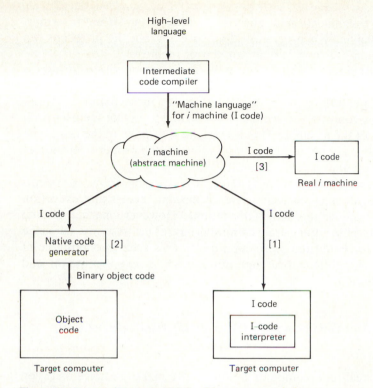

Figure 5.7 Intermediate-code high-level-language translation.

5.2.2 Systems Programming Languages and Applications Languages

When considering high-level languages for real-time applications, you will find it useful to make the distinction between *systems programming languages* and *applications languages*. A systems programming language is intended to produce programs for use in the computer system, such as operating systems, I/O device drivers, assemblers and compilers, and editors. Languages such as C and BCPL are commonly considered to be systems programming languages. An applications language, such as COBOL or APL, is intended for the user of a computer and is designed to allow the development of programs to solve problems in areas such as business, numerical analysis, or computer graphics.

A systems language is designed for maximum efficiency and will support data types and operators that are close to those supported by the machine. An applications language is designed to make the applications program easy to write and will support data types and operators that are useful in the application area. As an example, a system language such as C supports the array data type but supports operations only on individual elements of the array. An applications language for numerical analysis applications, such as APL, not only allows operations on elements of an array but also has operations such as assignment, multiplication, and addition of entire arrays. This makes numerical-analysis programs easier to write, but the programs run quite slowly and require a considerable amount of memory.

A systems programming language normally has some provision for access to physical features of the computer, such as the input/output system, since programs such as operating systems require access to these features. Applications languages, by contrast, are specifically designed to isolate the user from the physical aspects of the machine. The desire here is to prevent the user, who is not expected to be a computer expert, from affecting the proper operation of the computer or operating system.

Normally, for real-time computer systems, it is desirable to use a systems programming language, since considerations of machine access and execution efficiency are important in real-time applications. However, many real-time applications also require numerical computations. The systems language chosen in these cases should therefore support features such as floating-point calculations and basic mathematical operations such as exponentiation and trigonometric functions.

5.3 CHOOSING BETWEEN ASSEMBLY AND HIGH-LEVEL LANGUAGES

The designer of a real-time computer system must make a choice between using assembly language or a high-level language to implement the software portions of the system. In the past, designers have usually chosen to use assembly language, but the trend is now toward the use of higher-level languages. We believe that this trend will continue to the point where high-level language implementations will predominate.

5.3.1 Advantages of High-Level Languages

There are some very good reasons why a high-level language rather than assembly language should be used in a real-time computer system. The reasons can be summarized as

- Increased software productivity
- Increased software quality
- Increased software portability

Software productivity Software productivity refers to the cost in time and money for the development of the software portion of a real-time system. In applications where less than a few hundred systems are to be produced, the software-development costs are the predominant cost in the computer portion of the system. A common industry rule of thumb is that each completely debugged and documented program statement costs from $10 to $20. These figures apply whether the program statement is an assembly-language statement or a high-level-language statement. However, each high-level-language statement typically replaces four or five assembly-language statements. Each high-level-language statement is much more cost-effective than is each assembly-language statement. As an example, a relatively small application might require 2000 assembly-language statements costing $20,000 to $40,000 to develop. The equivalent program in a high-level language might require only 500 statements and cost only $5000 to $10,000 to develop. The high-level language program will also be completed sooner than the equivalent assembly-language program. This can be a significant advantage in a competitive market.

Software quality   Software quality refers to some of the design issues discussed in Chap. 4. In order to produce software that is modular, uses structured control constructs, and is easy to test and maintain, it is desirable that the programming language support program modules (functions or procedures) and structured control constructs and that it be understandable. If the language produces source statements that are understandable in the sense that they allow the reader to determine what overall purpose each statement is to accomplish, then the program will be easier to verify, test, and maintain. Assembly language is notoriously bad on all these points. Modules are not well-supported, particularly in terms of "hiding" (preventing access to) information within modules. At best, structured control constructs are supported in a clumsy fashion. Finally, assembly-language programs are very difficult to read in terms of determining what the program is supposed to accomplish. On the other hand, high-level languages are available which do support the writing of high-quality, understandable software.

Software portability   Software portability refers to the ability to run a program written in one computer language on different types of computers. As already discussed, assembly languages are so tied to particular machine architectures and instruction sets that they are almost completely nonportable. High-level languages are far more portable. While it is unfortunately rather common that differences exist among implementations of a language for different computers, the differences are normally small and the conversion process is relatively simple. Portable software allows a user to reuse expensive software when hardware is upgraded to new computers. Also, useful modules developed for one system can be used in another system based on a different type of computer.

A related point is that in a design team using high-level languages and a modular approach to the design, many of the designers won't have to spend a lot of time learning about the details of the instruction set and assembly language of every new computer that is used. Only the designers writing modules that directly make use of the machine-language instructions will have to learn these details.

5.3.2 Disadvantages of High-Level Languages

Given the advantages of high-level programming languages that we have just described, why would any designer of a real-time computer system choose to program in assembly language? The reasons given usually include some of the following points:

1. Programs written in high-level languages run too slowly for real-time applications.
2. Programs written in high-level languages require too much memory for program storage.
3. The machine-level operations (e.g., I/O) required can't be done from a high-level language.
4. No high-level language is available for the microcomputer used.

Execution speed penalty   As already discussed, high-level language programs that are interpreted run slowly because the program must be translated line by line every time it is executed. As a result, most high-level languages used in real-time applications are compiled. However, it is also true that the compiler for a high-level language will usually produce a machine-language program that is longer than that produced from the equivalent assembly-language program written by a good assembly-language programmer. Since the program is longer, it will run slower and take up more memory space.

Memory-space penalty   Compilers produce longer machine-language programs for two reasons. First, high-level language operations often bear little relation to the basic operations that can be carried out by the computer. As an example, the C programming language requires that all arithmetic operations be carried out on data that is at least 16 bits long. Thus, to produce machine-language code for the C version of the subtraction program discussed earlier in this chapter, the compiler must add machine instructions to expand the 8-bit data to 16-bits by adding leading zeroes. Then a 16-bit subtraction must be performed which requires several instructions on an Intel 8085 since it has only an 8-bit subtraction instruction. Finally, the result must be shortened back to 8 bits prior to storing it in memory. The net result is a machine-language program that is considerably longer than the one produced by the assembly-language version of the subtraction program.

The second reason why high-level languages produce longer machine-language programs is that the process of translating a program into machine language is complex, involving a number of decisions and compromises. A compiler must do this automatically and perform the translation for all correctly written programs. As a result, the decisions and compromises made on issues such as the allocation of memory or the registers that should be used may not be the best for a particular program. A programmer writing in assembly language, on the other hand, can make all these decisions based on a detailed knowledge of the program being written. Basically, compiler programs are not yet quite as clever as a good assembly-language programmer.

Mitigating factors Before we give up on high-level languages because they are too slow and require too much memory, we should consider a few points. First, the newer 16- and 32-bit microprocessors such as the Motorola 68000 family and the Intel 8086 family are considerably faster than their 8-bit predecessors. In addition, their instruction sets have been designed to allow more-efficient translation of high-level languages to machine language. Second, the cost of memory is declining rapidly, so the cost penalty of longer machine-language programs is decreasing.

The quality of compilers for high-level languages is also improving. Modern compilers incorporate optimization phases to improve the quality of the machine-language programs they produce. They may not be quite as clever as a good assembly-language programmer but some claim to be at least as clever as an average assembly-language programmer.

Because of these improvements, the speed of high-level language programs is more than adequate for a large number of real-time applications and the cost of the additional memory required is no longer a deterrent in all but the most cost-sensitive applications.

In many cases a greater improvement in execution speed can be obtained by designing a more-efficient program than by writing it in assembly language rather than a high-level language. The book *Writing Efficient Programs*, listed in the bibliography at the end of this chapter, provides some excellent guidelines.

Availability The criticism that high-level languages are not available for a particular computer or that the languages available are not suitable for real-time applications is also not as valid as it once was. For example, the C programming language, which is well-suited to real-time applications, is available on all the major 16-bit microprocessors and is also available for many of the 8-bit microprocessors. In fact, for some machines, C compilers are available from several different vendors. The same situation is true for other popular high-level languages such as Pascal and FORTRAN.

5.4 REQUIREMENTS FOR REAL-TIME HIGH-LEVEL LANGUAGES

5.4.1 Speed and Efficient Use of Memory

The basic requirement for a real-time high-level language is that it produce object code which executes quickly and does not take up too much storage space. How much of a speed and space penalty must be paid when using a high-level language rather than assembly language depends on several factors:

1. Choice of high-level language. Some high-level languages produce more efficient machine language than do others. Languages which are compiled to machine language will almost always run faster than languages which are interpreted or use an intermediate code. Different compilers for the same language also produce object code of widely varying efficiency.
2. Nature of program. Assembly language will have a greater advantage in short, simple programs where the programmer can take advantage of special features of the problem or machine. In bigger programming projects, the need for more structure and for modularization of the program reduces the gains possible by assembly-language "tricks."
3. Experience of programmer. Designing an efficient assembly-language program requires an experienced programmer who is familiar with the machine being programmed. An "average" programmer who is not particularly familiar with the computer may produce assembly-language code that is no more efficient than that produced by a good high-level language translator.
4. Type of computer. Modern 16-bit microcomputers have instructions which assist high-level language translators in producing efficient object code. Thus the advantage of assembly language can be expected to decrease as the new microcomputers are incorporated into real-time systems.

Test results indicate that assembly-language programs run two to three times as fast as the best high-level language programs for short programs running on 16-bit microcomputers. The speed advantage of assembly language reduces to less than two-to-one in large, complex programs.

5.4.2 Benchmarks

Unfortunately, it is difficult to obtain information about which languages and compilers produce the fastest and shortest machine-language programs. No standardized method of rating languages and compilers exists. As a result, conflicting claims are made regarding the quality of machine language produced by various languages and compilers. Test programs, called *benchmark* programs, have been developed to compare the quality of different languages and compilers. However, different benchmark programs produce different comparative ratings because the different programs have a different mix of instructions. For example, one benchmark program might emphasize mathematical operations on floating-point numbers and another benchmark program might emphasize operations on strings of characters. A particular language or compiler may be stronger in character handling than in floating-point math and as a result will perform better in one benchmark than in the other. Therefore, it is important to choose a benchmark program that has a mix of instructions similar to the actual programs that will be written. A programmer planning to develop a word-processing program should choose the string-handling benchmark rather than the floating-point-math benchmark, for example.

One source of information on high-level languages and compilers is magazines aimed at users of small computers. They often contain articles reviewing or comparing compilers. A series of articles in *Byte* magazine [1, 2] used a simple benchmark program to evaluate the performance of several languages and compilers on some popular microcomputers. User groups for different makes of computers also publish performance information on compilers in their newsletters. If the application warrants, it may be worthwhile to purchase several compilers and develop benchmark programs to test them. This also allows the quality of documentation, ease of use, and vendor support to be evaluated.

5.4.3 Interface to Assembly Language

Despite the advantages of programming in a high-level language, it is sometimes necessary to write at least part of a program in assembly language. This may be because for one part of the program speed is critical and it must execute as fast as possible or because some operation must be performed that is difficult or impossible to perform in the high-level language. Therefore, there must be some means of integrating the high-level and assembly-language portions of the program so that data can be passed back and forth.

If the high-level-language compiler produces assembly language in a form suitable for a relocating assembler, the problem is basically solved. The programmer accesses an assembly-language module by performing a function (subroutine) call. Since the compiler can't find a function with that name in the high-level-language program, it enters the function name as an EXTERNAL name in the assembly-language program it produces. The assembly-language program is assembled by the relocating assembler to produce a relocatable object module. The module is linked with other modules generated from assembly-language portions of the program to create the complete machine-language version of the program. During the linking operation, the EXTERNAL function names are replaced with the addresses of the assembly-language modules that are to be called.

In order to properly pass data from a high-level-language module to an assembly-language module, and vice versa, the programmer must know something about how data is represented at the machine level and how a high-level-language function call is translated into machine language. The documentation for the compiler must contain this information if the compiler is to be used with assembly language. We will discuss interfacing high-level language with assembly language in more detail in Chap. 7.

5.4.4 Compiler Utilities

Many high-level language compilers produce assembly language that is compatible with a popular relocating assembler. Other compiler packages include their own relocating assembler so that the user gets a completely integrated program-development system. Figure 5.8 shows a block diagram of such an integrated program-development system. The compiler accepts high-level language modules and produces *assembly-language modules*. The compilation of the program as a set of separate modules is called *separate compilation*. The ability to perform separate compilation is an extremely useful feature, since a programmer can compile and test a module prior to the completion of the entire program. The assembly-language modules produced by the compiler as well as assembly-language modules written by human programmers are assembled into relocatable object modules by the assembler. These modules can then be linked by the linker to produce the final machine-language program.

Many of these systems have a *library* facility. A relocatable object module representing an often-used function can be entered into a special library file by a *librarian* program. During the linking operation, the linker can be comman-

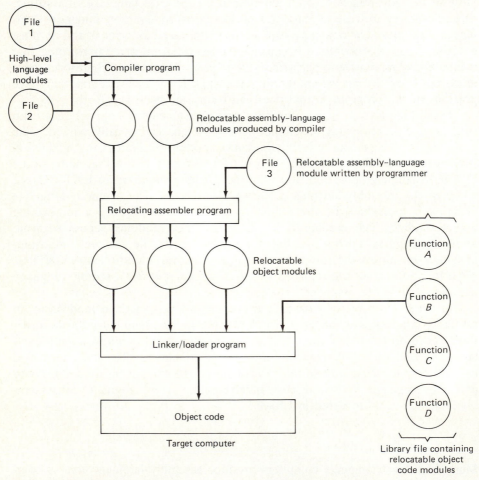

Figure 5.8 High-level-language-compiler system supporting separate compilation, inclusion of assembly-language modules, and library modules.

ded to search the library file for modules whose names match EXTERNAL names in the modules it is linking. If it finds a match, it links that module into the program. The programmer can thus call a function in the library from a high-level language program through a subroutine call. Library files containing standard mathematical and input/output functions are supplied with many compilers. The user is free to add extra library files containing functions useful to the particular application at hand.

5.4.5 Producing ROMable Programs

In microcomputer based systems the memory is commonly divided into read-only (ROM) and read/write (RAM) memory. These two types of memory will be discussed in detail in Chap. 8. The program is stored in the ROM portion of memory and the variables (data) are stored in the RAM portion of memory. The address ranges for the ROM and RAM will vary from system to system. High-level-language programs written for systems of this type must produce machine-language programs in which the program and data areas are separate and the addresses of these areas can be specified. Compilers which produce this type of machine-language program are said to produce ROMable code, meaning that the program can be stored in read-only memory. Compilers producing relocatable assembly language usually produce ROMable code, since relocatable assembly-language programs have separate program and data segments whose starting memory addresses are specified during the linking operation.

5.4.6 Access to the Computer's Hardware

Programs in real-time systems often directly control external devices connected to the computer. As a result, high-level languages used in real-time systems must have provision for access to the input/output system of the computer. This may consist of special instructions in the high-level language or library functions that can be called from the high-level language program.

In many systems, an external device can send a signal, called an *interrupt*, to the computer which causes the computer to suspend execution of the program. Another program, called the *interrupt service routine*, or *interrupt handler*, then executes. It normally carries out some task associated with the external device that sent the interrupt signal. It may accept some data from the device, for example. Once this routine is completed, execution of the suspended program resumes. The computer has methods of ignoring interrupt signals, responding only to selected signals, and determining which interrupt service routine will be executed when an interrupt is received.

A high-level language used in systems where interrupts occur must have some means to command the computer to carry out these functions. This is normally done through the use of library functions. We will discuss the design of functions to control computer interrupts in Chaps. 7 and 14.

5.4.7 Bit-Manipulation Operations

In a real-time system many operations depend on the status of individual bits. For example, the system may monitor a set of switches and represent them as bits packed within a word in memory. The real-time language must allow these bits to be accessed and manipulated. The necessary functions include bitwise logical operations (OR, AND, complement) and left and right shift operations (see App. A).

5.4.8 Reentrancy

A final requirement for a real-time language is that it produce *reentrant* programs. In a reentrant program, the execution of the program can be suspended in the middle of a function, a new program can begin to execute which uses the same function, and after this new program has completed execution, the original program can resume execution as if the suspension had not occurred.

In a real-time system, the execution of a program may be suspended by an interrupt from an external source. The interrupt service routine that then executes may use some of the same functions (subroutines) as the suspended program (see Fig. 5.9). If the functions are not reentrant, it is possible for the interrupt service routine to adversely affect the operation of the suspended program.

The basic requirement for a reentrant program is that variables within a function be stored in such a fashion that variables from one invocation of the function are kept separate from variables in another invocation of the same function. If this were not the case, the second invocation of the function would alter the data stored in the variables by the suspended first invocation of the function. When the first invocation resumed execution, the values of its variables would be different from the values before the suspension of execution and the function would not produce the correct result.

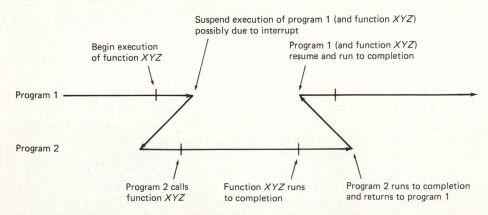

Figure 5.9 Example of a situation where reentrant programs are required.

Modern languages such as Pascal and C allow reentrant programs to be written with little difficulty.

5.5 SOME REAL-TIME PROGRAMMING LANGUAGES

5.5.1 BASIC

BASIC language began in the mid-1960s as a language for teaching computer programming to university students. It was designed to be easy to learn, simple to use, and inexpensive to run. BASIC has been transported from its original large main-frame environment to most minicomputer and microcomputer systems. It has become the "standard" language in personal-computer systems.

BASIC is actually more suited to applications programming than to systems or real-time programming. Despite this, it is one of the most widely used languages in real-time applications. There are several reasons for this:

1. BASIC is known or can be quickly learned by engineers involved in real-time-systems designs. Supposedly better languages are often not known by nonspecialists and are more difficult to learn.
2. BASIC interpreters or compilers were available for the minicomputers or microcomputers being used whereas compilers for better languages were not.
3. Extensions to the original BASIC language by vendors or real-time computer systems removed the most glaring deficiencies of BASIC in real-time applications.

The original version of BASIC did not support an engineering approach to software development. The support for subroutines was extremely primitive, making the development of modular programs difficult. The control structures were also very simple, encouraging the uncontrolled use of GO TO instructions to make up for missing control structures. BASIC did not allow the convenient definition of new data types or structures and did not enforce type conventions for types defined by the language.

Extensions have been introduced by many different vendors to make up for many of these deficiencies. Unfortunately, the extensions are not standardized, leading to a proliferation of noncompatible dialects of BASIC. A new ANSI standard for BASIC has been developed which includes a standard set of the most useful extensions. Until this standard is adopted by compiler vendors, portability of BASIC programs will continue to be a problem.

Extensions have also been added to BASIC to make it more useful in real-time programming applications. Access to memory locations is commonly done by commands called PEEK (read a location) and POKE (write to a location). Access to assembly-language subroutines is often implemented with the CALL command. The CALL command will accept a parameter specifying the starting address of the assembly-language routine and a list of parameters to be passed to the routine. Most versions of BASIC do not have bit-manipulation instructions. They must be implemented with assembly-language routines. These various extensions to aid real-time programming are nonstandard, although an IEEE working group is trying to develop a standard set of extensions.

BASIC has developed a bad reputation among academic computer scientists because of its poor support for good programming techniques. BASIC is not the best choice of languages for large programming projects in cases where a better language is readily available. However, in the real world, better languages are not always readily available. Also, a large number of systems are in the field which are using BASIC successfully. An attempt to change these systems to a new language would be expensive and unpopular. Pragmatism will dictate that BASIC will continue to be used in these systems.

5.5.2 FORTRAN

FORTRAN was developed as an applications language in the mid-1950s in order to replace assembly-language programming in scientific and engineering applications. The language has gone through an extensive evolution over the years and has built up a huge software base. Most engineers and scientists are taught FORTRAN during their undergraduate training and it remains the dominant language for engineering and scientific applications.

FORTRAN supports modular programming to the extent that it has a provision for named subroutines whose variables are local to the subroutine. Initially, FORTRAN had extremely primitive control structures consisting of the DO loop, the arithmetic IF, and GO TOs to statement numbers. The latest standard version of FORTRAN (FORTRAN 77) has added the IF-THEN-ELSE construct as well as the ability to form *blocks* of statements which are treated as a single statement by this construct. FORTRAN has a wide variety of data types suitable for scientific applications, including double-precision floating-point and COMPLEX data types. The language also has an extensive set of mathematical operators and functions. The support for processing nonnumeric data is not nearly as good.

As with BASIC, this language is not inherently well-suited for real-time applications. Access to memory locations must be done through nonstandard PEEK and POKE type operations. Bit manipulation is not supported in the language itself. Access to assembly language is done through the FORTRAN subroutine call facility. A linker program will link in the assembly-language module. This procedure will vary from system to system, since the method of passing parameters and linking modules is implementation-dependent (i.e., not standarized).

Despite its disavantages as a real-time language, FORTRAN is used in real-time applications. Some of the reasons why it might be used include:

1. A desire to use the same language for both real-time and data-processing applications.
2. A need for FORTRAN's excellent numerical processing capabilities. Signal processing and advanced process-control systems are examples of systems where such capabilities are required.
3. Cases where FORTRAN is the only language supported by the chosen hardware.

If these considerations do not apply, some other language should be chosen.

5.5.3 FORTH

The FORTH language was developed in the early 1970s as a systems programming and real-time control language for minicomputers. Its first applications were in the control of radio telescopes and related scientific equipment, but it has been applied to a wide variety of real-time tasks. Versions of FORTH are available for most popular microcomputers.

The language is unusual in its syntax and its implementation and is therefore difficult to describe in a few words. Basically the language consists of a series of machine-language *dictionary definitions* of the basic elements of the language. Each definition could be considered a special form of a subroutine. A FORTH program is written by building up statements based on these dictionary functions. The programmer can also extend and customize the language by adding new dictionary definitions based on already existing dictionary functions.

FORTH is basically an extremely modular language, since the most natural way to program is to define new dictionary items for any desired function. It also supports a top-down programming style, since dictionary functions are built up of previously designed dictionary functions. A program can thus consist of a few top-level dictionary definitions which are based on a hierarchy of lower-level dictionary definitions. FORTH supports most of the control constructs required for structured programming but requires considerable attention on the part of the programmer to ensure that the proper variable is being tested by the construct. The basic FORTH package supports only a few simple data types, such as bytes and integers. The basic package does not even support arrays. However, extended versions are available that supply array and floating-point data types. Type checking and other error checking is virtually nonexistent. FORTH's biggest disadvantage from the viewpoint of software engineering is that it encourages the writing of unreadable programs. The syntax of the language uses a reverse Polish notation similar to that on Hewlett-Packard calculators which is less easily understood than the standard type of syntax. Also, the nature of the language discourages the use of named variables which normally help the readability of a program.

Since FORTH was designed for real-time applications, it has the features required of a real-time language. Each FORTH implementation for a particular type of computer includes an assembler that is integrated into the FORTH-language system. Assembly-language code can be intermixed with FORTH statements and assembly language can be used to create new dictionary definitions. Thus all the hardware access features and bit-manipulation instructions required for a particular application can be very easily implemented.

FORTH has become a fairly popular language for real-time applications. It seems to be most popular in research laboratories where a minicomputer or microcomputer is used to control laboratory equipment. FORTH is interactive in the same way as an interpreted language, but its programs execute much faster than those written in standard interpreted languages. Users in a research environment where the control program for a laboratory setup will change frequently obviously find the interactive nature of FORTH an advantage. Program changes can be made and tested very quickly.

The FORTH language system is extremely compact. A usable program-development system can reside on a computer with only 16 kbytes of memory and very limited amounts of mass storage. FORTH is therefore popular with engineers who do not have a powerful software development system at their disposal. In many instances, a FORTH program can be developed on the computer system that will eventually be used in the real-time system. The fact that many FORTH software-development packages are extremely inexpensive is also a powerful inducement when development funds are limited.

For large real-time systems destined for industrial applications, the merits of FORTH diminish considerably. Benchmarks indicate that FORTH programs do not execute as rapidly as programs written in more conventional compiled languages. The advantages of FORTH's interactive nature and its minimal development system requirements are often outweighed by problems of documenting and maintaining FORTH's rather cryptic programs. In fact, an interactive language system often encourages quick and undocumented "fixes" of program bugs. Fixes of this type often cause more problems than they cure.

5.5.4 PL/M

PL/M is a systems and real-time programming language developed by Intel Corporation for use on its microcomputers. Independent vendors supply languages which are very similar for other types of microcomputers. Intel's preeminent position in the microcomputer business along with PL/M's history as the first high-level language implemented for microcomputers have made PL/M an important real-time language.

PL/M was designed in the early 1970s and shows the influence of the movement toward software engineering and structured programming. The nature of PL/M is such that a program is composed of *blocks* of program statements. Each block can be considered as a unit for control purposes. Blocks can be nested so that there are blocks within blocks. The net effect is that PL/M strongly supports a modular and top-down style of programming. PL/M has a very full set of control constructs including IF-THEN-ELSE,

DO-WHILE, and DO CASE (SWITCH). PL/M for the Intel 8080 microcomputer supports BYTE and 16-bit ADDRESS data types as well as arrays of these types. It also allows more complex data structures based on BYTE and ADDRESS variables to be defined. PL/M for the Intel 8086 adds floating-point numbers to these data types. The number of mathematical operators supplied is small, being limited to the basic arithmetic operations.

PL/M is very much a real-time and systems programming language. Its limited range of data types and its limited number of operations do not make it very suitable for programs requiring complex numerical calculations. The support for machine-level operations, on the other hand, is very good. Memory locations can be addressed by means of pointers. The I/O section on Intel microcomputers can be directly accessed from PL/M. PL/M includes a complete complement of bit-manipulation instructions. PL/M even has provision to control the interrupt system of the microcomputer. This last feature is rather machine-specific and nonportable.

The typical PL/M compiler system supports separate compilation of PL/M program modules. Inclusion of assembly-language modules into a PL/M program is also possible, as is the maintenance and use of a library of precompiled modules.

PL/M should be seriously considered as a real-time language by users of Intel microcomputers. The language is well-supported by Intel and is reasonably portable among Intel microcomputers. PL/M has the deficiency that not many versions exist for non-Intel computers and that Intel supports the PL/M compiler only on its own development systems.

5.5.5 Pascal

Pascal was developed in the early 1970s as a language for teaching programming to university students. It was designed to implement the concepts of structured programming which had been defined in the late 1960s. It was also designed to be a relatively "small" language which would allow the development of compilers producing efficient machine language. Over the past decade, Pascal has migrated out of the teaching environment into applications, systems, and real-time programming applications.

Like PL/M, Pascal is a *block-structured* language in which blocks of statements can be used as a unit by control structures. Pascal makes heavy use of subroutines which it calls procedures and functions. Pascal also supports both local variables known only within a procedure and global variables known throughout a program. These three features allow the programmer to write Pascal programs in a top-down modular style. Pascal has all the required control constructs to allow the writing of programs directly from a structured flowchart.

Pascal has integer, real, boolean (logical), and character data types. It supports arrays and *records* of these types and supports the definition of both new data types and new data structures. Pascal is a *strongly typed* language—

that is, each variable must belong to one and only one type and conversions between types are strictly limited. For example, a boolean variable (1 or 0) can not be treated as an integer by performing arithmetic operations on it. This strong typing is an aid to the production of correct programs but can be something of a hindrance in some real-time and systems-programming applications.

Pascal was not designed specifically as a real-time language. In fact, its origin as a teaching language makes the language somewhat inflexible in systems applications. For example, Pascal has pointers to memory locations but they cannot be explicitly set by the programmer to specific addresses. This makes sense in a teaching environment where access to the machine is not desirable but is an inconvenience in real-time and systems-programming applications. Pascal did not originally allow separate compilation of modules, which guaranteed that an entire program could be checked by the compiler for any intermodule errors in data types or parameter passing but made the development of large programs by a team rather difficult. These problems, as well as other deficiencies for real-time applications, have been dealt with by adding extensions to the basic Pascal language. Unfortunately, the extensions are nonstandard and affect the portability of Pascal programs.

The designer of Pascal, Niklaus Wirth, has designed a Pascal-like language called Modula-2 which is designed specifically for real-time programming. It solves most of the problems mentioned in the previous paragraph. Compilers for Modula-2 are only beginning to appear, so it is difficult to assess how much of an impact it will have on industrial applications. However, it has already generated considerable interest and deserves consideration once reliable and efficient compilers become available for it.

Pascal compilers are available for most 8- and 16-bit microcomputers. Many engineering organizations have adopted Pascal as their standard language for microcomputer applications, since the language strongly supports an engineering approach to software design, is well-supported by software vendors, and has proved flexible enough to support applications, systems, and real-time programming.

5.5.6 Ada

The Ada language has been under development since 1975 under the sponsorship of the U.S. Department of Defense. The Department of Defense wanted a common language for all its *embedded* computer systems. An embedded computer system is a computer system that forms an integral part of a larger system such as an aircraft or a missile defense network. In many cases an embedded computer system and a real-time computer system are one and the same.

The design of Ada has been strongly influenced by Pascal, so the languages look very similar in the areas of program structure and control features. However, Ada is designed for large software-development projects, so the language has been designed to support separate compilation of program modules while still being able to perform cross-module checking for errors. In addition, Ada has far more support for I/O operations than does Pascal.

Ada includes all the features required of a real-time language either within the language itself or as standard *packages* that can be linked into a program. Ada includes features not found in most other languages—for example, provision to handle exception conditions such as interrupts or arithmetic overflows.

Ada's wide range of capabilities is not achieved without cost. The language is quite "large" and requires some time to learn thoroughly. Ada has been criticized for being too "large." The critics claim that it will be very hard to design good compilers for Ada and that it will be virtually impossible to have complete error checking, since the language is so complex. The required development system for Ada is also quite large. At present a DEC VAX super minicomputer is required to run the complete Ada compiler system. The development of more powerful microcomputers during the 1980s will reduce the cost of the Ada development system.

Ada is a very new language and complete compilers for microcomputers are only now becoming available. As a result, there has been no field experience to indicate how successful Ada is as a real-time language running on microcomputer systems. However, Ada has the support of the U.S. Department of Defense and so is likely to be well-supported and widely used in the future.

5.5.7 C

C was developed at Bell Labs in the early and mid-1970s as a systems programming language. It has been used very successfully on minicomputer systems to produce operating systems (UNIX) and a wide variety of programming tools (including compilers for FORTRAN, Pascal, and C). C compilers are available for most microcomputers and the language is gaining in popularity very rapidly. We will use C in the programming examples given in this book because we believe it is one of the best choices for a high-level language to be used in real-time systems. An introduction to the C language is given in App. E.

5.6 CHOOSING A LANGUAGE

Recommending a particular programming language is always dangerous. Each language has a group of fanatic adherents who will be enraged if their language is ignored. A recommendation based on the theoretical appeal of a particular language may founder on issues such as the availability and quality of compilers for the computer system to be used. In many cases the choice will be based on such practical factors as familiarity with the language, previous use of the language within the organization, or availability of good software packages written in the language.

In cases where a high-level language is being used for the first time in an organization and the computers being used are the newer 16- and 32-bit microcomputers, C, Pascal, or Modula-2 should be considered first. Organizations which plan to develop very large systems or plan to sell to the United States military should begin to consider Ada even though the language is not yet well established. In many organizations, previous use of Intel microcomputers will give PL/M an edge in the choice of language.

In our opinion, BASIC and FORTRAN should be considered only in cases where there are criteria besides suitability for real-time programming to be considered. Some of these criteria are:

1. A large base of software already written in these languages must be used in future applications.
2. Systems or application software supplied by the vendor of the computer must be used and is written in FORTRAN or BASIC.
3. Programming must be done by programmers familiar with these languages and retraining of these programmers is not feasible.
4. FORTRAN or BASIC are already available in the organization and all real-time programming projects are so simple that the cost of a new language system cannot be justified.

FORTH should be considered in cases where an economical development system is important (e.g., small companies) or where an extremely interactive development environment is important (e.g., research and development laboratories).

SUMMARY

A computer can be programmed in its binary machine language, in its assembly language, or in a high-level programming language. One of the decisions a system designer must make is the choice of which language level to work at.

Programming in machine language is almost never justified, since it is very time-consuming and prone to error. Programming in assembly language is still at the level of the basic operations performed by the computer, but the programmer uses mnemonic names for these operations rather than their binary operation codes. In addition, memory locations can be assigned names so that binary memory addresses do not have to be used when writing the program.

Relocating, macro, and structured assemblers are assemblers with additional features to make the task of programming easier and more disciplined. However, programming in assembly language always requires close attention to the details of the computer's architecture and instruction set. Since assembly language is tied so closely to the computer's architecture and instruction set, the assembly language for each different type of computer is different. As a result, assembly-language programs cannot be moved directly from one computer to another (they are nonportable).

High-level languages, on the other hand, are portable. In addition, they reduce programming effort and allow a more disciplined approach to program design. Systems programming languages retain the access to the computer hardware that is one of the advantages of assembly language. The disadvantage of programming in a high-level language is that the resulting machine-language program will execute somewhat more slowly and require somewhat more memory than the machine-language program resulting from the equivalent program written in assembly language. In a large number of real-time applications these disadvantages no longer outweigh the many advantages of programming in a high-level language.

Modern structured high-level languages such as C and Pascal are available for most microcomputers. If possible, they should be chosen in preference to older less-structured languages such as BASIC or FORTRAN.

EXERCISES

5.1 The following program is written in Intel 8085 assembly language. Identify all the names defined by the programmer and list them in a symbol table. Then calculate the numerical value of each name in the symbol table.

```
;Program to add the bytes stored in VARA and VARB and store the result
;in SUM.
;
PROGRAM   EQU    200H
DATA      EQU    1000H
          ORG    PROGRAM
LOOP:     LDA    VARB
          MOV    D,A
          LDA    VARA
          ADD    D
          STA    SUM
          JMP    LOOP
          ORG    DATA
VARA:     DS     1
VARB:'    DS     1
SUM:      DS     1
          END
```

5.2 At what memory address will the preceding program begin when it is loaded into the target computer?

5.3 When the Intel 8085 microprocessor is turned on, it is reset so that its program counter contains 0000. Add the appropriate assembler directives and assembly-language instructions to the program in Prob. 5.1 so that the program will execute immediately after the computer is turned on. The existing portions of the program must remain in the same locations in memory.

5.4 A macro resembles a subroutine at first glance but the two are actually quite different. What are the differences? What are the advantages and disadvantages of each?

5.5 Why aren't there portable assembler programs that allow the same assembler-language program to run on more than one type of computer?

5.6 Take a high-level programming language with which you are familiar and attempt to answer the following questions:

    (*a*) Is it usually interpreted or compiled?

    (*b*) Are both interpreters and compilers available for it?

    (*c*) Is it a system programming language or an application language?

    (*d*) What data types does it support?

    (*e*) Does it have the structured control constructs?

    (*g*) Are there compilers or interpreters available for microcomputers?

    (*h*) How well does the language (in its microcomputer implementation) meet the requirements for a real-time programming language:

        (i) Production of efficient machine-language code

        (ii) Ability to use assembly-language functions from the high-level-language program

        (iii) Production of ROMable code

        (iv) Access to the computer's hardware

        (v) Bit-manipulation instructions

        (vi) Production of reentrant code

REFERENCES

1. Gilbreath, J.: "A High Level Language Benchmark," *Byte*, September 1981, p. 180.
2. Gilbreath, J., and G. Gilbreath: "Eratosthenes Revisited: Once More Through the Sieve," *Byte*, January 1983, p. 283.

BIBLIOGRAPHY

Assembly Language

Barron, D. W.: *Assemblers and Loaders*, Elsevier, New York, 1972.
Kane, G., D. Hawkins, and L. Leventhal: *68000 Assembly Language Programming*, McGraw-Hill, New York, 1981.
Leventhal, L. A.: *8080A/8085 Assembly Language Programming*, McGraw-Hill, New York, 1978.

High-Level Languages

Aho, A. V., and J. D. Ullman: *Principles of Compiler Design*, Addison-Wesley, Reading, MA, 1977.
Bentley, J. L.: *Writing Efficient Programs*, Prentice-Hall, Englewood Cliffs, NJ, 1982.
Brodie, L.: *Starting FORTH*, Prentice-Hall, Englewood Cliffs, NJ, 1981.
Feuer, A., and N. Gehani (eds.): *Comparing and Assessing Programming Languages*, Prentice-Hall, Englewood Cliffs, NJ, 1984.
Gehani, N.: *Ada: An Advanced Introduction*, Prentice-Hall, Englewood Cliffs, NJ, 1983.
Grogono, P.: *Programming in Pascal*, Addison-Wesley, Reading, MA, 1980.
Hubin, W. N.: *BASIC Programming for Scientists and Engineers*, Prentice-Hall, Englewood Cliffs, NJ, 1978.
Katzan, H.: *FORTRAN 77*, Van Nostrand Reinhold, New York, 1978.
Kernighan, B. W., and D. Ritchie: *The C Programming Language*, Prentice-Hall, Englewood Cliffs, NJ, 1978.
McCracken, D.: *A Guide to PL/M Programming for Microcomputer Applications*, Addison-Wesley, Reading, MA, 1978.
Zarella, J.: *Language Translators*, Microcomputer Applications, Suisun City, CA, 1982.

SOFTWARE DESIGN EXAMPLE: AN I/O DRIVER FOR A SIMPLE PERIPHERAL DEVICE

To show how high-level languages can be applied in a typical real-time application, we will describe the development of *I/O driver functions* for a keypad/display card which might be used in a small stand-alone system. The functions will handle the input of data from the keypad to a program and the output of data from a program to the display.

Functions of this type, sometimes referred to as *device handlers*, are often written in assembly language. In fact, simple assembly-language programs to control keypads and alphanumeric displays are commonly used examples in introductory texts on microcomputers. We will see that these functions can be written in a high-level language with all the resulting benefits discussed in the previous chapter.

The programs presented in this chapter and later chapters in this book are written in the C programming language. Readers who are unfamiliar with this language should study App. E before proceeding. The programs presented in this book don't make use of all the special capabilities of C, are heavily commented, and are accompanied by flowcharts. Therefore readers familiar with another high-level programming language shouldn't have much difficulty in learning how to read them. We have chosen the C language because it is widely used in real-time microcomputer programming and it offers the features required when a program must deal directly with computer hardware.

6.1 THE PRO-LOG 7303 KEYPAD/DISPLAY CARD

The 7303 keypad/display card, shown in Fig. 6.1, is manufactured by Pro-Log Corporation. It is designed to be used in small computer systems built with standardized circuit boards, called STD bus boards. The keypad/display card can be plugged directly into such a system. We will discuss the design of computer systems using standard circuit boards, such as STD bus cards, in Chaps. 13 and 14.

The keypad/display card can be used for simple data entry and display tasks and can replace a terminal or a special-purpose control panel. Figure 6.2 shows the layout of the card. The keypad contains 24 keyswitches numbered from hexadecimal 0 to hexadecimal 17 which can be read by the computer. The twenty-fifth keyswitch, marked RESET, is tied directly to reset circuits in the computer and does not affect our design. The two rocker switches and the LED display also do not enter into our design. The alphanumeric display is important to our design. It can display eight characters, each of which can be a capital letter, a numeral, or a punctuation symbol from the ASCII character set.

The keypad/display card has two I/O registers, or *ports*, that can be accessed by the computer. There is an I/O register called the control port used to control the display and another I/O register called the data port used to both send data to the display and read data from the keypad. I/O registers will be discussed in more detail in Chap. 9. The addresses of the ports can be selected by jumper wires on the circuit board, but the board is delivered with a

Main Features of the 7303 are:

- 8-position alphanumeric display with ASCII input
- 24 programmable keys plus reset
- Repairable keyboard and replaceable key labels
- 8-bit binary LED display
- 2 rocker switches
- Simple program control of displays and keys
- On-card I/O ports for processor control
- Single +5 V Operation

Figure 6.1 Pro-Log 7303 keypad/display card. (*Pro-Log Corp.*)

control port address of 0xD1 and a data port address of 0xD0. The card is usually used in computer systems with a separate I/O address space (see Sec. 2.5).

For the purpose of this example we will assume that the keypad/display card will be used in a computer system with a separate I/O address space and that the C compiler has two library functions, input( ) and output( ), that handle I/O operations to this address space. The input( ) function is called as follows:

$$\texttt{input(portaddress);}$$

The parameter portaddress is the address of the I/O register which contains the data that the input( ) function returns to the calling function. The output( ) library function is called as follows:

$$\texttt{output(portaddress, data);}$$

Again, the parameter portaddress is the address of the I/O register. The parameter data is the data that is to be sent to the I/O register.

C compilers for microcomputers with separate I/O address spaces, such as the Intel 8085, Zilog Z80, and the Intel 8088, 8086, and 80286 family, almost invariably have library functions of the type we have just described. Compilers for other high-level languages intended for use in real-time applications usually have similar facilities.

6.2 DISPLAY HANDLER

6.2.1 Programmer's Model of the Display

When developing a program to control an external device, it is a good idea to develop a simple model of the device as seen by the computer. The model should contain only the data essential to the proper control of the device by the program. Figures 6.3 and 6.4 show such a model for the display portion of the keypad/display card. Figure 6.3 shows the relationship of the two computer accessible ports to the display and Fig. 6.4 shows the function of each bit position in the two ports.

The low-order 7 bits of the data port contain the 7-bit ASCII code for the character to be sent to the display. The high-order bit (d_7) of the data port must be set to 1. The control port selects the position in the display to which the character will be sent and controls the transfer of the data from the data port to the display. The position of the character is specified by a 3-bit binary number stored in the three low-order bits of the control port $(d_2\, d_1\, d_0)$. The number corresponds to the display positions shown in Fig. 6.2. Transfer of data from the data port to the display is controlled by the WR bit in the control

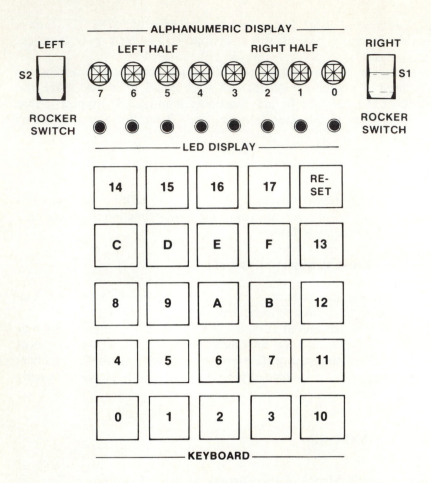

Figure 6.2 Keypad/display card layout. (*Pro-Log Corp.*)

register (d_3). Data is transferred when the WR bit is 1. When the WR bit is 0, the display ignores any data sent to it.

Data must be sent to the control and data ports in the proper sequence in order for a character to be sent to the display successfully. The character code and the binary number specifying the position of the character in the display must be valid before the WR bit is set to 1 and they must be held valid for a short period after the WR bit is reset to 0. The flowchart in Fig. 6.5 illustrates the sequence of operations required to meet these requirements.

6.2.2 Design of Display Handler Function

We now have sufficient information to design a display handler function. The handler function will be passed a pointer to a string constant by the calling function. The handler function will step through the string using the pointer

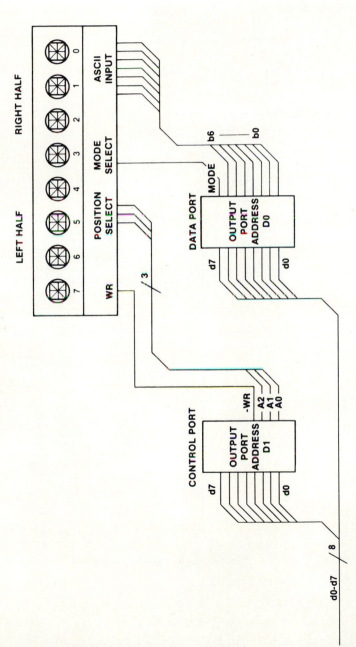

Figure 6.3 Display interface. (*Pro-Log Corp.*)

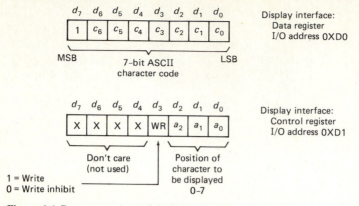

Figure 6.4 Programmer's model of the display interface.

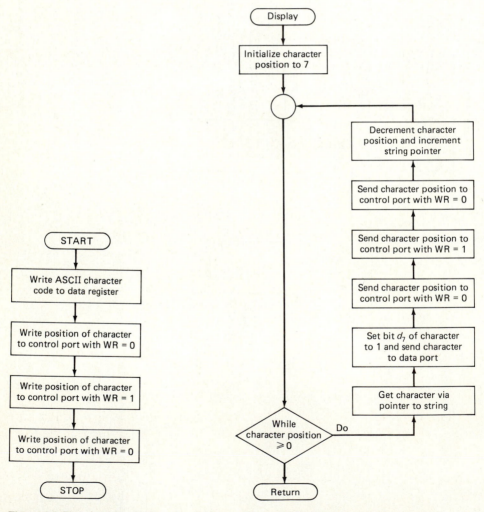

Figure 6.5 Flowchart for display of a single character.

Figure 6.6 Flowchart for display handler.

and send the first eight characters in the string to the display. The design of the handler function is illustrated by the flowchart in Fig. 6.6.

Writing a C program from the flowchart is quite straightforward. The library function output( ) is used to perform the actual output operations to the ports on the keypad/display card.

Display Handler Function

```
#define DATAPORT      0xD0    /* address of data port */
#define CONTROLPORT   0xD1    /* address of control port */
#define BIT7MSK       0x80    /* mask to set bit 7 of character to 1 */
#define WRON          0x8     /* mask to set WR bit to 1 */
#define WROFF         0x7     /* mask to set WR bit to 0 */

/*display( ) is passed a pointer to a string. It sends the first 8
  characters in the string to the alphanumeric display in the Pro-Log
  7303 Keypad/Display card. The calling routine must ensure that the
  string is formatted so that the message appears correctly on the
  display. In particular, display( ) does no checking that the string is
  the proper length.                                                   */

display(stringptr)
char *stringptr; /* pointer to string of characters to be displayed */
{
int position;        /* display position */
position = 7;
while(position >= 0)

    {
    output(DATAPORT, *stringptr|BIT7MSK);
    /* send character to data port with bit 7 set to 1 */
    output(CONTROLPORT, position & WROFF);
    /* send character position to control port with WR = 0 */
    output(CONTROLPORT, position|WRON);
    /* send character position to control port with WR = 1 */
    output(CONTROLPORT, position & WROFF);
    /* send character position to control port with WR = 0 */
    position--;  /* decrement display position */
    stringptr++;/* increment pointer to point to next character */
    }
}
```

The function makes use of the bitwise logical operators to alter individual bits in the control ports. For example, bit 7 in the data port is set to 1 by the bitwise OR of the 7-bit ASCII character code sent to the data port and the mask BIT7MSK which has the binary pattern 10000000. The result of the OR operation is an 8-bit value in which the most-significant bit (bit 7) is set to 1 and the remainder of the bits are set to those in the 7-bit ASCII character code. This value is then sent to the data port by the output( ) function. The two masking operations which set the WR bit to 1 or 0 act in a similar fashion. Refer to App. A for more information on bitwise logical operations and the use of masks.

6.3 KEYPAD HANDLER

6.3.1 Programmer's Model

The programming model for the keypad interface is shown in Fig. 6.7. Only the data port is used in the interface to the keypad. The data port actually consists of two separate ports, an input port and an output port, sharing the same I/O address. When the computer reads data from the data-port address, the data comes from the input port. When the computer writes data to the data-port address, it is sent to the output port. The data port is also used to send data to the display. However, since the display reads the data-port lines only when the WR line from the control port is set to 1, the data port can be used to read the keypad without affecting the display.

The keyswitches on the keypad are arranged in a 4×6 *matrix*. Bits 0 to 3 of the output port are connected to the four columns of the matrix, while bits 0 to 5 of the input port are connected to the six rows of the matrix. The circles on the keypad side of the two ports indicate that the ports invert data sent through them. Thus a 1 sent to a bit position in the output port register will result in a low-level voltage on the output line corresponding to that bit position. Similarly, a low-level voltage applied to one of the lines on the keypad side of the input port will result in a logic 1 in the corresponding bit position in the port register.

6.3.2 Keypad Scanning

The keypad is read by *scanning* the keypad a column at a time. Each column is driven to a low voltage in turn by the output port while the other three columns are driven high (about 5 V). A keyswitch closure in the column that is driven low causes the row connected to the keyswitch to go low. A row which has no keyswitch closed, or has one closed in a column which is driven to a high level, will remain at a high level. As each column is driven low in turn, the input port is read by the program to see if any of the rows is driven low. If so, the keyswitch which is depressed can be determined, since the program keeps track of which column is being driven and the row value is determined by the bit position in the input port.

The keypad handler must do more than simply scan the keypad to detect a key depression. A code must be returned to the calling function corresponding to the key that is depressed. In our handler we will return an integer corresponding to the hexadecimal number on the key. The handler function must also ensure that it responds to each key closure only once. This can be achieved by implementing the ASM diagram shown in Fig. 6.8. The keypad handler function is entered on the assumption that the previously decoded key is still depressed. The function repeatedly scans the keypad until it determines that all keys are released. It then makes the transition to the next state, where it scans for the next key depression to decode and return to the calling

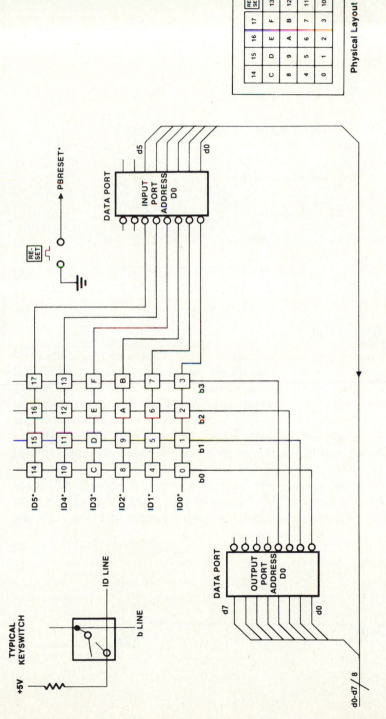

Figure 6.7 Keypad interface. (*Pro-Log Corp.*)

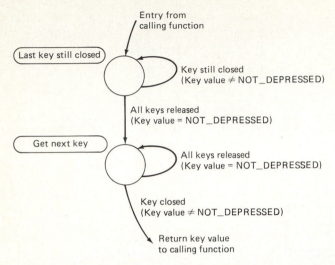

Figure 6.8 ASM diagram for keypad handler function.

function. This ensures that each key depression is detected, decoded, and returned to the calling function only once.

6.3.3 Keypad Debouncing

A final task of the keypad handler function is to deal with the problem of *contact bounce* when a keyswitch is opened or closed. When a mechanical switch is opened or closed, the switch contacts bounce, opening and closing many times before coming to rest. A switch typically bounces for several milliseconds. The switches in the keypad are specified to reach a stable state within 15 ms. If the keypad handler repeatedly scans the keypad while a keyswitch is bouncing, it will appear that the switch has been depressed several times instead of only once.

One method of "debouncing" the keypad is shown in Fig. 6.9. When the program detects that a keyswitch has been closed, it waits for a period longer than the maximum time for the keyswitch to stop bouncing and then scans the

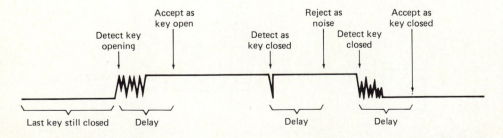

Figure 6.9 Keypad debouncing.

keypad again to verify that the keyswitch was actually depressed. This deals with contact bounce and also protects against electric noise pulses that may be interpreted as switch closures. The same approach is used when checking for all keys released.

6.3.4 Keypad Handler Design

A flowchart for the keypad handler can now be developed which implements all the requirements we have discussed. It is shown in Fig. 6.10. The C language function can then be written using this flowchart as a guide.

```
#define NOT_DEPRESSED      -1    /* code returned by scankeys( ) when
                                    no key is depressed */

#define DEBOUNCETIME       20    /* value passed to wait( ) function
                                    so it waits 20 milliseconds */

/*rdkbd( ) reads the keypad on the Pro-Log 7303 keypad/display card. It
   ensures that each key entry is read only once and debounces the keys.
   It returns an integer corresponding to the hexadecimal number on the
   key pressed. rdkbd( ) makes use of two functions; scankeys( ) which
   performs the actual scanning and decoding of the keypad and wait( )
   which is a delay routine used in the debouncing operation */
rdkbd( )
{
int firstvalue, secondvalue, keyvalue;
/* First ensure that the previous key is released */
do
    {
    keyvalue = scankeys( );    /* check keypad for key closure */
    if(keyvalue = = NOT_DEPRESSED)
        {
        wait(DEBOUNCETIME);    /* delay and then check again */
        keyvalue = scankeys( );
        }
    }
while(keyvalue ! = NOT_DEPRESSED)
/* Now scan for a new key closure since all keys have been released */
do
    {
    firstvalue = scankeys( );    /* check keypad for key closure */
    if(firstvalue ! = NOT_DEPRESSED)
        {                               /* if closure detected - delay and */
        wait(DEBOUNCETIME);    /* check again */
        secondvalue = scankeys( );
        if(firstvalue = = secondvalue)
            keyvalue = secondvalue; /* otherwise keyvalue remains
                                        NOT_DEPRESSED */
        }
    }
while(keyvalue = = NOT_DEPRESSED)
return(keyvalue);
}
```

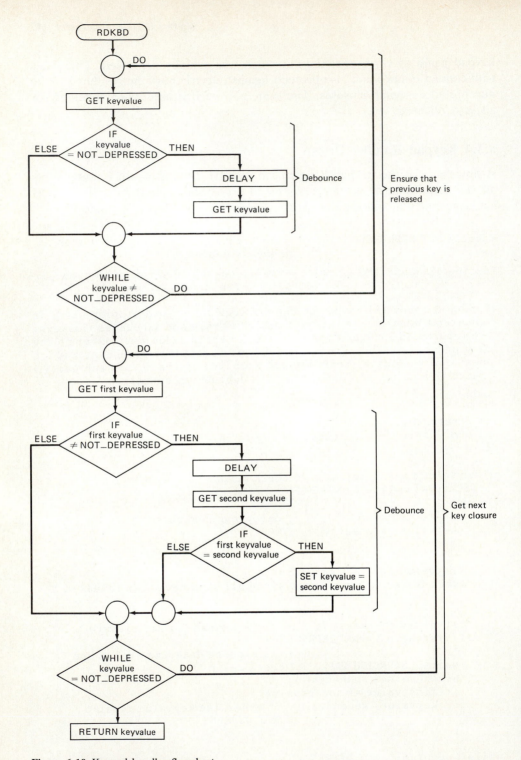

Figure 6.10 Keypad handler flowchart.

148

The keypad handler function rdkbd( ) calls a low-level function, scankeys( ), to actually scan and decode the keypad. Scankeys returns the integer value of the key if a key is depressed and returns -1 (NOT_DEPRESSED) if no key is pressed. The flowcharts describing the scankeys( ) function are shown in Figs. 6.11 and 6.12.

In order to drive each column of the keypad to a low voltage level in turn, a process called *strobing*, a variable (colstrobe) is initialized to 1. This strobe variable is sent to the keypad output port, where it causes column b_0 to go low since the port inverts data sent to it. The strobe variable is repeatedly

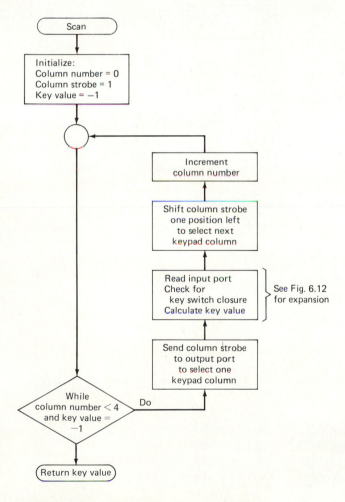

Figure 6.11 Flowchart for keypad scanning function.

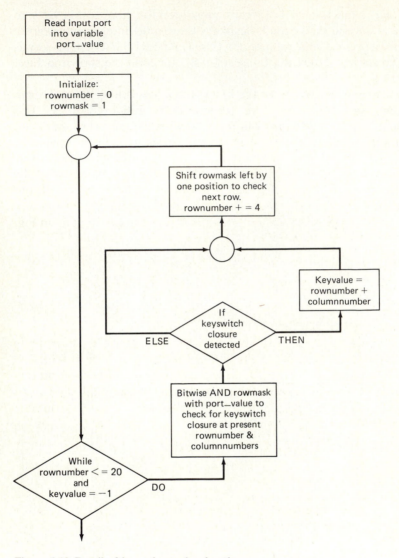

Figure 6.12 Detail of keypad scanning function.

shifted left and sent to the output port. This causes the following bit patterns in the output port:

$$
\begin{array}{ll}
00000001 & \texttt{colstrobe} = 1 \\
00000010 & \texttt{colstrobe} < < = 1 \\
00000100 & \texttt{colstrobe} < < = 1 \\
00001000 & \texttt{colstrobe} < < = 1
\end{array}
$$

As a result, each column in the keypad is driven low in turn.

In order to check each row of the keypad for a key depression, another variable (rowmask) is also initialized to 1 and repeatedly shifted left. The rowmask variable is bitwise ANDed with the data from the keypad input port to mask off the data for one row alone (the row corresponding to the bit position where rowmask has a 1). By strobing the keypad columns in an outer program loop with colstrobe while scanning the keypad rows with row-mask in an inner loop, it is possible to test each key in the keypad for a closure.

The decoding of the keypad to return an integer corresponding to the depressed key could be done by creating a rectangular array with dimensions equal to the number of keypad rows and columns. The array would contain integers corresponding to the keys in the physical array represented by the keypad. This technique allows for an arbitrary assignment of integer codes to key locations. However, in this case we can calculate the integer code directly. Examining the logical keypad layout (different from the physical layout) in Fig. 6.7, we see that the number on a key can be calculated by

$$\text{Keyvalue} = \text{column number} + \text{row number}$$

where column number = 0, 1, 2, 3
       row number = 0, 4, 8, C, 10, 14 (hexadecimal)

In the scan function the column number and row number are initialized to 0. The column number is incremented by 1 every time the column strobe is shifted left. The row number is incremented by 4 every time the rowmask is shifted left. When a keyswitch closure is detected, the column number and row number are added to create the integer corresponding to the depressed key.

The C code for the scankeys( ) function which follows was written directly from the flowcharts describing the design of the function.

```
#define KEYPAD        0xDO    /* Address of keypad ports */

/*scankeys( ) is a keypad scan routine for the Pro-Log 7303 keypad/
    display card. It performs the actual scanning of the keypad and deter-
    mines which key has been depressed. The function returns an integer
    corresponding to the key which is depressed or a -1 (NOT_DEPRESSED)
    if no key is pressed.    */

scankeys( )
{
int colstrobe, colnumber, keyvalue, rownumber, rowmask, port_value;

colnumber = 0;
colstrobe = 1;
keyvalue = NOT_DEPRESSED;

while((keyvalue == NOT_DEPRESSED) && (colnumber < 4))
    {
        output(KEYPAD, colstrobe); /* strobe a column */
        port_value = input(KEYPAD); /* read the rows */
        rownumber = 0;
```

```
        rowmask = 1;
        while((keyvalue = = NOT_DEPRESSED) && (rownumber < =20))
                {
                if((rowmask & port_value) ! = 0)
                        keyvalue = colnumber + rownumber;
                rowmask < < = 1;/*shift rowmask left*/
                rownumber += 4;
                }
        colstrobe < < = 1;/* shift colstrobe left */
        colnumber++;
        }
return(keyvalue);
}
```

The rdkbd( ) function also calls the function wait( ), which creates the delay required in debouncing the keypad switches. This function is a simple loop calibrated to produce delays in increments of 1 ms. Since the timing is done by software, this function is extremely system-dependent. A change in the C compiler used, the type of computer used, or the clock frequency of the computer will require recalibration of the function. The following C program is one example of a wait( ) function:

```
/*wait(time)  produces  a  delay  of  ''time''  millisesconds  where
  ''time''is an unsigned integer passed to the function. The function
  must be calibrated for different systems by adjusting the constant
  MAGICNUMBER so that the function call, wait(10000), produces a delay
  of ten seconds.                        */

#define MAGICNUMBER 84

wait(time)
unsigned int time;
{
unsigned int i, j;
for(i = 0; i < time; i++)
        for(j = 0; j < MAGICNUMBER; j++)
                ;                               /* do nothing statement */
}
```

SUMMARY

The design of the keypad and display handlers illustrates the design process described in Chap. 4. Each handler constitutes a self-contained module. Information on the hardware details of the interface to the display or the keypad is hidden within the module. A programmer using these modules need only know what parameters must be passed to the modules and what values the modules will return. If a change occurs in the hardware (for example, a change in the port addresses), changes in the software are limited to the device handler modules. Even if the computer used in the system were to be changed, C's high degree of portability ensures that the reprogramming effort to adapt the program to the new computer would be modest. The structure of a C program as a set of functions encourages modular design. The use of local variables within a function and the passing of function parameters by value also helps to prevent unplanned interaction among modules.

This design example also shows that most of the work in program design is in the stages prior to that of actually coding the program. In these examples, most of the design time was spent in understanding the details of the hardware interface, determining exactly what was required of the functions, and designing the structured flowcharts of the functions. The actual translation from structured flowchart to C language program was relatively easy. The availability of the control constructs used in structured flowcharts in the C language makes the translation process particularly easy. The ability to group statements into self-contained blocks also helps make the translation process easier.

Finally, the example shows that programs which control hardware interfaces must typically do a considerable amount of bit manipulation such as masking and shifting. The bitwise logical operators and shift operators in the C language are invaluable for such operations. The direct access to memory locations and I/O ports provided by the C language and its I/O library is also of great benefit when designing programs of this type. In the past, device drivers of this type have usually been written in assembly language because of the requirements for bit manipulation and access to the I/O system, but C is now a good replacement for assembly language in all but the most speed- or memory-critical applications.

EXERCISES

6.1 Using the keypad and display handler functions developed in this chapter, write a C program that allows a programmer to load machine-language programs into the computer. The program must do the following:

1. Accept the starting memory address for the machine-language program (in hexadecimal) from the keypad.
2. Accept each byte of machine-language code (in hexadecimal) from the keypad and store it in the correct memory location. Assume that each memory location contains 1 byte.
3. Display the current memory location and the value being loaded into it (in hexadecimal).
4. Allow the programmer the option to examine the contents of memory locations without having to load new values into the locations.

BIBLIOGRAPHY

Pro-Log Corp.: *7303 Keyboard/Display Card User's Manual*, 2411 Garden Road, Monterey, CA, 93940.

PART
THREE

LOWER-LEVEL
CONSIDERATIONS

SEVEN

INTEGRATING
ASSEMBLY-LANGUAGE COMPONENTS

Assembly language was first introduced in Sec. 3.5.2 as a convenient way of writing machine instructions. The format of assembly-language statements and the translation of the assembly-language instructions into machine instructions using an assembler program was later described in Sec. 5.1. While the particular mnemonics that are used to represent each machine instruction and the spectrum of assembler directives available are determined by the assembler's developer, the order of the fields in the assembly-language format is usually the same:

label | mnemonic | operand | comment

While most application software can be developed in high-level language, there are certain applications where assembly-language program components may be required. For example, the manipulation of internal CPU registers cannot be done explicitly in a high-level language such as C.

In this chapter we discuss two particular topics:

1. The information that one must gather to write a program in assembly language
2. Some important applications of assembly language used in conjunction with high-level-language programs

7.1 PROCESSOR BACKGROUND INFORMATION REQUIRED

For a given microprocessor architecture (Secs. 2.1 to 2.3) with an assembler program (Sec. 5.1) that will run on the chosen development system (Sec. 3.5.3 and Chap. 15), a certain amount of microprocessor-specific information must be gleaned from the data books describing the microprocessor and the documentation for the assembler. In this section, the necessary information for the Motorola MC68000 family (MC68000, MC68008) and the Intel 8086 family (the 8086, 8088, 80186, 80188, and 80286) will be categorized and described in detail. For most references to the common features of all the listed Intel products we will use the term "Intel family" rather than list all the processors. We will also use the term "Motorola family" for the Motorola processors.

The following list of important information must be assembled about any processor that you are considering for design purposes.

- The programming model of the CPU
- The representation of data in memory
- The representation of instructions in memory
- How instructions are represented in assembly language
- The addressing modes
- The instruction-set description

7.1.1 Programming Model of the CPU

From the point of view of a programmer, only the internal CPU registers that can be examined, tested, or modified are of interest. These registers constitute the programming model of the CPU (see also Sec. 2.3.2). The majority of the instructions in the processor utilize the registers in this model in some way. Figure 7.1 shows these registers for the MC68000 and Fig. 7.2 for the Intel 8086 family.

Motorola MC68000 family   Aside from the status register, all data and address registers are 32 bits long. The MC68000 family *data registers* can be used for byte, word (16 bits), or longword data (32 bits). The MC68000 family *address registers* can be used only for word or longword data.

The *program counter* "points to" (contains the address of) the next instruction to be executed. In the MC68000, only 24 bits are output to memory, giving an active address range of 16 Mbytes.

The *status register* has

- Five bits dedicated to displaying results of CPU operations on data (bits 0 to 4).
- Three bits (bits 8 to 10) that show the current interrupt priority level (to be discussed in Sec. 7.7).
- One bit (bit 15) that indicates whether the CPU can be "single-stepped" one machine instruction at a time by an external pulse used for debugging. This is called *trace mode*.
- One bit (bit 13) that indicates whether the CPU is in *user* or *supervisor* mode. The states and levels displayed in bits 8 to 15 can be altered only by the MC68000's "protected instructions," which affect the status register only when the machine is in supervisor mode.

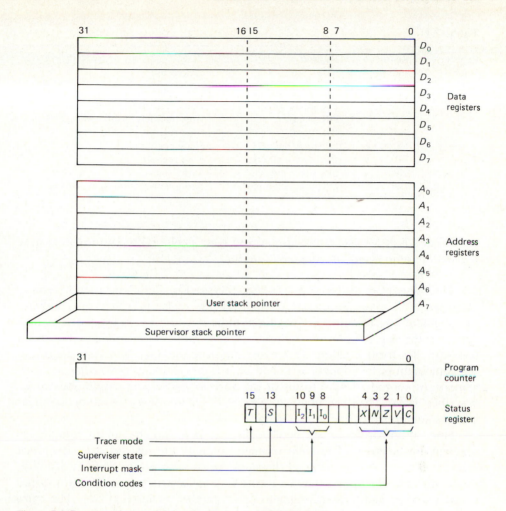

Figure 7.1 Programming model for the Motorola MC68000 family.

The supervisor mode, or "state," is privileged in that certain instructions can be executed only in this state. When user programs run in user state, for example, an error condition in a program, such as the presence of an illegal instruction, will cause a transition to the supervisor state (to be discussed in Sec. 7.7.2) which can run a program to report the error and take other appropriate actions. The supervisor state can then provide smooth handling of the error. Thus the supervisor state provides a degree of security in a system (if desired). The MC68000 has the protected instructions to prevent a user program from managing to sneak into supervisor state without going "through the door"—i.e., without using a TRAP instruction to get into the supervisor state (see Sec. 7.7.2).

Intel 8086 family   All registers in the Intel 8086 family are 16 bits long. The general registers are divided into data and pointer-index (i.e., address regis-

Table 7.1

Register number	Register address	Register name	Special use
0	000	AX (accumulator)	Word multiply, word divide, I/O
1	001	CX (counter)	Count for strings, shifts, rotates
2	010	DX (data)	Word multiply, word divide, I/O
3	011	BX (base)	Base register in addressing
4	100	SP (stack pointer)	Stack operations
5	101	BP (base pointer)	Base register
6	110	SI (source index)	String source and index register
7	111	DI (destination index)	String destination, index register

ters). Each register has some special use (see Table 7.1). The uses of these registers in addressing are discussed in Sec. 7.1.5.

The segment registers—CS, DS, ES, SS (see Fig. 7.2)—are used to extend the addressing space beyond the 16-bit address registers' range of 64K byte. The segment registers can be considered to hold a 16-bit *segment address* which becomes the top 16 bits of a 20-bit *segment address component* (bottom 4 bits = 0). To form a 20-bit (1 Mbyte) physical memory address, the 20-bit segment address component is added to a right-justified 16-bit *offset address component* computed as a result of the various addressing calculations to be described later. Thus the physical address is greater than the 20-bit segment address component by an amount up to 64K byte.

Since there are four segment registers, there can be up to four active segments in memory. The *code segment* contains the instructions. The *stack segment* contains a region called the *stack*, whose use is described in Sec. 7.2. The *data segment* contains data referenced by the instructions that is not stored in the stack or in the extra segment or as part of an instructioon. The *extra segment* is always used in the computation of the physical address of a string destination. The segments can be nonoverlapping, partially overlapping, or completely overlapping to conserve memory when the full 64K byte is not required in any segment. When completely overlapped, the contents of the segment registers are equal (CS = SS = DS = ES).

The program counter in the 8086 family is called the *instruction pointer* (*IP*).

A *status word* contains 4 bits (overflow, carry, zero, sign) that carry the same results of CPU operations as the condition codes in the MC68000. One bit (auxiliary flag) is required by the 8086 family for its binary-coded decimal arithmetic instructions. An extra bit (the parity bit) in the 8086 can be used to indicate errors in data received by the processor (see Chap. 10). The status word also has a trace mode bit, called the *trap flag*, that can be set when the CPU is to be single-stepped; a single *interrupt enable bit* that recognizes external interrupts to the CPU when set; and a *direction flag* that when set

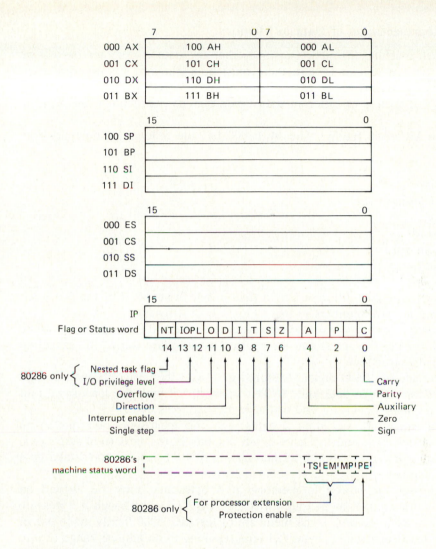

Figure 7.2 Programming model for the Intel 8086 family.

autodecrements the index register during a string instruction, and when reset autoincrements the index register.

There is no supervisor state available on the 8086, 8088, 80186, and 80188 processors. However, there is a supervisor state on the 80286 called the *protected state* which contains an extra register, called the *machine status word*, shown in Fig. 7.2. This word contains 4 bits, one of which (protection enable, PE) when set indicates the active protected state. The remaining 3 bits (MP, EM, and TS) are used to coordinate with a processor extension such as a numeric coprocessor.

7.1.2 Representation of Data in Memory

Since each machine instruction assumes that the data it operates on will be available in a particular format, there must be a standard storage format for each data type in memory. One measure of the "power" of a processor is the number and variety of different data types that its instructions can process.

Motorola MC68000 family   The MC68000 has the following data types (see Fig. 7.3):

- Bit
- BCD digit
- Byte
- Word
- Longword
- Address

A data type of a different length than a byte is accessed by the processor by reading integral numbers of bytes (1, 2, 3, or 4 bytes).

Bit data in memory is accessed by reading a byte and then operating on the correct bit. Both the access and the operation take place during the execution of a single instruction.

Although two BCD digits are stored in 1 byte, they are also accessed and processed as a single data unit. BCD data is stored with the least-significant digit at the least-significant end of the highest-address byte.

Data *words* are stored in 2 bytes with the most-significant bits in an *even-addressed* byte and the least-significant bits in the next (odd-addressed) byte. *Longwords* must also be aligned on even-byte boundaries and take up 4 bytes altogether.

Addresses are stored in memory in 4 bytes and must be aligned on even-byte boundaries as in longwords (see Fig. 7.3). Although 4 bytes are assigned to each address, some members of the MC68000 family make use of only the least-significant 3 bytes (24 bits) resulting in an address range of 0 to 16 Mbytes, while other processors have 32-bit addresses (0 to 4 Gbyte addressing range) but still allowing software developed for the 16-Mbyte MC68000 to run unmodified.

Intel 8086 family   The Intel 8086 family has the following data types (see Fig. 7.4):

- *Unpacked BCD digit*. A digit contained in a byte.
- *Packed BCD digit*. A digit stored with least-significant (LS) digit in LS nibble of the lowest address byte.
- *String*. A contiguous sequence of 1 to 64K bytes ordered from low- to high-memory addresses.

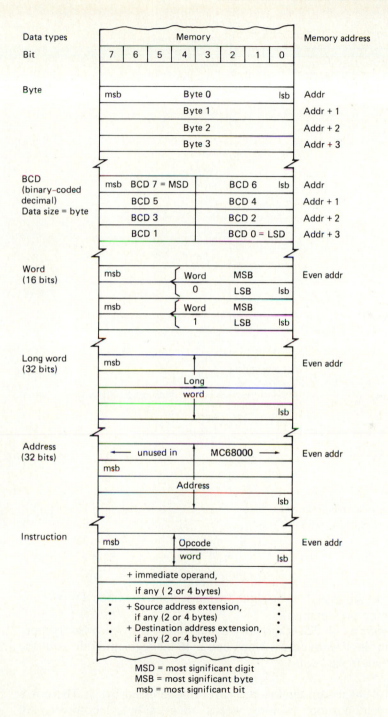

MSD = most significant digit
MSB = most significant byte
msb = most significant bit

Figure 7.3 Data organization in memory of the Motorola MC68000 family.

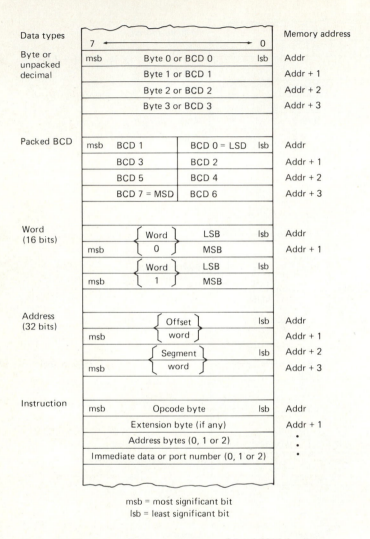

msb = most significant bit
lsb = least significant bit

Figure 7.4 Data organization in memory of Intel 8086 family.

- *Byte*. A signed or unsigned binary numeric value.
- *Word*. A 16-bit signed or unsigned binary numeric value (twos complement representation).
- *Pointer* (*address*). A 32-bit quantity composed of a 16-bit offset component in the lower-address word and a 16-bit segment address in the higher-addressed word.

Words can be aligned on either odd- or even-addressed bytes. There is a performance penalty for not aligning words on even bytes, however. All memory accesses take place one word at a time over the 16-bit bus of the 8086. The least-significant byte of the 16-bit data comes from an even address and the most-significant byte comes from an odd address. Thus if a data word starts on an odd address, only one-half of it is obtained on the first 16-bit memory read (the most-significant half) and a second memory cycle is required to get the second half.

It is important to note that *Intel and Motorola have chosen opposite conventions for the storage of data in memory*: Motorola stores the most-significant byte (MSB) of a multibyte data type (i.e., address, longword, or word) in the *lowest* memory address and Intel stores the most-significant byte in the *highest* memory address of the corresponding data type.

7.1.3 Representation of Instructions in Memory

In general, a machine-language instruction can contain several types of information:

1. The operation code ("opcode")
2. The data type (or size)
3. Data source addressing information
4. Data destination addressing information

The operation requested by the operation code determines the need for further information. For example, nonmemory reference instructions do not require any data of the type 2, 3, or 4 just described.

Motorola MC68000 family   The MC68000 has *variable-length instructions*. Each instruction has at least one word containing an operation code. The number of subsequent words that will be fetched depends on the operation code and the addressing modes specified in the first word. The machine-language instructions all start on even bytes and the operation code word is 2 bytes in length (see Fig. 7.3).

In MC68000 instructions, this information is encoded in up to four fields holding

1. An instruction word including a variable-length operation code
2. An immediate operand (optional)
3. A source address extension (optional)
4. A destination address extension (optional)

The contents of the operation code field determine the presence of any remaining instruction fields. The remaining fields are in the order shown in Fig. 7.3. For short instructions, such as MOVEQ ("move quick"), all the information required is contained in a single 16-bit word. MOVEQ moves 1 byte of *immediate* data (data stored in the instruction), nnnnnnnn, into data register number ddd. The bit pattern in memory for the MOVEQ instruction is

MOVEQ

op. code	reg

Contents of even address: 0111 ddd 0
Contents of next address: nnnnnnnn

Instructions can refer to data that is longer than 1 byte and the immediate data can thus be 1, 2, or 4 bytes long. Since the MOVEQ instruction is implicitly a byte operation, the immediate data is only 1 byte long.

Data manipulated by the instructions can be also be located in CPU registers or memory. The next field in Fig. 7.3 can contain a *source address extension*. The use of this information will become evident in the subsection describing *addressing modes* (Sec. 7.1.5). The length of this field can be either 2 or 4 bytes.

Similarly, the next field is the *destination address extension*, which can be either 2 or 4 bytes long.

The exact instruction formats for each instruction can be obtained from the Motorola user's manual listed in the bibliography of this chapter. The maximum instruction length is 10 bytes.

Intel 8086 family   Instructions in the Intel 8086 family are also of variable length. The operation code, however, is contained in a single byte and thus the minimum instruction length is 1 byte by comparison with the 2-byte minimum instruction length of the MC68000.

When more than the operation code is present the order becomes

1. Operation code (1 byte)
2. Operation-code extension (1 byte if present)
3. Address component (2 bytes if present)
4. Immediate data or port number (1 or 2 bytes if present)

The maximum instruction length is 6 bytes; for example, "MOV immediate to memory" is composed of the combination

Operation code 1 byte	Operation extension 1 byte	Destination address 2 bytes	Data 2 bytes

For the exact instruction format for each instruction one should consult the Intel user's manual listed in the bibliography of this chapter.

7.1.4 Instruction Format in Assembly Language

The same information that is required for machine code can be represented in assembly language in more easily readable form for humans. A set of mnemonic equivalents to the machine instructions for a particular microprocessor can be defined by anyone. The semiconductor manufacturer generally defines one set in the data books describing the microprocessor. When a software development system is used (see Chap. 15) that is not developed by the chip manufacturer, a different set of mnemonics is often used in the assembler for the development system (e.g., Tektronix 8560 or Hewlett-Packard 64000 development systems). It is important to be aware that such differences can exist.

Also, when a high-level language compiler is used that has not been developed by the chip manufacturer, the assembler that goes with it may use different mnemonics. For example, the MC68000 assembler mnemonics that are used in this chapter were defined by Whitesmiths Ltd., an independent software vendor who wrote the C compiler used for the C examples in this book. The Whitesmiths' assembler does not use capitals in the mnemonics, and several instruction mnemonics differ from the Motorola mnemonics. For example, the mnemonic for the Motorola assembler command "move (data at the effective address) to the status register" is

$$\text{MOVE} \langle \text{effective address} \rangle, \text{ SR}$$

whereas the Whitesmiths' command is

$$\text{mtsr} \langle \text{effective address} \rangle$$

The use of the term "effective address" will be described in Sec. 7.1.5.

Although details of Whitesmiths' assembler will not be explicitly given here, there is sufficient context so that the meaning of each instruction will be evident.

7.1.5 Addressing Modes

It is advantageous for instructions to be able to refer to the locations of data in a variety of ways, called addressing modes, first introduced in Sec. 2.2.1. Each microprocessor family has its own repertoire of addressing modes selected by the microprocessor architects. One of the reasons for having a variety of addressing modes is to reduce the number of bytes associated with each instruction. One way that this can be accomplished is by indirect addressing (see later in this section). Another goal is to efficiently access data in *data structures* such as lists and stacks (also to be described).

The MC68000 MOVE instruction and the similar Intel 8086 family MOV instruction will be used throughout the following discussion. This instruction moves data from a source location to a destination location. Each MC68000 addressing mode will be described first and then the equivalent Intel 8086 addressing mode(s) will be discussed.

The format of the Motorola MOVE instruction will be given in both its machine representation and its assembly-language representation (Whitesmiths' notation for MC68000). In the figures that follow, the MC68000 instructions are shown as words for convenience rather than for their memory representation as a byte-addressed array (see Fig. 7.3). Registers will be represented in 32 bits as they are in the machine.

Since the minimum Intel instruction length is 1 byte, the Intel instructions will be shown in units of 1 byte (as addressed in memory).

Motorola operation code   Let's first look at the fields that Motorola allocated to the MOVE instruction (Fig. 7.5). The unique portion of the operation code is only 2 bits, both 0. The operand size (ss) can be byte (ss = 01), word (ss = 11), or longword (ss = 10). In assembly language the operand size is specified by appending a ".b" (byte), ".w" (word), or an ".l" (longword) to the mnemonic for the operation code. Thus a MOVE instruction which moves a 16-bit word would be written as "move.w". The remaining bits are divided into *source address* and *destination address* components. In each of these, there is a "register" and a "mode" field. The *mode field* and *register field* together are used to determine which address mode the processor must use.

There are 12 addressing modes defined for the MC68000 CPU but there are only 3 mode bits (giving only eight possible codes). Since five of the addressing modes do not use the data or address registers, one of the eight codes in the mode bits (mode 111) uses the 3 register bits to specify one of the five modes that do not require a data or address register to be specified.

How each mode is specified and operates will be described next. The name used for each mode was given by Motorola. For the purposes of the following discussion, the destination will always be data register 1 (register = 001, mode = 000, and called "d1" in assembly language—see Fig. 7.6). The source address will be determined by the mode under discussion. Also, the size of the data will be assumed to be "longword" (ss = 10).

Intel operation code   Rather than dividing up the 1-byte operation code into consistent fields for the MOV instruction, Intel uses seven different operation codes to specify the different groups of addressing modes. The only fields which appear in certain opcode bytes are

1. A 1-bit *width field*, "w," has w = 1 when the data is a word (16 bits) and w = 0 when the data is a byte.

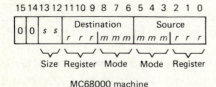

MC68000 machine

Figure 7.5 MC68000 machine-instruction format for MOVE.

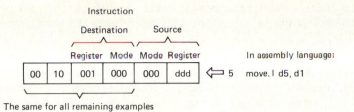

Figure 7.6 MC68000 data register direct addressing.

2. A 3-bit *register field*, "reg," that can be set to any register number from 000 to register 111. If w = 1, then all eight 16-bit registers can be used (AX, CX, DX, BX, SP, BP, SI, DI). If w = 0, then only the 8-bit data registers are valid (AL, CL, DL, BL, AH, CH, DH, BH).
3. A 1-bit *direction field* "d" when set to 1 indicates that the data destination is a field called "reg" in the operation code extension byte; otherwise "reg" is the source.

For all the 8086 examples to follow we will assume that the destination is the accumulator register (reg = AX = 000) and the source will vary depending on the addressing mode.

The operation-code byte and the following extension for the MOV instruction are

$$\text{Operation code} = 100010 \text{ d w} = 100010 \text{ 1 1}$$
$$\text{Extension} = \text{mod reg r/m} = \text{mod 000 r/m}$$

The resulting bit pattern is shown, where
d = 1, reg = destination register. (If d = 0, reg = source register.)
w = 1, move a word. (If w = 0, move least-significant byte) and mod, r/m depend on which addressing mode is selected (to be discussed).

a. Register addressing In register addressing, an operand is located in a CPU register.

Motorola data register direct (mode 000) In the data register direct addressing mode, the location or *effective address* of the source operand is the data register whose binary-coded number appears as "ddd" in bits 0 to 2 of the instruction (Fig. 7.6).

An example of a MOVE instruction using data register direct addressing would be represented in assembly language by

$$\texttt{move.l d5, d1}$$

The meaning of this is "Take the longword data in data register 5 (ddd = 101) and put it in data register 1." This mode simply takes the data *directly* out of the data register.

Motorola address register direct (*mode 001*) In address register direct addressing, the effective address of the source operand is the address register whose binary-coded number appears as "aaa" in bits 0 to 2 of the instruction (Fig. 7.7).

An example of a MOVE instruction using address register direct addressing is represented in assembly language by

$$\texttt{move.l a5, d1}$$

The meaning of this is "Take the longword data in address register 5 (aaa = 101) and put it in data register 1."

This mode is equivalent to mode 000, except that the address register is being used. Since the address registers cannot be used for byte data, the size must not be specified as byte (i.e., ss must not be 01).

Intel register operand mode (*mod = 11*) For register operand addressing, the extension byte is coded appropriately:

$$\text{Extension} = \text{mod reg r/m} = 11\,000\,\text{src}$$

The resulting bit pattern is shown for moving a word of data from a register source (src = any register from 000 to 111) to the accumulator destination (reg = AX = 000), where

$$\text{mod} = 11 \text{ for register operand mode}$$

When mod = 11 and d = 1 (in operation-code byte), then r/m = source register and reg = destination register (note that r/m = destination register and reg = source register if d = 0).

A separate operation code (10001100) is required to MOV data from a segment register to memory or a register. The extension byte defines the destination.

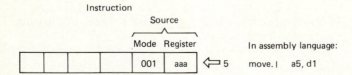

Figure 7.7 *Address register direct* addressing.

b. Register indirect addressing In register indirect addressing, the *effective address* of an operand is contained in an address register. Since the *address* of the operand is contained in the register rather than the operand itself, the term *indirect addressing* is used.

Motorola address register indirect (mode 010) In Fig. 7.8 the effective address of the source operand is contained in address register number "aaa". For the example,

move.l (a5), d1

the meaning of this instruction is "Use the contents, mmm . . . m, of address register 5 as the address of longword data in memory, fetch the data, and deposit it in data register 1."

This mode is very useful in that it allows a single-word instruction to address the whole of memory. By comparison with the "absolute long address" mode to be described later, it saves two instruction words.

Motorola address register indirect with postincrement (mode 011) Here the effective address is calculated as in address register indirect (mode 2) (steps 1 and 2 shown in Fig. 7.9). After accessing memory (step 3), the address register is incremented by 1, 2, or 4 bytes to get past the present byte, word, or longword, respectively, to point to the next operand in memory (step 4).

An example of the MOVE instruction using this mode is

move.l (a5)+, d1

This mode is useful for accessing data in lists. After each indirect access, the address register is automatically incremented by the size, in bytes, of the data item being moved and so is ready to read the next element of the list, and so forth. This mode can also be used to get data from a data structure called a *stack*. This will be discussed in detail in Secs. 7.3 and 7.4.

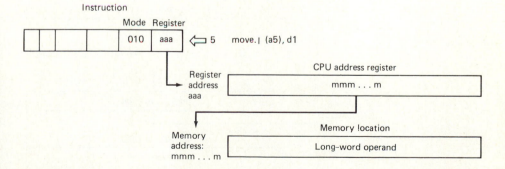

Figure 7.8 *Address register indirect* addressing.

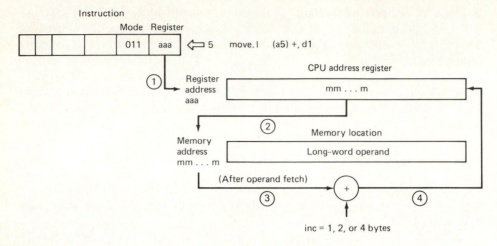

Figure 7.9 *Address register indirect with postincrement* addressing.

Motorola address register indirect with predecrement (*mode 100*)   In this mode the address found in the address register (pointed to by the instruction source register field) is decremented by 1, 2, or 4 bytes, depending on the size of the operand, *before* it is used (step 3 in Fig. 7.10) to fetch the contents of memory.
    In assembly language, an example MOVE instruction would be

$$\texttt{move.1 -(a5),d1}$$

This mode is useful for reasons similar to those given for the previous mode.

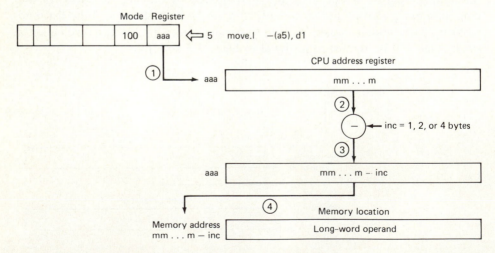

Figure 7.10 *Address register indirect with predecrement* addressing.

Intel register indirect mode (mod = 00) If the destination register is the accumulator (d = 1, reg = 000) and a word is being transferred from memory, the extension for this mode is

$$\text{Extension} = \text{mod reg r/m} = 00\ 000\ 100 = \text{SI}$$
$$101 = \text{DI}$$
$$011 = \text{BX}$$

Note that mod = 00 means "register indirect" and no displacement field follows the extension byte. The r/m field contains the register number of SI, DI, or BX registers. The r/m field can be set = BP, but then a 2-byte displacement field (set to zero for this mode) must follow the extension.

There is no equivalent autoincrement-autodecrement mode in the Intel 8086 family.

c. Based addressing This mode can be used when the location of an operand is known only relative to a *base address* before assembly of the program. The relative location is stored in the *displacement field* of the instruction, and the base address in the address register is set up at run time. This addressing scheme is called *based addressing*.

Based addressing can be used when the base address to an array is passed to a subroutine and the subroutine then accesses a certain element of the array. Thus the offset into the array is known by the assembler/linker, but the base address is determined at run time.

Motorola address register indirect with displacement (mode 101) A 16-bit displacement (+ or −) can be added to the address in the CPU address register in this mode. Before the displacement is added to the 32-bit contents of the CPU register, the displacement is correctly represented in 32 bits (step 3 in Fig. 7.11) by replicating the sign bit (bit 15) in each of the bits from 16 to 31. An example of this addressing mode is

```
move.l -12(a5),d1
```

Intel base addressing (mod 01 or 10) Based addressing in the Intel 8086 family adds a base address in the base register (BX) to either a short (8-bit) or a long (16-bit) displacement (stored after the extension byte in the instruction) to produce an effective address in the data segment. The base pointer (BP) can also be used as the base and when this is used, the stack segment is addressed.

The format for the extension byte for this form of addressing is

$$\text{Extension} = \text{mod reg r/m} = \begin{matrix} (01 = \text{short disp.}) & (110 = \text{BX}) \\ 000 \\ (10 = \text{long disp.}) & (111 = \text{BP}) \end{matrix}$$

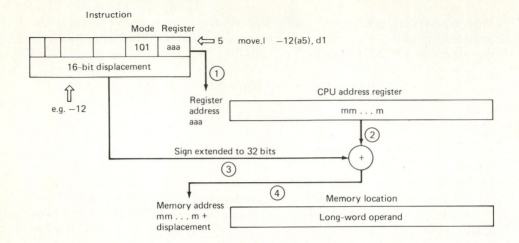

Figure 7.11 *Address register indirect with displacement* addressing.

d. Indexed addressing In indexed addressing, the base address is stored along with the instruction at assembly time and an offset is calculated at run time and is stored in a register (the *index register*).

Indexed addressing is useful for addressing a data structure such as a list where the location of the bottom of the list is determined by the assembler/ linker and the particular list elements are addressed relative to the base address at run time.

Motorola There is no indexed addressing mode in the MC68000 that corresponds to the definition just given. The indexed mode can be approximated by using the based indexed mode (discussed next) and storing the base in one register and the offset in another register.

Intel indexed addressing (mod = 10) If a 16-bit number is stored in the displacement word following the extension and this displacement is regarded as a *base address*, then this mode becomes an indexed addressing mode. Note that the operation code and extension bits are the same as in the based addressing mode with a 16-bit displacement; only the programmer's interpretation differs. The SI and DI registers can be used as the index registers. Thus the extension byte becomes

$$\text{e? } \quad \text{Extension} = \text{mod reg r/m} = 10\ 000\ 100 = \text{SI}$$
$$101 = \text{DI}$$

e. Based indexed addressing Based indexed addressing has both the base address and the offset address stored in registers, allowing both to be computed at run time. This could be used, for example, in a subroutine to access the entries to an array whose base address is passed as a parameter (see Sec. 7.4.2).

Motorola address register indirect with index (mode 110) The contents of the address register containing the base, mm..m, can be modified by two sources in this mode to produce an effective memory address (see Fig. 7.12):

1. The first is an 8-bit displacement contained in the second word of the instruction (step 3).
2. The second is the contents of an *index register* (step 4).

This mode can be used to implement a form of indexed addressing by storing a base address in the address register aaa, and an offset, called an "index," in any of the remaining general registers iii.

Note that in conventional indexed addressing, the base address would be determined at assembly time and stored in the word(s) following the instruction.

For mode 110, a displacement can be also be added to the sum to allow addressing in which only the displacement must be known prior to run time.

An example of this in assembly language is

$$\texttt{move.l 4(a5,d3.l),d1}$$

where "d3.l" adds a 32-bit index from data register 3.

Intel based indexed mode (mod = 10) The based indexed mode uses both the base registers BX and BP and the index registers SI and DI. As with the MC68000, a displacement (8 or 16 bits for 8086) stored in the instruction can be added to the sum of the base and index registers in the computation of the effective address.

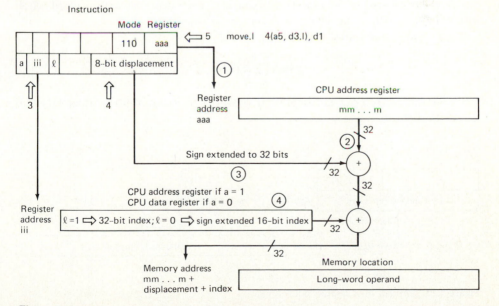

Figure 7.12 *Address register indirect with index* addressing.

If a 16-bit displacement is also added (mod = 10) and the destination register is the accumulator (reg = 000), the extension byte becomes

$$\text{Extension} = \text{mod reg r/m} = 10\,000\,000 = \text{BX} + \text{SI} + \text{disp.}$$
$$001 = \text{BX} + \text{DI} + \text{disp.}$$
$$010 = \text{BP} + \text{SI} + \text{disp.}$$
$$011 = \text{BP} + \text{DI} + \text{disp.}$$

The 2 bytes containing the displacement follow the extension byte.

Note here that the mod bits are again the same as the based addressing and the indexed addressing modes but that the r/m bits are different in order to encode the sums of registers and displacement.

f. Absolute addressing In absolute addressing, the instruction carries the full effective address of the operand.

Motorola absolute short address (mode 111, reg 000) In this mode, the second word of the instruction contains the address of the operand, giving an address within the first 32K bytes of memory (sign = 0), or the top 32K bytes of the memory when the sign bit of the word is set (see Fig. 7.13). For example, if the address is −1 (i.e., all bits are set in the 16-bit word), then the address will be the top address in memory when the word is sign-extended, regardless of the number of address bits used in the target system memory, provided that its size is a power of 2.

This mode can be advantageous since it saves one instruction word when the data is at either the top or the bottom of memory.

An example assembly-language instruction is

```
move.l 1024,d1
```

This instruction fetches the longword contents of location 1024 (decimal) and deposits them in register d_1.

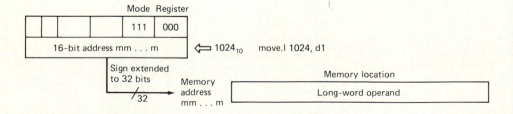

Figure 7.13 *Absolute short address* addressing.

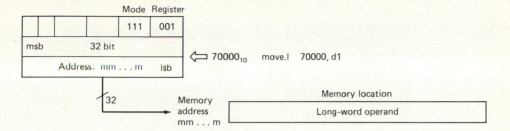

Figure 7.14 *Absolute long address* addressing.

Motorola absolute long address (*mode 111, reg 001*)   This mode is the same as the previous mode except that the address can be 24 bits (see Fig. 7.14).
An example is

$$\texttt{move.l 70000,d1}$$

In real-time systems it is very common for one to require direct access to a particular memory location, because in memory-mapped I/O, the device registers appear as locations in memory.

Intel absolute addressing (*no mod bits*)   There is a separate operation code for MOV's absolute addressing mode in the 8086 family. The operation code for the movement of memory data to the accumulator (the only register allowed to receive absolute-addressed data) becomes

$$\text{opcode} = 101000\,w$$

If we wish to transfer a word of data (w = 1), then the full instruction becomes

$$101000\,1 = \text{byte 1 (opcode)}$$
$$\text{low add.} = \text{byte 2}$$
$$\text{high add.} = \text{byte 3}$$
$$\left.\right\} \text{source address}$$

g. Relative addressing   Relative addressing is similar to based addressing except that the base register is the program counter.
The purpose of this mode is to allow a programmer to access data that is *n* bytes away from the present value of the program counter. This could be used, for example, to read an operand stored along with the program.

Motorola program counter with displacement (*mode 111, reg 010*)   The register used in this mode is the program counter, but otherwise this mode behaves like "address mode indirect" (see Fig. 7.15).

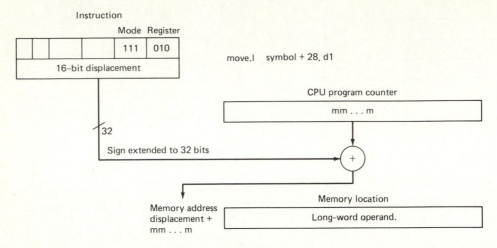

Figure 7.15 *Program counter with displacement* addressing.

In assembly-language programming, one of the ways to locate an operand is to express it with respect to a defined symbol. The assembler can then figure out what the address is relative to the program counter and insert the correct displacement.

An example is

```
symbol: move.l a3,a4 *define location called ''symbol''
        .
        .
        .
        move.l symbol +28,d1
```

The assembler will figure out how far the source operand will be away from the address in the program counter upon execution of the MOVE instruction. The displacement calculated is then placed in the source address extension word of the instruction. Assembly-language programs that make use of program counter relative addressing modes are inherently relocatable (or *position-independent*), so that they don't require a special relocating assembler.

Intel 8086 family   There is no relative addressing mode for data movement in the Intel 8086 instruction set. This mode does appear, however, in jump, branch, loop, and call instruction addressing.

h. Relative indexed   The relative indexed mode is similar to the based indexed addressing mode except that the base address is in the program counter. The index from the base address is stored in another register.

This mode can use the contents of the index register to step through a table that is displaced *n* bytes from the present address in the program counter.

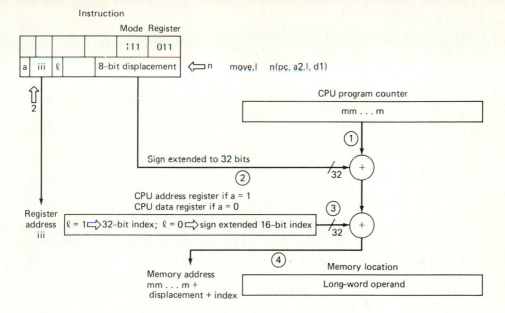

Figure 7.16 *Program counter with index* addressing.

Motorola "program counter with index" (mode 111, reg 011) This mode behaves like mode 110, with the exception of the program counter replacing the address register (see Fig. 7.16).

In this case the program counter must be explicitly identified in the assembly language:

$$\text{move.l } n(pc, a2.1), d1$$

Intel 8086 family There is no equivalent addressing mode in the Intel 8086 family.

i. Immediate In immediate addressing, the operand is stored following the operation code in the instruction. This is convenient for storage of constants in the program.

Motorola immediate data (mode 111, reg 100) In this mode the operand follows the instruction word. Byte, word, and longword operands are allowed. Single bytes are stored in the least-significant byte of the word following the instruction word.

An example of a longword move is

$$\text{move } \#73652, d1$$

where "#" indicates "immediate" (see Fig. 7.17).

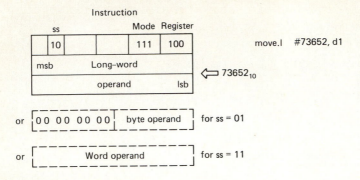

Figure 7.17 *Immediate data* addressing.

Intel immediate operand An immediate operand can be transferred to any register numbered in the "reg" field of the operation code. The data can be either byte (w = 0) or word (w = 1). The operation code for carrying this out is

$$\text{opcode} = 1011 \text{ w reg}$$

To transfer a 16-bit word to the accumulator, the instruction thus becomes

$$\text{opcode} = 1011 \text{ 1 000}$$

$$\begin{array}{ll} \text{immed.} & \text{low data} \\ \text{data} \quad = \\ & \text{high data} \end{array}$$

j. String addressing The Intel 8086 family has a group of instructions that manipulate strings of bytes or words. The index registers SI and DI are used to hold the source address and the destination address, respectively. When a MOVS instruction is carried out on byte-length data, for example, the SI and DI addresses are autoincremented to point to the next byte.

There are no equivalent instructions in the MC68000.

k. I/O Addressing As we discussed in Chap. 2, there are two methods of addressing I/O ports: memory-mapped I/O and I/O-mapped I/O. Both techniques put addresses on the address bus but in I/O-mapped, a status pin on the microprocessor informs the system that the address is an I/O address during the execution of an I/O instruction.

Motorola The Motorola MC68000 uses memory addresses to obtain I/O addresses. This is called memory-mapped I/O, which is described in Sec. 2.5.

Intel   There are two ways of addressing I/O ports in the Intel 8086 family. The first is by memory-mapped and the second is I/O-mapped input/output, also described in Sec. 2.5.

The Intel instructions that carry out the I/O-mapped I/O require a port address. The port address can be either in a byte (256 possible port addresses) in the instruction or in register DX (64K possible port addresses).

Table 7.2

Functional operations				
Arithmetic	Logical	Shift	Test	Alter
add	and	asl	cmp	s_{cc} -- *
adda -- *	andi	asr	cmpa	ext
addi			cmpi	
addq			cmpm	
addx				
abcd				
sub	or	lsl	tst	tas
suba --*	ori	lsr		
subi				
subq				
subx				
sbcd				
neg	eor	rol	btst	bchg
negx	eori	ror		bclr
nbcd				bset
muls	not	roxl		
mulu		roxr		
divs	clr			
divu				

Movement and control						
PC	Address only	Data only	Data/address	SR	USP	System
jmp	lea	movep	move	mtsr	mtusp	stop
jsr	pea	moveq	movem	mfsr	mfusp	reset
rts	movea	swap	exg	mtccr		illegal
rtr	link	ext				chk
rte	unlk					nop
bra						trap
bsr						trapv
b_{cc}						
db_{cc}						

--* = no condition codes are set in these operations.

7.1.6 Instruction Set Description

In processors such as the Intel 8086 and 8088 and the Motorola MC68000, there are many more instructions than in the earlier microprocessors, and the sophistication of the instruction set has also increased. It is necessary to organize the instruction set in your mind in some way that makes sense to you. The "best" way to do this depends on the microprocessor's instruction set as well as on the user.

The difficulty of getting familiar with an instruction set is increased by the number of apparent inconsistencies in the instruction set. For example, in the MC68000, one can shift an operand by only 1 bit if the operand is in memory, whereas multiple-bit shifts can be specified for operands in registers. Some instructions are restricted in the size of operands that they will handle or by the addressing modes that can be used with them. Some of the apparent inconsistencies are understandable when the limitations of the machine architecture are considered, but some depend on choices made by the designers for reasons ranging from the philosophical to the pragmatic.

One way of partitioning the instruction set is to view the instructions as principally

1. *Functional operations* on data or addresses
2. *Movement and control instructions* that mostly involve the movement of information between a source and destination

Almost all functional operations affect the status register bits.

Control operations such as program branch operations can be considered to be a special case of information movement operations since they involve movement of addresses into the program counter.

The MC68000 instruction set partitioned as described is shown in Table 7.2.

The meaning of the various instructions is given in App. C.

7.2 ASSEMBLY-LANGUAGE PROGRAMMING

The structure of an assembly-language program for the Intel 8085 was given in Sec. 5.1. A simple program for the subtraction of two numbers, written in assembly language for the Motorola MC68000, is given here (see Fig. 7.18).

For this assembler, any text between an asterisk "*" and the end of the line is considered to be comments and is ignored. The term ".text" is an assembler directive to identify the code segment of the program. The directive ".data" identifies the data segment of the program. The directive ".even" ensures that instructions and data words are located on even-byte boundaries, as required by the MC68000.

You may have noticed that the program has no absolute memory addresses

```
*Program to calculate diff = var2—var1
                .text                   *''instructions to follow''
                .even                   *''align them on even bytes''
program:                                *a label = address of next byte

          .
          .
          .
                move.w      var2,d1     *d1 = var2
                subr.w      var1,d1     *d1 = d1 − var1
                move.w      d1,diff     *diff = d1
          .
          .
          .
                jmp         program     *do it again

                .data                   *''data to follow''
                .even                   *''on even byte boundaries''
var1:           .word                   *one word of data
var2:           .word                   *another word
diff:           .word                   *and another
```

Figure 7.18 Example assembly language program.

specified. This is because the program is relocatable and the final location is determined by the linker, as described in Chap. 5. This program can be assembled and loaded into the target MC68000 system as a stand-alone program.

Many software projects, however, require a number of subroutines, will have parameters to pass to these subroutines, and will have both high-level and assembly-language routines. When subroutines originally written in assembly language are called by a high-level-language routine, the assembly-language subroutine must transfer control back to the high-level-language routine in the way expected by the compiler of the high-level-language calling routine. Furthermore, if the assembly-language routine requires parameters from the calling routine, these can be accessed only if the assembly-language programmer knows the compiler's parameter-passing conventions. These topics are the subject of the next few sections.

7.3 SUBROUTINE CALLS AND RETURNS

An assembly-language *subroutine* is an instruction sequence that can be executed at any point in a program without requiring the instruction sequence to be stored at every location that it is needed. Where required, the subroutine is *called* (executed) by the program. Upon completion, the subroutine *returns* to the calling program.

The scenario shown in Fig. 7.19 allows an instruction sequence to be "called" by issuing a "jump" instruction which transfers control to the instruc-

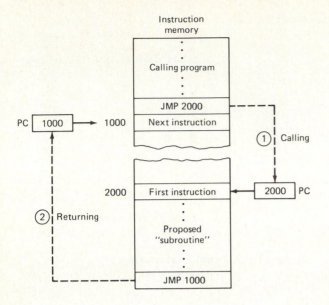

Figure 7.19 Unworkable subroutine linkage.

tion in location 2000. The return jump to location 1000 (the *return address*) is fixed and therefore the instruction sequence cannot be used at any other point in the calling program.

One way around the problem is to store the return address somewhere in the subroutine whenever the subroutine is called and then use this address when returning from the subroutine to the calling program. This used to be the principal calling mechanism in minicomputers with magnetic core memories. When the program and subroutines are stored in read-only memories, however, a return address cannot be written there at run time.

The most common method of calling a subroutine uses a separate memory area called a *push-down stack* (or simply "stack") in read/write memory to store the return address. Stacks have the property that the last data item stored on the stack is always the first item retrieved from the stack. For this reason they are sometimes called last-in-first-out stacks. The operation of storing an item on the stack is called *pushing* an item onto the stack. Conceptually, an item is pushed on "top" of all items already stored in the stack.

```
item3                          item4
item2      ''push item4''       item3
item1                          item2
                               item1
```

        Stack before   Stack after

The operation of removing an item from the stack is called *popping*, or *pulling*, an item off the stack. Conceptually, the item at the top of the stack is removed from the stack. The push and pop operations cause the stack to operate in a last-in-first-out fashion.

```
item4                     item3
item3      ''pop''        item2
item2                     item1
item1
```

Stack before   Stack after

Pushing data onto a stack involves filling the stack from high memory (the "bottom" of the stack) to low memory (the "top" of the stack), or vice versa. Stack growth in either direction is supported by the autoincrement and decrement addressing modes in the Motorola MC68000, but the Intel 8086 stack instructions (push, pop) support only stack growth toward low memory.

In the MC68000, data is pushed onto the stack through the use of a "register indirect with predecrement" addressing mode. The register used is called the *stack pointer*. The contents of the stack pointer always point to the top item on the stack. Any address register can be used as a stack pointer, and thus up to eight stacks can be manipulated at one time in the MC68000. Address register A_7, called the *system stack pointer* (SP), is actually two registers. When the system is in user mode, a reference to A_7 will cause the *user stack pointer* (USP) to be used. When the system is in supervisor mode, the *supervisor stack pointer* (SSP) will be used. In this way a supervisor system stack can be kept separate from a user system stack. The system stack is automatically used by certain instructions, such as subroutine calls and returns, for example.

Thus when calling the subroutine using the MC68000 JSR (jump to subroutine) instruction, the contents of the program counter are first pushed onto the stack (step 1 in Fig. 7.20). The system stack pointer is decremented by 4 bytes (step 2), by a "register indirect with predecrement" addressing mode that is implicitly used by this instruction. Since the program counter contains the address of the instruction immediately after the JSR instruction, the stack now contains the correct return address.

After the contents of the program counter are pushed onto the stack, the program counter is loaded with the address of the subroutine (the address is the effective destination operand of the JSR command) and execution continues at that address.

At the end of the subroutine, an RTS (return from subroutine) instruction is executed. This instruction loads the program counter with the contents of the location pointed to by the stack pointer—the contents of the top of the stack (step 4). The stack pointer is now incremented by 4 (step 5) to point to the next item on the stack, since the top item has been "removed." This process is called "*popping* the program counter (contents) off the stack." The stack and stack pointer are now as they appeared before the subroutine, and execution continues at the instruction after the JSR instruction.

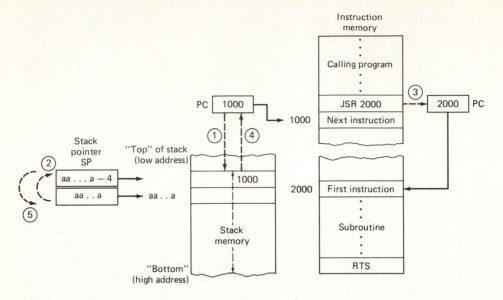

Figure 7.20 Subroutine linkage using the stack.

It should be noted here that the contents of a stack implemented with any of the eight address registers can be popped into a destination other than the program counter by using the MOVE instruction with "address register indirect with postincrement" addressing mode.

With this subroutine linkage, there is no reason why a subroutine could not call another subroutine, since the stack returns to its previous state after each RTS.

7.4 SUBROUTINE PARAMETERS

A piece of information passed to a subroutine is called a *subroutine input parameter*. Subroutines can also return parameters to the calling program, called *subroutine output or return parameters*.

Although a subroutine that receives no parameters could be used to output an internal constant to the real world each time it is called, most subroutines operate on data passed to them and return data to the calling program (henceforth to be called the "caller"). There is a wide variety of acceptable protocols for doing this, and the best technique depends on the job.

7.4.1 Parameters in Registers

The simplest, the fastest, but not necessarily the best, way to pass a parameter is to put it in a data register and then call the subroutine. The subroutine can use the same register while operating on the data, and any result can also be

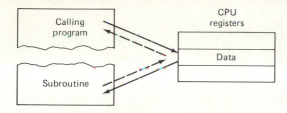

Figure 7.21 Parameter passing using registers.

stored in a register for return to the caller (Fig. 7.21). This method is fast, because internal CPU registers can be accessed faster than memory.

It is necessary to be very careful when using the registers in conjunction with high-level-language routines, since the high-level language may use some registers to store data that must remain unchanged. If the high-level-language program calls an assembly-language subroutine which uses one or more of these registers to pass parameters to another subroutine, the data stored by the high-level-language program will be wiped out. The compiler documentation for the high-level language will normally list the registers whose contents must be preserved when an assembly-language subroutine is called. It is still possible to use these registers in an assembly-language subroutine as long as the registers are saved immediately upon entry to the subroutine and are restored just prior to return from the subroutine. One easy way to do this is to push the register contents onto the stack upon entry to the subroutine and then pop them off the stack just prior to the return instruction.

Registers can also be used to return parameters to the calling routine. Many implementations of the C language make use of the registers to return a value from a C subroutine (function). For example, Whitesmiths' implementation of the C language on the MC68000 uses data register d_7 to return a value from a function to the calling function. Obviously, an assembly-language subroutine called by a C program can use the same technique to return a value to the C-language calling program.

7.4.2 Parameters in Dedicated Memory

Another technique for passing parameters is to use a dedicated region of memory called a *parameter area* (Fig. 7.22). The calling program deposits the parameters in the parameter area and the subroutine reads them. Return parameters can also use this same mechanism. The dedicated memory can be defined as a part of the calling program's data area, the subroutine's data area, or outside of any routine in a *global* area.

Some problems can occur when a subroutine is being executed and a program interrupt occurs from an external device. Although interrupts are discussed later (Sec. 7.7), what happens is that the CPU saves certain registers immediately and enters an *interrupt service routine* (a subroutine) written by the user. If the interrupt service routine calls the same subroutine that was just interrupted, the new parameters passed by the interrupt service routine to the

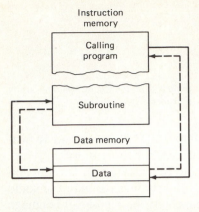

Figure 7.22 Parameter passing using dedicated memory.

subroutine must not overwrite the old parameters. If they do, then when the interrupt service routine is completed and execution continues in the interrupted subroutine, its parameters will have been corrupted. A compiler that allows such parameter corruption does not produce *reentrant code* (see Sec. 5.4).

A reentrant method that allows parameters to be stored in the data area of the calling program is to have the calling program deposit the parameters in the data area and then deposit the base address of the parameters (i.e., the address of the first parameter) into a CPU register. As long as the subroutine knows the relative location of each parameter, it can access them by adding the relative address to the base address. If an interrupt occurs, the base register is saved on the stack. If the interrupt service routine calls the subroutine, it passes the base address of its own data area in memory, which is distinct from the area used by the original calling program. When execution returns from the interrupt service routine, the previous base address register is popped off the stack and the subroutine can continue to access the original parameters which have been preserved.

7.4.3 In-Line Parameter Area

Another way to pass parameters when the program is in read-write memory is to store the parameters "in-line" after the instruction (JSR) that calls the subroutine (see Fig. 7.23).

Since the "return address" points to the parameters, the subroutine can use them and then increment the return address past the parameters to point to the true return address.

In a similar way, the addresses of the parameters can be stored after the call instruction, and the subroutine then uses this address to fetch the parameters. The use of the address-pass instead of the parameters themselves is called *call by reference*.

A compiler using this technique would not produce ROMable code (see Sec. 5.4), since the program must be stored in read/write memory.

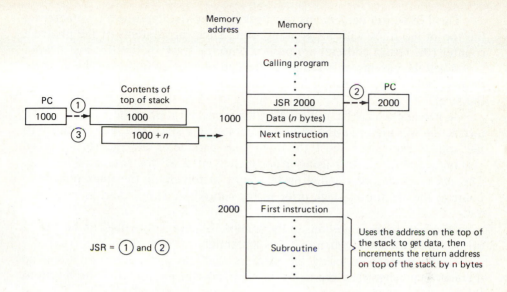

Figure 7.23 Parameter passing in-line.

7.4.4 Parameter Passing on the Stack

The user stack can also be used to pass parameters as well as to store a return address (see Fig. 7.24). Before the JSR instruction is issued, the parameters are pushed onto the stack. This is accomplished on the MC68000 by using a MOVE instruction with a predecremented stack pointer (SP) for destination addressing.

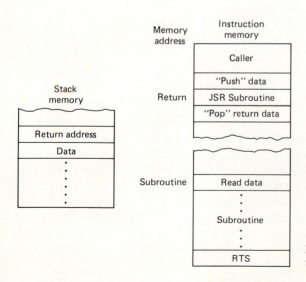

Figure 7.24 Parameter passing on the stack.

This instruction decrements the stack pointer and deposits a parameter on the top of the stack with SP pointing to it. Then the contents of the program counter (the return address) are pushed onto the stack by the JSR instruction.

When the parameter is required, the subroutine can address it by a 4-byte displacement (+ve) from the SP to jump over the PC contents on the top of the stack.

A parameter can be returned in the same location if desired. An RTS instruction will then *pop* the return address off the stack and return control to the caller. The caller must then pop the parameter off the stack (called "cleaning up the stack") using a postincremented SP for source addressing. The stack can also be cleaned up by simply incrementing the stack pointer to point to the old top of stack prior to the subroutine call if there is no return data.

Of course there is no limit to the number of parameters that can be passed on the stack except for memory size constraints.

One advantage of using the stack is that the parameters are using up memory only when they are being used by the subroutine. After the return to the main program, the storage reservation disappears. This is not the case when dedicated memory is used.

Stack usage can be taken one step farther by means of a *stack frame* to allow not only storage of the parameters of a subroutine on the stack but also storage of all subroutine "local" (automatic) variables. Use of this technique can lead to subroutines that are both reentrant and ROMable.

The stack frame   A *stack frame* can be used for three purposes when calling a subroutine:

- Storing the return address
- Passing the subroutine parameters
- Providing temporary (called *dynamic* or *automatic*) variable storage for the subroutine

Thus there is no permanent (*static*) storage necessary unless the subroutine requires some variables to be held constant between invocations of the subroutine.

To set up a stack frame, an extra pointer, called the *frame pointer*, is required. The subroutine's dynamic variables and parameters lie between (i.e., are "framed" by) the address in the frame pointer and the address in the stack pointer. References to the dynamic variables and parameters are made with respect to the frame pointer.

The process of setting up a stack frame is shown in Fig. 7.25.

Assume initially that the caller has a stack frame in which the frame pointer (FP) points to the "bottom" of the frame at location zz . . z. The stack pointer (SP) points to the "top" of the frame at location yy . . y.

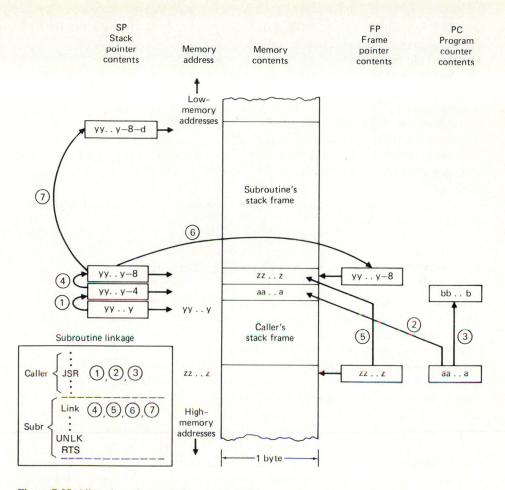

Figure 7.25 Allocation of a stack frame with LINK.

The calling program executes a JSR instruction which decrements SP by 4 bytes (step 1 in Fig. 7.23), deposits the return address stored in the program counter (PC) onto the stack (step 2), and puts the destination address of the subroutine into PC (step 3).

The first instruction of the subroutine will be a LINK instruction in the MC68000. This has three steps:

- Push FP onto the stack (steps 4 and 5).
- Copy SP into FP (step 6).
- Allocate d bytes of frame storage by decrementing SP by d (step 7).

Thus the state at the end of the LINK instruction is such that the old program counter and the old frame pointer are stored on the stack. This information is used by the UNLK instruction to exactly reconstruct the state of the stack prior to the subroutine call.

The UNLK instruction, placed at the end of the subroutine, reverses the effect of the LINK. The UNLK instruction

- Copies FP into SP
- Pops the old frame pointer zz . . z back into FP

An RTS instruction will then pop the old program counter's contents aa . . a back into the PC and increment the SP in the process to the original value yy . . y.

The stack-frame construction has also been used in C-language translation, as will be seen in Sec. 7.5.

7.5 LINKING HIGH- AND LOW-LEVEL PROGRAMS

Oftentimes an assembly-language routine must be written and linked to a high-level-language calling program (see Sec. 5.4). For example, if a special task requires high speed and the equivalent C program is too slow, a more efficient assembly-language routine might be desired.

Conversely, an assembly-language program might require a C-language library routine. In either case, parameters must be transferred between assembly-language subroutines and C functions.

One of the benefits of high-level language is that details of passing parameters are hidden from the user; but when a high-level–low-level interface is desired, all these details must be exposed by referring to the compiler documentation. For example, in Whitesmiths' version of the C language, a parameter is passed to the subroutine on the stack, as just described. However, when two or more parameters are listed, as in the following function call contained in a C program,

```
z = function(x, y); /*line from program*/
```

.

the parameters are pushed on the stack from right to left. Thus y is pushed and then x. An assembly-language program must retrieve these parameters in the correct order from the stack. It should be pointed out that the conventions used by various compiler-writers may be similar but not identical. You should always study the documentation for the compiler that you are using to see exactly how it handles parameter passing.

In addition to reading the compiler documentation, a good way to learn how to link assembly-language subroutines to high-level-language programs is to compile a simple program that calls a subroutine and study the assembly-language linkages set up by the compiler for the high-level-language subroutines.

We have done this in Fig. 7.26 for a very trivial program in C. The program passes one parameter "a" and returns one parameter "b". The program defines only *automatic* (i.e., *dynamic*) variables.

The first thing to notice about the translation of the program into assembly language is that the main program has a return at the end (RTS). Thus even the main program is a subroutine. The program that calls the main program will be discussed in Sec. 7.6.

The compiler uses an underscore (such as in _main) to indicate that the symbol is to be known as an external (global) symbol.

The status of the stack after the first LINK instruction has been executed is shown in Fig. 7.26. It can be seen that the FP and SP differ by 12 bytes. Each "int" in Whitesmiths' C for the MC68000 is 32 bits long (4 bytes). There are two integers plus one integer parameter that get passed to the subroutine, making three integers, or 12 bytes in all.

Note that the C parameter passed to the subroutine is the value of the parameter, not its address. Parameters are thus *passed by value*, as opposed to the *pass by reference* method described earlier (Sec. 7.4).

The second line and third lines of the program take the integer "1" stored as an immediate byte in the MOVEQ instruction and deposit it in the automatic variable "a" on the stack. The fourth line transfers the contents of "a" to parameter storage for the subroutine called by a JSR.

The subroutine has a 4-byte automatic variable "b" created by LINK, and the parameter "a" is transferred into it by the next MOVE.

The return parameter is seen to be passed in register d_7 as opposed to being passed on the stack. Thus Whitesmiths' C implementation uses the stack frame for input parameters and automatic variables but returns parameters via the registers.

If an assembly-language program is being written to interface to a C program, it must obey the parameter-passing conventions of the compiler being used. Details on the conventions should be sought in the compiler documentation. Other C implementations may have different conventions. Other high-level languages are very likely to have different conventions.

7.6 START-UP ROUTINES

As pointed out in Sec. 7.5, the main C program is also a subroutine, so it is necessary as a minimum to have a short routine that transfers control to the main program and accepts control from the main program upon completion. This start-up routine may also have to initialize some processor registers such as the stack pointer and may set the processor interrupt system to the desired state. The exact implementation of this start-up routine depends on the target system.

When a computer is turned on or reset, it starts executing a program located at some point in memory determined by the contents of the program counter. In some computers the program counter is set to some predetermined

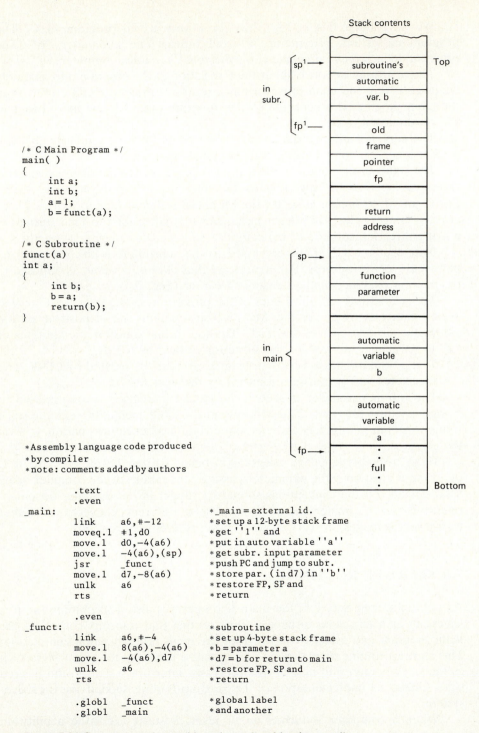

```
/* C Main Program */
main( )
{
    int a;
    int b;
    a = 1;
    b = funct(a);
}

/* C Subroutine */
funct(a)
int a;
{
    int b;
    b = a;
    return(b);
}
```

```
*Assembly language code produced
*by compiler
*note: comments added by authors

        .text
        .even
_main:                                    * _main = external id.
        link     a6,#-12                  * set up a 12-byte stack frame
        moveq.l  #1,d0                    * get ''1'' and
        move.l   d0,-4(a6)                * put in auto variable ''a''
        move.l   -4(a6),(sp)              * get subr. input parameter
        jsr      _funct                   * push PC and jump to subr.
        move.l   d7,-8(a6)                * store par. (in d7) in ''b''
        unlk     a6                       * restore FP, SP and
        rts                               * return

        .even
_funct:                                   * subroutine
        link     a6,#-4                   * set up 4-byte stack frame
        move.l   8(a6),-4(a6)             * b = parameter a
        move.l   -4(a6),d7                * d7 = b for return to main
        unlk     a6                       * restore FP, SP and
        rts                               * return

        .globl   _funct                   * global label
        .globl   _main                    * and another
```

Figure 7.26 C program and assembly code produced by the compiler.

value (normally zero) when the computer is turned on or reset. In other computers, the program counter register is loaded with the contents of some predetermined memory locations. In the first case, the computer will start execution at the value (address) loaded into its program counter. The start-up routine (or a JUMP to the start-up routine) should be placed there. In the second case, the computer will start execution at the address stored in the predetermined memory locations. The address of the first instruction in the start-up routine should be stored in these locations.

For example, if the target system is totally stand-alone with its program in nonvolatile memory, an MC68000 will start executing in the location whose address is stored in the 4 bytes starting at location 4 in memory (see Sec. 7.7). The first instruction of the start-up routine can be located here. This start-up routine could call an operating system, if any, which can, in turn, call a C program such as "main" or the startup routine could call "main" directly.

The program segment shown next is an example of a start-up routine that calls "main" directly.

```
.text
.even
movea.1  #OX7FFE,a2   *7FFE (hex)=highest even RAM
                      *address
mtusp    a2           *set user stack pointer to 7FFE
andi.w   #OXDFFF,sr   *put the system in user state
jsr      _main        *call function main( ) in user
                      *program
.
.                     *instructions to handle returns
.
```

It is assumed here that the target memory (in this case the Motorola M68000 Educational Board), has 32K bytes of RAM memory. The stack starts at the top of memory on an even address.

The ANDI instruction sets status bit 13 to 0 while leaving the other bits alone just before transferring control to the main program.

Since stacks are filled from high memory to low memory, a high even address must be selected for the "bottom" of the user stack and this is placed in USP, the user SP. The system is then placed in user mode by forcing the supervisor bit to be zero. Finally, the program "main" is called. If main returns to the start-up routine, instructions must follow the JSR to handle it.

7.7 COMPUTER INTERRUPT SYSTEMS AND SERVICE ROUTINES

A very important aspect of real-time computing systems is the capability of handling interrupts. As introduced in Sec. 2.3.4, an *interrupt* is a signal to the processor from an external source that can cause the processor to be switched from its present instruction sequence to execute an appropriate program called an *interrupt service routine*, or *interrupt handler*. The microprocessor's facilities that support interrupts are called the *interrupt structure*, or *interrupt system*. The combination of the interrupt structure and the interrupt service routines are useful in synchronizing computer activities with external real-time events.

Because of the differences in interrupt systems between different microprocessor types, interrupt handling software is very machine-specific and nonportable. The programmer must learn the details of the interrupt system of every microprocessor for which interrupt handling software is to be written. In most cases, portions of the interrupt handling software must be written in assembly language, since machine-specific operations are required which are not available in common high-level programming languages. There are, however, enough similarities between interrupt systems that some general comments are possible for dealing with interrupts and the programming of interrupt service routines.

7.7.1 Interrupt Systems and Events

Interrupt input signals Interrupts from external devices are initiated by a logic level signal applied to an interrupt input to the microcomputer. One of the first pieces of information that the designer should extract from the documentation for the microcomputer is whether the interrupt is caused by the signal going from a high to a low level (a *low-true* signal; see App. B) or by the signal going from a low to a high level (a *high-true* signal). The designer must ensure that the interrupt signal from the external device makes the proper transition when an interrupt is supposed to occur.

The designer must also check to see whether the interrupt inputs are *edge-sensitive* or *level-sensitive*. An edge-sensitive interrupt input has an internal flip-flop which is set when the interrupt signal makes a transition from low to high (rising edge-sensitive) or from high to low (falling edge-sensitive). Such an input is useful when the interrupt signal is a short pulse that might be missed if interrupts are momentarily disabled while the signal is active. The internal flip-flop holds the information that an interrupt has been requested even though the external signal is no longer active. A level-sensitive input has no internal flip-flop to latch the interrupt signal. If the interrupt signal becomes inactive before the microcomputer's interrupt system recognizes it, the interrupt will not occur.

Resetting interrupt request signals If it is important to ensure that the interrupt signal lasts long enough to be recognized by the microcomputer, it is equally important that the signal not last too long. Upon responding to an interrupt, the microprocessor locks out any further interrupt request signals. After completing the interrupt service routine, the interrupt system is usually programmed to respond to interrupts again. If the same interrupt signal is still active, the microprocessor will respond again. One interrupt request can thus erroneously generate several response cycles from the microcomputer. To avoid this, the designer must ensure that the interrupt signal is reset to an inactive state during the interrupt service routine prior to reenabling the interrupt inputs.

In the case of edge-sensitive interrupt inputs, the interrupt request is usually reset automatically. When the microcomputer responds to the interrupt, the internal flip-flop which latches the interrupt signal is reset to the inactive state. In the case of level-sensitive interrupt inputs, the interrupt signal may be reset by one of the signals (an interrupt acknowledge signal) generated automatically by the microcomputer as it responds to the interrupt. The interrupt signal may also be reset as a side effect of the interrupt service routine. For example, in the real-time-clock interrupt service routine to be discussed in Chap. 14, the act of reloading the counter in the Intel 8253 counter-timer IC automatically resets the interrupt signal. However, it is also possible that the programmer may have to explicitly reset the interrupt signal by writing data to a control register in the interrupting device. The only general advice that can be given is that the designer should check carefully to see how the interrupt request signal is to be reset.

Interrupt identification and dispatch   In many real-time systems there will be several external devices that can generate interrupt signals. The interrupt system must have some means of identifying the source of the interrupt and branching to the appropriate interrupt service routine. The two techniques commonly used are called *polling* and *vectoring*.

Polling   An interrupt system that makes use of polling to identify the source of the interrupt usually uses only one interrupt input line. The interrupt input is commonly low-true (see App. B) and the interrupting devices are tied to the interrupt line by *open collector gates* (see Sec. 13.1 and App. B). Any one of the sources can generate an interrupt by driving the interrupt line low. Such a system, using a Motorola 6809 microprocessor, is shown in Fig. 7.27. Upon receiving an interrupt, the starting address of the interrupt service routine is loaded into the microprocessor's program counter, causing execution to transfer to the interrupt service routine. This starting address is usually stored in predetermined locations in memory. In the Motorola 6809 system shown, the 16-bit address is stored in locations FFF8 and FFF9. The programmer must ensure that the correct address is stored in these locations.

After being entered, the interrupt service routine must first determine the source of the interrupt. This is where the polling comes in. Each interrupting device contains a status register with 1 bit indicating whether the device has requested an interrupt. The interrupt service routine polls the possible interrupt sources by addressing and reading the contents of each status register and checking to see if the bit indicating an interrupt request is set. Once the interrupting device is identified, the appropriate operations to service the interrupting device can be performed and the interrupt request signal from that device can be reset.

Interrupt systems using polling are simple and don't require much special circuitry in the microprocessor. However, the response time to an interrupt is

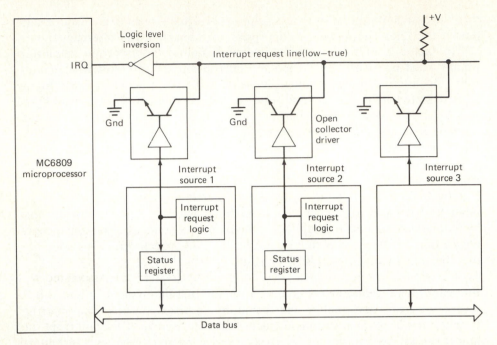

Figure 7.27 Polled interrupt system.

lengthened by the time required to execute the instructions which poll the potential interrupt sources to determine the actual source of the interrupt. Since interrupts are used to provide quick response to some external event, a faster method of identifying the interrupting device and branching to the correct interrupt service routine is often required.

Vectoring Interrupt systems using vectoring provide a faster response. In a vectored system the interrupting device identifies itself directly to the microprocessor, eliminating the need for the microprocessor to poll all the possible sources. Execution branches directly to the interrupt service routine associated with the device generating the interrupt request.

Before we begin discussing ways in which this identification can be performed, we should clarify some terminology. The term "interrupt vector" is often used rather loosely. It can refer to the code sent to the microcomputer by the interrupting device to identify the source of the interrupt; it can refer to the memory address where the starting address of the interrupt service routine is stored; or it can refer to the actual starting address of the interrupt service routine. In this chapter, we consider the interrupt vector to be the starting address of the interrupt service routine and the identifying code to be the interrupt vector code. However, the manufacturer's manuals for a particular microprocessor may use the terms differently.

In some systems, such as the Intel 8085 based system shown in Fig. 7.28, each interrupting device is given its own interrupt request line. Each interrupt

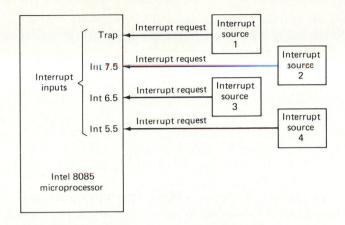

Figure 7.28 Vectored interrupt system with individual interrupt request lines.

request line has a unique starting address, or interrupt vector, associated with it which is loaded into the program counter when an interrupt occurs on that line. Such systems are sometimes referred to as *autovectored* interrupt systems. Identification of the interrupt and dispatch to the interrupt service routine are automatic and do not require any action on the part of the interrupting device other than the generation of an interrupt request signal. However, the number of interrupt sources is limited by the number of available interrupt request lines. Microprocessor ICs normally can accommodate only four or five separate interrupt request lines.

Priority encoder   One way to accommodate more interrupt request lines is to use a separate IC to encode the interrupt requests. As shown in Fig. 7.29, the individual interrupt request lines from the interrupt sources are fed to the encoder IC. It generates a binary code corresponding to the interrupt line which is requesting an interrupt. In the system shown, eight interrupt request lines can be encoded into a 3-bit binary code. This code is sent to the

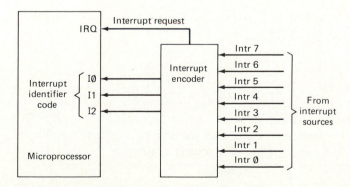

Figure 7.29 Vectored interrupt system with encoded interrupts.

microcomputer along with an interrupt request signal. The code is used within the microcomputer to determine the starting address for the interrupt service routine associated with the interrupt source.

The IC that encodes the interrupt requests must deal with situations where two or more interrupt requests occur simultaneously. Therefore, the IC is normally a priority encoder which encodes only the highest priority interrupt request if more than one request is received simultaneously. Depending on the design of the encoder, the priority of an interrupt request line may be fixed, or it may be possible for the microprocessor to control the priority of an interrupt request line by writing a control code to the priority encoder IC.

Vector code generation An alternative way to achieve interrupt vectoring is for the interrupting device to use the data bus to deliver an identifying code. The device normally first generates an interrupt request signal. When it receives an interrupt acknowledge signal from the microcomputer indicating that the interrupt has been recognized, it then places its identifying code onto the system data bus, where it is read by the microcomputer. The code is used by the microcomputer to determine the starting address of the interrupt service routine.

Figure 7.30 shows such a system. The interrupt sources use open collector drivers (see App. B) to generate active low interrupt requests on a single interrupt request line. Upon receiving an interrupt request (step 1), the microprocessor sends an interrupt acknowledge signal down the interrupt

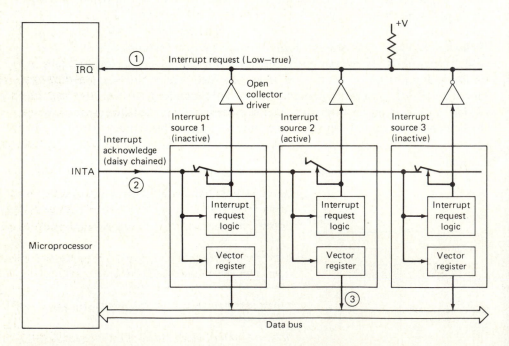

Figure 7.30 Vectored interrupt system where interrupt sources generate identifying codes.

acknowledge line (step 2) which is connected to all the interrupt sources. This signal causes the source which generated the interrupt request to place an interrupt vector code stored in a vector register onto the data bus (step 3) where it is read by the microprocessor. The interrupt acknowledge signal will also cause the interrupt source to deactivate its interrupt request signal.

A system such as this must also have a way of dealing with simultaneous interrupt requests from more than one source. A *daisy chain* is used to interrupt the acknowledge line. In a daisy-chained system, the interrupt acknowledge line is passed through each interrupt source. Internal logic within each source prevents the interrupt acknowledge signal from being passed on to other sources farther down the line if the source is requesting an interrupt. Thus the first source along the daisy chain to request an interrupt is the only one to receive the interrupt acknowledge signal and so is the only one which will place its interrupt vector code onto the data bus. Obviously, in a daisy-chained system, an interrupt source's priority depends on its position along the interrupt acknowledge daisy chain.

One problem with the system just described is that every interrupt source must have the necessary logic to handle the daisy chaining of the interrupt acknowledge line and to send a vector code to the microprocessor. Many modern microprocessors have special interrupt control ICs which are designed to reduce the need for special circuits in the interrupt sources. For example, the National Semiconductor NS16202 interrupt control unit IC is designed to work specifically with the interrupt system of the National Semiconductor NS16032 microprocessor. As shown in Fig. 7.31, the NS16202 can accept up to 16 individual interrupt request lines. It handles the arbitration of simultaneous interrupt requests and sends a single interrupt request to the NS16032 micro-

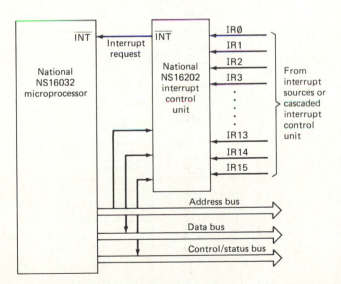

Figure 7.31 Vectored interrupt system using special interrupt control integrated circuit.

processor. The microprocessor then acknowledges receipt of the interrupt and the interrupt control unit places an 8-bit interrupt vector code on the microcomputer data bus. The system can be expanded to accept up to 256 different interrupt sources by cascading interrupt control units.

Interrupt vectors in memory In any vectored interrupt system, the vector code sent to the microcomputer is used to determine the starting address of the interrupt service routine to be executed. The code could be used directly as the address. However, in most microprocessors using vectored interrupts, an indirect method is used. In the scheme shown in Fig. 7.32, when an interrupt occurs, the previous contents of the program counter are saved (step 1), usually on a stack, so that the interrupted program can resume execution after the interrupt service routine is complete. The interrupt vector code from the device (step 2) is used to calculate an offset which is added to the base address of a table of interrupt service routine starting addresses (the interrupt vectors). The result (step 3) is the address of the memory location where the starting address (the interrupt vector) of the interrupt service routine is stored. In systems where the memory word width is less than the width of memory addresses, following memory locations are used to store the remainder of the

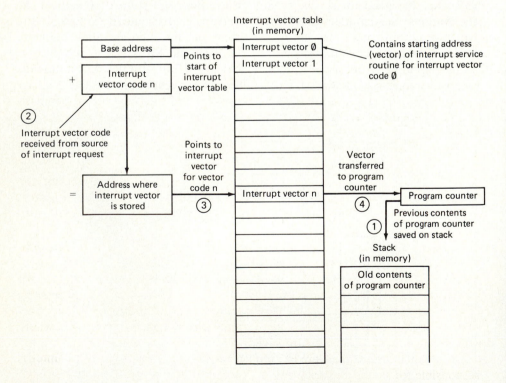

Figure 7.32 Interrupt service routine dispatch in a system with vectored interrupts.

address. This starting address is transferred to the program counter register (step 4), causing execution to branch to the interrupt service routine. The previous contents of the program counter are saved, usually on a stack, so that the interrupted program can resume execution once the interrupt service routine is complete. The advantage of this indirect scheme is that the interrupt vector code need be only long enough to accommodate the maximum number of interrupt sources expected in the system rather than being the length of a memory address.

In a vectored interrupt system such as the one just described, the programmer must ensure that the starting addresses of the interrupt service routines are stored in the interrupt vector table. Also, the interrupt vector codes may have to be stored in the interrupt sources or in the interrupt control IC. In any event, the programmer must have a clear understanding of how interrupt sources are identified and how execution is transferred to the interrupt service routine. This can be accomplished only by careful study of the manuals for the microcomputer, since the details vary considerably from system to system.

Saving and restoring the system state   One of the requirements in responding to an interrupt is that the execution of the interrupt service routine should have no unplanned effects on the program that is interrupted. In order to ensure this, the contents of any CPU registers used both by the interrupt service routine and the interrupted program must be saved at the beginning of the interrupt service routine and restored and at the end of the interrupt service routine. This is called *saving* and *restoring* the *system state*, or *context*.

As mentioned before, the contents of the program counter register are saved automatically (usually on the stack) when execution branches to the interrupt service routine. The program counter contents are restored automatically when execution transfers back to the interrupted program from the interrupt service routine. Whether any other registers are saved and restored automatically depends on the microprocessor used. Some microprocessors do not save any other registers automatically. Others save the status or flag register automatically, since it will almost certainly be altered during the interrupt service routine. Finally, some microprocessors save all the registers automatically or switch to an alternative set of registers for the interrupt service routine.

The penalty for saving all the registers automatically is that the response time to an interrupt is lengthened by the time it takes to save and restore all the registers. In many cases the interrupt service routine alters the contents of only a few of the registers, so saving all the registers is not actually necessary. Some microprocessors make provision for two types of interrupts—slow interrupts where all the registers are saved automatically and fast interrupts where only a few essential registers are saved.

In systems where registers are not saved automatically, the programmer is responsible for saving and restoring the contents of those registers used by both the interrupt service routine and the interrupted program. When programming in a high-level language, it is not always possible to know which registers will be used, so the only safe procedure is to save all the registers.

Registers are usually saved by pushing them onto the stack at the beginning of the interrupt service routine and are restored by popping them off the stack in reverse order at the end of the interrupt service routine. If the stack is used during the interrupt service routine, the programmer must make sure that the stack pointer is pointing to the proper location when the registers are restored by popping them off the stack.

Normally registers cannot be accessed directly by high-level-language instructions. Therefore the saving and restoring of the system state must be done in assembly language. However, the bulk of the interrupt service routine can usually be written in a high-level language if no direct access to the registers is required. The procedure is to write the high-level-language portion of the interrupt service routine as a subroutine which is called by the assembly-language portion of the interrupt service routine after the registers have been saved. When the high-level-language subroutine has completed execution, it returns to the assembly-language portion of the interrupt service routine, register contents are restored by popping them off the stack, and program execution is transferred back to the interrupted program.

Enabling and disabling interrupts   There are times when the microcomputer must ignore interrupt requests. For example, an interrupt request from a device should be ignored while its interrupt service routine is being executed in order to ensure that the interrupt service routine isn't repeatedly interrupted. As another example, spurious interrupt requests, which must be ignored, may be generated while the system is initializing itself after power is applied.

As a result of the need to ignore interrupt requests at certain times, microcomputers have provision to *disable* and enable interrupts. Sometimes interrupts are said to be *masked* rather than disabled. In any case, it must be understood that the interrupt request signals are still generated by the interrupt sources when interrupts are disabled, but the microcomputer does not respond to them.

There are two basic methods of controlling the enabling and disabling of interrupts. In the simpler of the two, all interrupt inputs are either enabled or disabled with the exception of *nonmaskable* interrupt inputs which are always enabled. The nonmaskable inputs are normally used to indicate an important event which requires fast response. An example is a warning that the ac power to the computer has failed. The computer must react quickly to carry out an orderly shutdown while there is still sufficient stored energy in the power supply to run the system.

When interrupts are enabled and disabled as a single block, the interrupt system is automatically disabled when an interrupt is recognized by the microprocessor. The system is reenabled at the end of the interrupt service routine. An interrupt request from another interrupt source will therefore be ignored until the interrupt service routine is completed. In many cases this does not pose much of a problem, since interrupt service routines are typically quite short and many devices can accept a short delay in responding to their interrupt request. If one interrupt source does require extremely fast response, it can be connected to the nonmaskable interrupt input.

However, some real-time systems may have devices which require relatively long interrupt service routines as well as some devices which require very fast response to their interrupt requests; in which case, an interrupt system where the interrupt inputs are organized in a *hierarchical priority structure* is required. In such a system, interrupt inputs can be selectively enabled or disabled. The interrupt sources requiring the fastest response are connected to the interrupt inputs with the highest priority. When an interrupt occurs, only the interrupt inputs with a priority less than or equal to the priority of the interrupt input that caused the interrupt are disabled. Thus an interrupt request from a high-priority device can interrupt the interrupt service routine of a lower-priority device.

Many modern 16-bit microcomputers have a priority-based interrupt structure based on the concept of a *priority level*. This priority level is incorporated as part of the status, or flag, register so that it can be monitored or changed under program control. The microprocessor will accept interrupt requests only from those inputs having priority higher than its current priority level. When an interrupt is accepted, the priority level is automatically changed to be equal to the priority of the interrupt source, thereby disabling any interrupts from any sources at the same level of priority or lower. When execution of the interrupt service routine has finished, the priority level is usually restored to the level prior to the interrupt.

In many cases the programmer will wish to enable or disable interrupts within a program. Some microprocessors have special machine-language instructions which allow the programmer to do this. The Intel 8085 microprocessor has the EI instruction to enable the interrupt system and the DI instruction to disable the interrupt system. In microprocessors where the current interrupt priority level is stored in a bit field of the status register, interrupt priority can be changed simply by altering the priority level bits of the status register. By setting the priority level to its highest value, all interrupt inputs except those that are nonmaskable are disabled. By setting the interrupt level to its lowest value, all interrupt inputs are enabled.

The method used to enable and disable interrupts under program control is machine-specific and will therefore not be available as part of a high-level language. However, the programmer can write short assembly-language subroutines to enable and disable the interrupt system. These subroutines can then be called from high-level-language programs.

Interrupt processing in the CPU From the processor's point of view then, there is a sequence of tasks to be carried out upon reception of an interrupt:

1. The processor (or external logic) determines whether the interrupt is of high enough priority to be recognized by the CPU.

2. If the interrupt is recognized, the processor acknowledges the interrupt with a signal back to the I/O devices. The processor may also lock out other interrupts until the present interrupt is serviced.
3. This signal will cause the interrupting device to put an interrupt code or interrupt vector code on the data bus, which the processor can use to find the interrupt service routine for the device in memory.
4. The processor *saves its state*—a combination of important registers such as the program counter and status register—on the system stack so that it can resume its program where it left off, after it services the interrupt. This process is also called *context saving*.
5. The processor then starts execution of the interrupt service routine at the address stored in the interrupt vector by placing that address into the program counter.
6. When execution of the service routine is complete, the processor pops the registers off the stack and resumes execution at the place it left off to service the interrupt.

The exact details of interrupt processing vary from processor to processor, but the main features—priority resolution, interrupt lockout while interrupt processing, the saving of the current state, execution of an interrupt service routine, and return to the previous state—are usually present in any system. In the more modern processors, most of the task can be carried out with few machine instructions.

7.7.2 Example of an Interrupt System

MC68000 interrupt structure   The MC68000 interrupt structure is embedded in a more general context called *exceptions*. Exceptions are events that modify the normal flow of instruction execution. These include not only device interrupts but also other external events such as bus errors (e.g., parity errors) and hardware reset requests. Internal events also produce exceptions, such as addressing errors, divide by zero, illegal instructions, and several special instructions (TRAP, TRAPV). The various exceptions are shown in Fig. 7.33.

The priority of the various exceptions and the point at which the processing of the exception takes over from the normal instruction sequence is also shown in Fig. 7.33.

If the CPU is in the *user state*, all exceptions result in a transition from user state to *supervisor state* in the MC68000. The automatic transition to an uncorrupted supervisor state upon an error condition allows the processor to recover from the error condition smoothly. The execution of an exception is, in fact, the only way to get into the supervisor state from the user state in the machine.

The supervisor state has several dedicated capabilities:

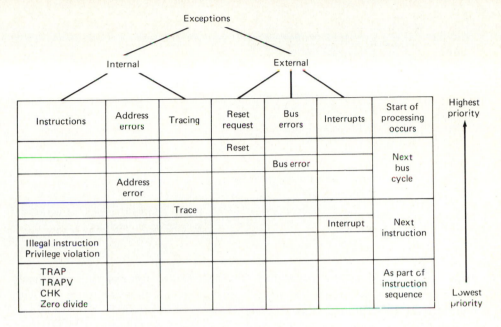

Instructions	Address errors	Tracing	Reset request	Bus errors	Interrupts	Start of processing occurs	Highest priority
			Reset			Next bus cycle	
				Bus error			
	Address error						
		Trace					
					Interrupt	Next instruction	
Illegal instruction Privilege violation							
TRAP TRAPV CHK Zero divide						As part of instruction sequence	Lowest priority

Figure 7.33 Types and priority of MC68000 exception processing.

- The status register, which contains the bit that records the CPU supervisor or user state, cannot be modified by any instruction in user state.

- The supervisor stack pointer (SSP) becomes the system stack pointer (SP) when the CPU is in supervisor state and the user stack pointer (USP) becomes SP when in user state. The USP can be MOVEd, however, when the CPU is in supervisor state by a privileged instruction (MOVE USP). This allows the supervisor program to store the user's "state" in the supervisor memory area.

- Any instructions that could modify the supervisor bit in the status register are allowed only while in supervisor state. (In user state the processor cannot be reset or halted and MOVE USP is not allowed.)

- The processor outputs a function code (on pins FC_0 to FC_2 of Fig. 7.34) which specifies whether the system is in user state or supervisor state and these signals can be used to partition the memory into completely separate supervisor or user memory if desired.

For this discussion we will concentrate on the events that occur during device interrupt processing. The other exceptions do not carry out the steps 1 to 3 described in Sec. 7.7.1.

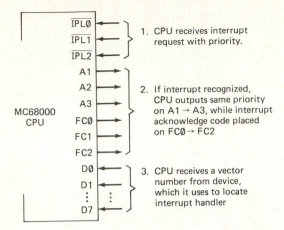

Figure 7.34 CPU/peripheral device signals during an interrupt.

Interrupt vectors in the MC68000 The MC68000 follows the previously described processing sequence. The important signals that communicate with the peripheral device are shown in Fig. 7.34.

The processor receives an interrupt request over three lines (IPL_0 to IPL_2) called the *interrupt priority lines*. When the peripheral device wishes to interrupt the processor, it encodes its priority onto these lines with 001 being the lowest level and 111 being the highest level. Otherwise the peripheral device sends 000 to the CPU. The priority level that a device is allowed to use is usually selected by jumpers on the interface printed circuit board.

If the interrupt level on IPL_0 to IPL_2 is greater than the processor priority stored in bits 8 to 10 of the status register, then the interrupt will be recognized. The existing status register is copied internally to the CPU and the supervisor bit is set and the processor priority is set to the level of IPL_0 to IPL_2. This locks out other interrupts at this level.

The processor informs the interrupting device that the interrupt has been recognized by placing the interrupt level code onto the address lines A_1 to A_3 when the function code on lines FC_0 to FC_2 is set to *interrupt acknowledge code* (111).

The third step in the MC68000 is to receive over the data lines an 8-bit *vector number* from the interrupting device. The vector number uniquely identifies the interrupting device. Since there are 8 bits in the number, 256 numbers are possible. The CPU effectively multiplies by 4 the vector number received, to produce a *vector address*. The vector address then becomes the address in memory from which the CPU fetches an *interrupt vector* after receipt of a vector number from the peripheral device. The interrupt vector is the starting address of the interrupt service routine. The user is responsible for placing the correct interrupt vector at the vector address generated by the device's vector number.

Because each vector number is multiplied by 4, the corresponding vector addresses occur at 4-byte intervals in memory starting at location 0 (i.e., 0, 4,

8, . . . , 1020). Also, since there are 4 memory bytes per vector number, a 4-byte interrupt vector can be stored starting from each vector address.

Some vector addresses are reserved for exceptions other than interrupts. Vector address 0 contains the RESET stack pointer. Vector address 4 contains the RESET interrupt vector (i.e., the contents of the program counter upon start up). Vector addresses between 8 and 96 can be automatically generated by the CPU by exceptions.

User device interrupt vectors can lie from memory location 256 (decimal) to location 1020.

It is possible to automatically select one of the seven interrupt vectors in locations 100 to 124 depending on the interrupt level shown on lines IPL_0 to IPL_2. This means of interrupt vectoring, called *autovectoring*, does not require a vector number from the peripheral device. The acquisition of an interrupt vector is CPU initiated and requires no handshaking with the interrupting device. The autovectored mode is selected by the device by asserting a special signal line to the CPU (the VPA line).

The interrupt vectors for any vector addresses used in the system from vector address 0 to 1020 must be user programmed. Also, the contents of memory locations 0 to 7 must contain the initial stack pointer and program counter contents.

7.7.3 The Interrupt Service Routine

An interrupt caused by a device is either a request for service from the interrupting device or, in some cases, an indication that another task must be carried out. For example, a clock device may interrupt the processor at a rate of 60 Hz. Each time an interrupt occurs, the processor's interrupt service routine will add an increment to the "time of day" stored in memory. Thus the exact time can be kept by the processor. In this situation, the clock device itself does not have any information to transfer other than the interrupt event.

The programmer is responsible for writing the interrupt service routine and placing the address of the interrupt service routine in the interrupt vector location so that when the CPU puts the interrupt vector into the program counter, the service routine will be executed.

If the CPU registers are to be used in the service routine, the existing register contents should be saved (usually on the stack). This must be done in assembly language. Then a high-level-language function can be called to carry out the body of the service routine. The high-level-language function must then return to the assembly-language service routine where the original registers are restored.

At the end of the service routine, an RTE ("return from exception") must be executed by the MC68000 that restores the status register and the program counter to their previous values. If the status register's supervisor bit is reset, then the processor effectively returns to user state. The equivalent Intel 8086 instruction is IRET, which restores the flags, instruction pointer, and code segment register.

Assembly-language interrupt service routines and C programs Very often the programmer may wish to write a dedicated service routine in assembly language. Since the service routine is never actually called by the C program and the C program is never called by the service routine, the question arises about how the assembler/linker will be forced to link the service routine.

This can be accomplished by *declaring* the name of the service routine as an *extern* in the body of the C function "main." This will force the linker to load the service routine (see Chap. 5), since the name of the service routine is a symbol that is *defined* as the address of the first line of executable code in the assembly-language program.

This is illustrated by a program skeleton shown in Fig. 7.35 for the MC68000.

The next problem is how to get the starting address of the interrupt service routine, "handler," into the memory location in low memory that holds it. This can be accomplished in the C program by initializing a *pointer* to the address (called the vector address) that holds the interrupt vector. Recall that when using vectored interrupts on the MC68000, the address that holds the interrupt vector for a device is automatically calculated by the CPU as 4 times the vector number delivered by the device in response to the CPU's "interrupt acknowledge" signal.

At run time, the pointer can then be used to insert the address of the service routine (i.e., the interrupt vector) into the correct low-memory vector address (i.e., the address that is 4 times the vector number delivered by the device). The service routine address cannot be inserted until run time, since you do not know where the interrupt service routine will be located until the program is linked.

The third problem is how to return a parameter from the service routine to the main program. This can be done by defining an *external* variable outside the main C program block. This will create (define) a global variable that can be referenced by the assembly-language subroutine.

The pointer ∗vect is initialized to a value that is the address of the interrupt vector. Also, the parameter var, which can be used by the service routine to send data to (or receive data from) the main program is defined as "external" (and allocated storage) by the second line of the C program.

Inside the main program, the service routine is declared as an external integer function, which means that the linker/loader will look elsewhere for its definition. Since the name of the service routine is its starting address at run time, the line

```
∗vect = handler;
```

can be translated as, "Let the contents of the vector address become the address of the interrupt service routine (i.e., the address of the handler)." If, for example, the interrupt vector for a particular device is located in hexadeci-

```
/* main program */
   int *vect = ⟨add. from 100 to 3FF⟩;
   int var;    /* external variable for communication with handler*/
   .           /* other variables you need */
   .
   .
   main( )
   {
           extern int handler( ); /*handler declaration*/
           *vect = handler;       /*put handler add. in vector*/
           .
           .                      /*your main program*/
           .
   }



*assembly language handler
           .globl   _handler     *external name of the handler
           .globl   _var         *external variable for communication
                                 * to C program
           .text
           .even
_handler                          *your handler code starts next line
           instructions
           .
           .                      *instructions in handler
           .
           rte                    *return from exception
```

Figure 7.35 C program with assembly-language interrupt service routine.

mal vector address 100 (OX100 is the lowest user vector address), then the contents of the vector address (the contents of location OX100) will contain the handler address after the execution of this instruction at run time.

The service routine's name is a defined symbol that is declared as a global for use by the C program. The parameter is not defined in the service routine but is "published" as a global so the linker/loader must look for it in the C program.

SUMMARY

The basic information that is needed for assembly-language programming is

1. The programming model of the CPU
2. The representations of data types, addresses, and instructions in memory
3. The representation of the assembly-language instructions and address modes
4. A decription of the instruction set

Since most large tasks are more easily broken up into a number of subroutines, the methods for calling a subroutine and returning from it are important. This is best accomplished by using the stack.

Most subroutines accept parameters from and deliver parameters to their calling program. There are a large number of ways to do this. C uses a stack frame for automatic input parameters and a register for returns. Both are compatible with reentrant and ROMable code.

Assembly-language subroutines to be linked with high-level programs must obey the calling and passing conventions of the high-level language.

Interrupt service routines receive control from the CPU after the recognition of an interrupt. These routines are often written in assembly language for speed and they can be linked with high-level-language routines for loading.

EXERCISES

7.1 Choose one 8-bit microprocessor, such as the Intel 8085. Using a data book from the manufacturer, summarize the important information by category as outlined in Sec. 7.1. Assume that the manufacturer's assembler will be used. List and number the addressing modes, but do not describe each in detail. Categorize the instruction sets into the two categories used in Sec. 7.1.

7.2 The instructions of the MC68000 may allow certain categories of addressing modes, depending on the instruction being executed and whether the operand is source or destination. For example, in a JMP operation, only "control addressing modes"—modes that might be useful in forming a jump address—would be allowed. The addressing categories are

Category	Allowed modes
Data	All but mode 1
Memory	All but mode 0 and 1
Control	2, 5, 6, 7-0, 7-1, 7-2, 7-3 allowed
Writable	All but 7-2, 7-3, 7-4

Note that destination of data is not allowed to be program-relative in a "writable" mode since the program may be in read-only memory (ROM).

List the categories (or in some cases modes that do not fall into the categories above) that are allowed for source and/or destination addresses of all the instructions listed as "functional operations" in Sec. 7.1 for the MC68000. Use a manufacturer's data book for reference.

7.3 Using either Whitesmiths' or Motorola's mnemonics, write an assembly-language program for the MC68000 that counts from 0 to 1 million. Upon reaching 1 million, the program toggles the most-significant bit of location 12345 (decimal), resets the "counter" to zero, and starts over again. The "counter" is to be a CPU register.

You can imagine that this program is carrying out the function of timer and the memory location is actually an input/output register (memory-mapped). Thus the peripheral device would see a square wave generated on the most-significant bit of the I/O register corresponding to the memory location addressed.

Using the instruction timing tables in the data book, figure out how long it takes your program to perform one pass through the timing loop?

7.4 Write a C program that defines three "int" variables—a, b, and c— and calls a MC68000 assembly-language subroutine that you will also write.

The assembly-language program adds "a" and "b" and passes the sum back to the C-language calling program.

The calling program puts the result in "c".

7.5 Repeat the task in Prob. 7.4, but this time arrange an assembly-language calling program and a C-language subroutine that actually performs the addition, passing the result back to the assembly-language routine.

7.6 Assume that your target system has a peripheral device such as the Motorola MC68230L8 timer chip as a component. Write a short C program to set up the timer chip's registers, to start it operating, and to type a message every time an interrupt occurs.

In the mode selected below, the timer chip decrements its counter register from a preloaded value to zero, whereupon a CPU interrupt is generated at priority level 2. The interrupt number that the device will deliver to the CPU when interrupted must be set up by the user in location 10023 (HEX).

The control register in the chip (location 10021) must be set up to A_1 (HEX) to cause the timer to operate in the desired mode.

The counter preload register must be set up to a suitable number to produce the desired time interval between interrupts. For the purposes of this problem, set the counter preload register to all 1s. The device's control registers of interest are memory-mapped to memory locations as follows:

Control/status register	*Memory address* (HEX)
Control register	10021
Interrupt number	10023
Counter preload register (most-significant byte)	10027
Counter preload register (middle byte)	10029
Counter preload register (least-significant byte)	1002B
Status register	10035

The least-significant bit of the status register, which indicates whether an interrupt has occurred, must be reset by writing a "1" to the byte before the next interrupt would be expected to occur. If this bit is not reset, the next interrupt will not happen.

Also write an interrupt service routine (that can be assembled and linked along with the C program) that clears status flag as described and informs the main program that an interrupt has occurred.

7.7 Identify the four fields in the following line of an MC86000 assembly-language program and briefly describe the purpose of each field:

```
startf2: move.l (a6),d0 *move the longword pointed to by
                        * register a6 to register d0.
```

7.8 The data in Table P7.1 shows the *initial state* contents of the registers and a portion of the memory for an MC68000 computer. Show the changes in register and memory contents that would take place if each of the following assembly-language instructions is executed. Use the same initial state for all instructions.

```
1. move.l d1,d0
2. move.l (a1),d7
3. move.l (a2),(a3)
4. move.l 8(a7),d4
5. move.l (a6)+,d2
6. move.l d5,-(a5)
```

Table P7.1

Register contents	Memory contents	
	Address	Data
	0FFC	00
	0FFD	00
	0FFE	00
	0FFF	00
$D_0 = 0000 \quad 0000$	1000	23
$D_1 = 4AB7 \quad 62F9$	1001	4B
$D_2 = 0001 \quad 0F60$	1002	CD
$D_3 = 0000 \quad 0000$	1003	67
$D_4 = 0000 \quad FFFF$	1004	FF
$D_5 = AB9A \quad 00DE$	1005	FF
$D_6 = 0000 \quad 100F$	1006	FF
$D_7 = 0000 \quad 09F0$	1007	FF
	1008	00
	1009	00
$A_0 = 0000 \quad 1000$	100A	00
$A_1 = 0000 \quad 1004$	100B	00
$A_2 = 0000 \quad 1008$	100C	3F
$A_3 = 0000 \quad 100C$	100D	88
$A_4 = 0000 \quad 1010$	100E	75
$A_5 = 0000 \quad 1014$	100F	00
$A_6 = 0000 \quad 1018$	1010	E1
$A_7 = 0000 \quad 0FFC$	1011	22
	1012	65
	1013	79
	1014	AB
	1015	C9
	1016	23
	1017	EE
	1018	04
	1019	A7
	101A	0F

7.9 Using illustrations of the stack, stack pointer, and program counter, describe what happens when the MC68000 calls a subroutine (JSR) and when it returns from a subroutine (RTS).

7.10 Get the documentation for a compiler other than a C compiler (e.g., a Pascal compiler) and describe how parameters are passed to a subroutine and returned from a subroutine.

BIBLIOGRAPHY

Wakerly, J. F.: *Microcomputer Architecture and Programming*, Wiley, New York, 1981.
(This excellent book covers aspects of both high-level (Pascal) and assembly-language programming. The assembly-language topics are treated both in general terms and in specific applications to example microprocessors.)
*Liu, Y-C, and G. A. Gibson: *Microcomputer Systems*: *The 8086/8088 Family—Architecture, Programming, and Design*.
(This book gives a good coverage of many topics related to the Intel processors. The book concentrates on assembly language, interfacing, and I/O considerations.)

Microprocessor data books

Motorola, Inc., *MC68000; 16-Bit Microprocessor User's Manual,* 3d ed., Prentice-Hall, Englewood Cliffs, NJ, 1982.

Motorola, Inc., *Motorola Microprocessor User's Manual,* 2d ed., Publication DL-120, 1981.

Intel Corp., *iAPX 86/88, 186/188 User's Manual—Programmer's Reference,* Publication 210911-001, Santa Clara, CA, 1983.

EIGHT

FUNDAMENTAL TECHNOLOGICAL
ALTERNATIVES

As discussed in Chap. 3, the basic problem of design is to meet the design specifications within the constraints of available technology, existing corporate resources, and the time available for completion. This chapter will focus on a discussion of the available technologies for design.

What does available mean? In a rapidly evolving field, one is tempted to design around the latest products announced by the semiconductor manufacturers in order to achieve better performance at the same cost or to obtain lower costs through reduced integrated circuit component counts. The potential problems that can arise if a newly announced component is selected are

- The part may not become available on schedule.
- The manufacturer may decide to never release it.
- The manufacturer may be the sole supplier. If there is a heavy demand for the product, delivery times may become long or cost of the device may remain high. It is also possible for the product to be withdrawn if it is not profitable.

Thus the announcement of a component does not qualify it as available technology. In general, the safest approach is to utilize components that are available on short delivery from multiple manufacturers.

Although this discussion is based on component availability, the same situation can arise at the board level and the subsystem level.

In this chapter we will discuss the characteristics of the basic technologies for transistor and logic circuits. In addition, utilization of these technologies at various levels of integration is discussed showing the design tradeoffs involved.

8.1 SEMICONDUCTOR TECHNOLOGIES

The basic semiconductor technologies are the foundation for the analog and digital components in a microcomputer system. The selection of an appropriate technology as the basis for a design may be an important factor in the success of the project. In the subsections which follow, the general differences between semiconductor technologies shown in Fig. 8.1 are discussed.

Semiconductor devices rely on the use of *n-type* and *p-type* semiconducting materials. An *n*-type semiconductor results when an intrinsic valence-4 semiconducting material such as silicon or germanium is doped with a valence-5 type of *doping impurity* such as arsenic. In *n*-type material, majority charge carriers (electrons) can be easily produced in proportion to the doping density.

A *p*-type semiconductor is produced by doping the intrinsic semiconductor with a valence-3 impurity such as boron or gallium. In this case the majority charge carriers are holes (absences of electrons).

The two types of transistors used in the fabrication of integrated circuits today are the *bipolar junction transistor* (usually called the *bipolar transistor*) and the *metal oxide semiconductor field effect transistor* (usually called the *MOS transistor*).

8.1.1 Field Effect Transistors

The MOS transistor (see Fig. 8.2) is conceptually the simpler of the two. An *n-channel enhancement* device consists of two *n*-type semiconductor *source* and *drain* "wells" separated by a *p*-type semiconductor over which a conductive *gate* lies. The gate is insulated by a very thin layer of silicon dioxide. Current (in the form of electrons) passes from the source to the drain under the gate when the gate is positive. The positive voltage on the gate attracts electrons which form (*enhance*) a conductive path, called the *channel*, through which the electrons flow from source to drain. On the other hand, when a negative (or zero) voltage is placed on the gate, the channel is no longer "enhanced" and

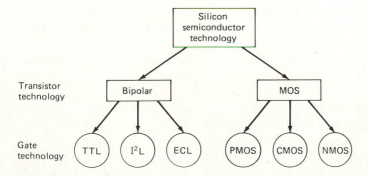

Figure 8.1 Silicon semiconductor technologies.

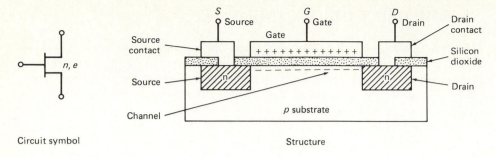

Circuit symbol   Structure

Figure 8.2 MOS transistor structure: *n*-channel enhancement type.

current stops flowing through the device. Thus the device acts like a switch controlled by gate voltage.

In a modified device called an *n-channel depletion* MOS transistor, an *n* channel is formed during the fabrication of the device. Thus current can flow from source to drain at zero voltage on the gate. In order to shut off the flow, the gate must be made sufficiently negative so that all charge carriers (electrons) are repelled from the channel.

In a *p-channel enhancement* device, current in the form of holes flows from the source to the drain through a *p* channel. The *p* channel is formed by placing a negative voltage on the gate. Holes can then flow from source to drain. If the gate voltage is reduced to zero or made positive, the channel will disappear and current flow will stop.

In a *p-channel depletion* MOS transistor there is a *p* channel that exists even at zero voltage on the gate and a positive voltage must be applied to the gate to turn off the device.

8.1.2 MOS Logic Circuits

Figure 8.3 shows an *inverter* made from *n*-channel MOS transistors. An inverter is a logic device for complementing a logic signal. If the input signal *X* is *true* (logic 1), the output signal will be *false* (logic 0), and vice versa. Logic devices such as inverters, ANDs, ORs, etc., are called *gates*. This use of the term "gate" should not be confused with the gate that controls the current flow through an MOS transistor.

The inverter is shown implemented using the two most common MOS logic families:

- *n*-channel MOS (NMOS): Fig. 8.3.
- Complementary MOS (CMOS): Fig. 8.4.

The NMOS inverter has two *n*-channel transistors. The top transistor is depletion mode and the bottom is enhancement mode. The power supply is a positive voltage. For the purpose of this discussion, assume that there is no load attached to the output.

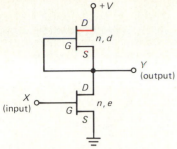

Figure 8.3 NMOS inverter gate.

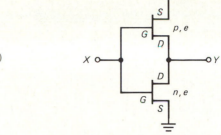

Figure 8.4 CMOS inverter gate.

When the input voltage X is 0 V (logic 0), the bottom transistor's channel is not enhanced and hence it acts as an open switch between source and drain. On the other hand, the top transistor has zero volts between gate and source, and hence this depletion transistor is on and its source must be the same potential as its drain ($+V$). Therefore the output voltage Y is equal to the supply voltage $+V$ (logic 1).

When the input voltage is positive (logic 1), the top transistor is still on, but the bottom transistor is also turned on by the positive voltage on the gate. A steady-state current then flows through the transistors between the supply and ground. The output voltage becomes the voltage drop across the lower transistor which is close to 0 V (logic 0) provided that the input voltage applied to its gate is sufficiently higher than its source voltage ($X > +V/2$).

Thus when the input is a 1, the output is 0, and vice versa, and the device is an inverter.

In the CMOS device shown in Fig. 8.4, when the input is 0 V (logic 0), the lower transistor is off just as in the NMOS lower transistor. If the gate of the top transistor is at 0 V and the source at $+V$ volts, then the gate is negative with respect to the source, which enhances the channel. Thus the output is at $+V$ volts (logic 1) and no current is flowing.

When the input is at $+V$ volts (logic 1), the top transistor is off and the bottom transistor is on. The output is then 0 V (logic 0) and no steady-state current flows.

From the discussion just presented, two important facts should be stressed.

1. Because of the very thin oxide layer under the gate, and because the gates are inputs to the circuit, electrostatic charges due to handling MOS and CMOS circuits can easily damage the oxide layer. Integrated circuits are usually stored in conductive foam. Assembly is usually carried out with special grounding precautions.
2. Current can flow in the steady state (i.e., when no inputs are changing) in NMOS circuits. In CMOS circuits, current flows only in the brief interval that the device changes state, with the exception of small "leakage" currents. CMOS circuits consume at least an order of magnitude less power. The power consumption is a function of clock speed in a clocked CMOS device such as a microprocessor.

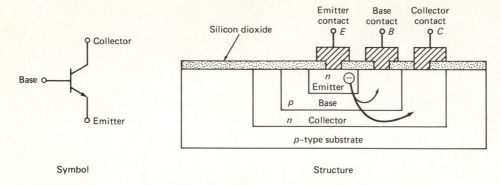

Symbol   Structure

Figure 8.5 Bipolar transistor structure: *npn* structure.

8.1.3 Bipolar Transistors

The bipolar transistor is shown in Fig. 8.5. The device shown is called a *npn transistor* corresponding to the type of semiconductor material in the *emitter*, *base*, and *collector*, respectively.

If a negative voltage is applied to the emitter with respect to the base, a net electron current will flow from the emitter into the base, where some electrons will contribute to current in the base lead. Provided that there is a positive voltage applied to the collector with respect to the base, the emitter electrons that did not contribute to a base current will flow into the collector and become a current in the collector lead. The collector current is approximately a fixed multiple of the base current.

8.1.4 Bipolar Junction Transistor Logic

In Fig. 8.6, the bipolar transistor is connected in a simple circuit. When the input voltage X is high, there will be a large collector current and low output voltage. The transistor is then said to be in *saturation*.

Conversely, when the input voltage is low, there will be very little collector current—a situation known as *cutoff*—and the output voltage will be high. Thus the circuit acts like an inverter. More complex junction transistor circuits

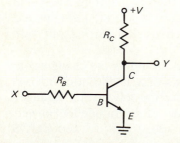

Figure 8.6 Bipolar junction transistor inverter.

are used in practice to form two very common types of logic using bipolar junction transistors.

- *Transistor-transistor logic (TTL)*
- *Integrated-injection logic (IIL)*

Integrated-injection logic has the advantage that no resistors are required in the logic circuits. Also, IIL can provide either low speed and low power or high speed and high power using the same device.

8.1.5 Other Technologies

The two remaining silicon technologies shown in Fig. 8.1—PMOS and ECL—are not widely used in microprocessors or support circuitry today.

PMOS stands for *p*-channel MOS and was the technology used in the first microprocessors. PMOS has been replaced by the faster NMOS process.

ECL stands for emitter-coupled logic. This is one of the highest speed technologies today. However, ECL also dissipates a large amount of heat and thus only a relatively few gates can be integrated on one chip without provision for cooling.

A new high-speed *gallium arsenide* technology is emerging that has about a 6:1 speed advantage and lower power requirements than silicon technologies. Integrated circuits have been fabricated and used in instrumentation where speed is important. It is at present more costly and has not achieved the same level of integration as silicon technology but it looks promising for high-speed signal processing and analog-to-digital conversion systems, for example.

8.1.6 Comparison of Main Technologies

Figure 8.7 shows the relative speed and power consumption of the main competing technologies today. From this figure, TTL is seen to exceed NMOS in speed and power consumption, while IIL and CMOS are similar over a wide range of speed or power. In fact, the power-delay product is approximately constant and is the same for IIL and CMOS.

PMOS is clearly at a serious disadvantage. NMOS continues to be a major technology, however, because of the high packing density of gates on a chip that is possible with NMOS.

A factor that is often important in selecting a technology for industrial application is the *noise immunity*—the ability of a device to be unaffected by voltage spikes on the signal and power lines. Since the logic threshold voltages (the voltages at which the device considers that a transition has taken place from one logic level to the other) are farther apart in CMOS than in MOS, the circuits will tolerate a greater amount of noise on the lines.

One advantage of bipolar technologies over MOS and CMOS is the capability of bipolar transistors (because of their low output impedance) to

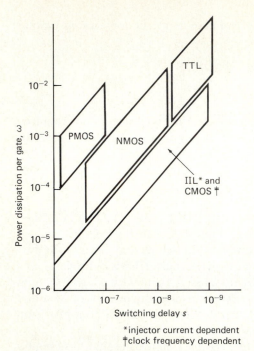

Figure 8.7 Power vs. delay for various technologies.

rapidly charge large capacitive loads. A data bus consists of a number of parallel wires with capacitance to ground and the capacitances of all the input gates tied to each wire. Thus each bus wire looks like a capacitor across the output of a gate that is driving the bus. TTL gates are better than MOS gates at driving long data buses. MOS transistors are also quite limited in their dc current drive capability; often only one equivalent "TTL load" (one TTL gate) can be attached to the MOS circuit output pin (see Chap. 13 for a discussion of ac and dc loads).

The technologies that are applicable in low-power battery applications are CMOS and, potentially, IIL. At the moment very few microprocessors offer extremely low power (e.g., RCA 1804) where the microprocessor can be placed in a "standby" mode with the clock stopped for long-term data-logging applications.

8.2 CHOICE OF INTEGRATION LEVEL

The second set of design alternatives focuses on the level of system integration to be employed in the design. Figure 8.8 illustrates the various stages in the construction of a complete turnkey system.

The stage, or *integration levels*, can be categorized into six levels, as shown in Table 8.1.

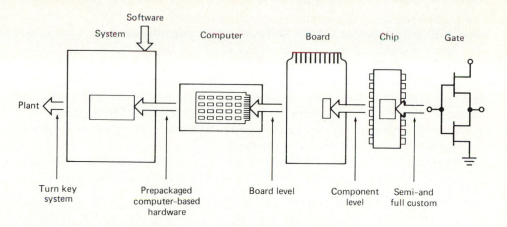

Figure 8.8 Stages in the development of a turnkey system.

Table 8.1

Level	Development time	Design cost	Cost per unit	Flexibility	Risk
Turnkey	Shortest	Least	Most	Lowest	Least
Prepackaged hardware					
Board level					
Component level					
Semicustom					
Full custom	Longest	Most	Least	Highest	Most

8.2.1 Turnkey Systems

A turnkey system is one that ideally can be purchased, moved to the installation site, and connected up and started without any design taking place by in-house personnel in the plant. Thus the development time, in-house design costs, and technical risks will be minimal, but the cost per unit will be high, reflecting the costs and profit at all other levels of integration in the table. The flexibility of the system to the unique requirements of an application may be limited, however, when different assumptions go into its design than apply in the particular application site.

8.2.2 Prepackaged systems

The term "prepackaged system" is used here to identify the application of a computer system purchased from a computer vendor to a particular real-time control task. The vendor supplies computer systems with interfaces and peripheral devices specified from its product line by the customer. The vendor usually supplies an operating system and some computer languages in which the users write their application programs. In this situation, the development time and design costs reflect the software development required.

There is some technical risk here. For example, the vendor may claim that the analog-to-digital conversion rate is 40,000 bytes/s and the disk transfer rate is 1,000,000 bytes/s and the user may assume that a program can be written that will sample the A/D at its maximum rate and store the data on the disk at that rate. But it may in fact turn out that such a program is not feasible using the support utilities in the operating system and that a much larger software development is required than was planned.

The cost per unit (e.g., $10,000 to $50,000) is not so high as a turnkey system and the system is more flexible, since the user is developing software tailored to the application.

8.2.3 Board-Level Design

As introduced in Chap. 1, board-level design refers to the purchase of available system functions as printed circuit boards, compatible with a particular system bus structure. This approach is more flexible than are previous approaches since boards can be selected from multiple vendors with a wider variety of function combinations per board.

Because of the wider variety of choices, this approach is more easily adapted to a particular application. The unit cost will be lower than that for prepackaged systems. Card prices for STD bus cards, for example, are $150 to $500, although special function cards can exceed that range.

There is a higher technical risk because of the possible incompatibilities between boards. For example, the manufacturer of an arithmetic processor board may claim that the product is "STD bus compatible," but the board may in fact work only with Intel 8085 and Z80 processors but not with Motorola 6809 STD bus cards. This is a general problem; compatibility problems should be investigated carefully.

The need to perform the *system integration* (assembly and debugging of the system hardware and software, including backplane, power supply, chassis, and cabinet) increases the development time and design cost. However, the sophisticated design or debugging tools required for the design levels below board level in Table 8.1 are not normally required at this level.

The remaining chapters in this book will assume that this level is the one the designer has selected, although much of the material applies regardless of the particular level selected.

8.2.4 Component-Level Design

Component-level design involves the selection and interconnection of individual integrated circuits to implement all the functions assigned to a board. Component-level design requires more development time and larger development costs. For example, the development time for an STD bus card can take 6 to 12 weeks. This is not all engineering time. There can be considerable delivery times involved (see Prob. 8.2). The corresponding costs can be $2500 to $4000. The unit cost, however, decreases significantly with the number of units produced. In quantities of greater than 100 units, the approach is most likely to produce lower unit costs than those for a board-level approach. Since the circuit can be designed to meet any special application requirements, the approach is also very flexible.

Because there is increased uncertainty about when a working prototype can be completed and at what cost, there is a significantly greater technical risk here, however.

There are a number of stages through which a component-level product must pass. These can be categorized into two classes:

- Development stages that have *fixed costs* (F) which must be recovered by the sale of N units
- Production stages that have *variable costs* (V) which depend on the number of units manufactured

The cost per unit, C, for N units can then be computed as

$$C = \frac{F}{N} + V$$

The development stages for a printed circuit board can be either carried out by in-house personnel and facilities or contracted out to experienced board designers. These stages are

1. *Circuit design.* Starting with a complete specification for the board, this step includes the circuit design, ordering of prototype parts, testing of a wire-wrapped (or equivalent) bench prototype circuit, production of schematic diagrams, circuit descriptions, and parts lists.
2. *Board layout.* This step involves positioning the various integrated circuits within board boundaries and drawing the conductive traces between pins in such a way that no two lines cross one another on the same side of the board unless there is to be an electric connection at the intersection. It is possible to have *multilayer circuit boards* that can have several interconnection planes. The board design with the printed circuit-board pattern is called the *artwork*. The artwork is usually drawn twice as large as the final board. The board layout step also includes the generation of an assembly diagram showing the position of each part on the board but not the interconnections.
3. *Photography.* The artwork is photographically reduced by one-half and a negative is prepared.
4. *Printed circuit-board etching.* The negative is used to etch the conductive traces onto copper-clad circuit board. Holes in the circuit board are drilled to the proper size to accept the integrated circuits, connectors, etc.

5. *Board debugging*. The printed circuit board is then assembled and tested. Major changes may result in the need to return to step 2 (or even step 1).
6. *Parts ordering*. Sufficient parts must be ordered to manufacture all the boards.

Once a viable prototype has been generated, the production stages can take place. The production methods selected depend on the number of boards to be built. The production stages are

1. *Printed circuit-card production*.
2. *Printed circuit-card assembly*. The integrated circuits and other parts are inserted on the board either by hand or by automatic parts inserters. The parts are then either hand-soldered onto the boards or *wave-soldered*—a continuous process for large numbers of boards.
3. *Testing*. Each board is then tested for functionality. This can be done automatically on a special test jig; or for small-quantity production, the cards can be tested in the final systems.
4. *Board repair*.

Generally speaking for large production, parts and bare boards should be tested before parts are inserted into the boards, since it is less costly to find faults before assembly.

8.2.5 Semicustom Integrated-Circuit Design

The last two design levels involve the design of the integrated circuits themselves. There are several reasons why a custom integrated circuit might be worthwhile in a particular application.

In the first place, integrated circuits with only a small number of logic gates per chip can be replaced with fewer, denser chips, resulting in more logic per board.

Provided that the quantity of boards is high enough, the development costs will be covered and the resulting costs per board will be lower with a higher level of integration. The minimum number of units (N) that must be manufactured can be computed by setting the cost per unit with custom circuits equal to the cost per unit with noncustom components, where

$$C(c) = C(n)$$

$$\frac{F(c)}{N} + V(c) = \frac{F(n)}{N} + V(n)$$

$$N = \frac{F(c) - F(n)}{V(n) - V(c)}$$

where $C(c)$ = cost per unit with custom circuits

$\quad C(n)$ = cost per unit with noncustom components

$\quad F(\ )$ = fixed costs

$\quad V(\ )$ = variable costs

$\quad N$ = "break-even" number of units.

An additional advantage to the custom circuitry is that it makes it more difficult for a competitor to copy your design (but not impossible).

Semicustom approaches, called *gate arrays*, provide arrays of gates on a single chip. The designer need only specify to the manufacturer the interconnections between the gates. The manufacturer can then complete the final metalization process steps on the chip, which make the connections that you have specified. The chip is then packaged.

Another approach, called the *standard cell* approach, uses CAD systems and libraries of standard logic cells that the designer can select to implement the design. This does not have the flexibility of a full custom design and requires all the semiconductor processing steps. The design can be carried out more rapidly than the full custom approach but less rapidly than the gate-array approach.

The number of transistors per chip for the above techniques is about 10 to 15,000 compared with a potential of about several hundred thousand for a full custom chip. The development time can be 9 to 12 weeks at a development cost of $5000 to $10,000. Compared with the same logic inserted as multiple integrated circuits in logic boards, the reliability will be superior for the single integrated component.

8.2.6 Full Custom Design

In a *full custom* design, the position of each of the transistors in each logic gate in the circuit is determined by the designer. In a microprocessor containing 400,000 transistors, this task would be impossible without the assistance of computer-aided design (CAD) tools and "library cells" containing logic modules such as registers, adders, etc., that are known to work.

It has been estimated that this type of design for a single chip may require one worker-year to complete at a design cost of $60,000 to $100,000, assuming that it is carried out by contract.

Since designers have control of the most basic technology, they do not have to buy from intermediate suppliers and thus the per-unit costs should be the lowest.

Because any circuit function is possible at this level, the circuit can do whatever is needed for the application and hence have the highest flexibility rating in the table.

Figure 8.9 shows a comparison between the various custom approaches and a component-level approach where the standard medium-scale integrated circuits (MSI) are replaced by one of the three custom methods.

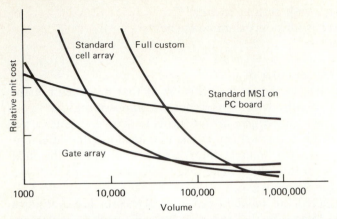

Figure 8.9 Custom design approaches compared with component level. MSI integrated circuit components are replaced by a custom integrated circuit. (*Adapted from* [1].)

8.3 MEMORY TECHNOLOGIES

The memory that is directly addressed by the CPU for both program and data is called the "main memory." There are other types of memory (e.g., disk memory) which we will discuss in the next chapter.

Main memories used to be primarily constructed from small ferrite cores that could be magnetized in one of two directions, thus storing a 1 or a 0 in each core. Although the method retains the magnetization when power is shut down, and memory cells can be both read and written to, it is relatively expensive to produce and not very dense by comparison with semiconductor memory. In most systems today, semiconductor memory is used. The main forms of semiconductor memory are shown in Fig. 8.10.

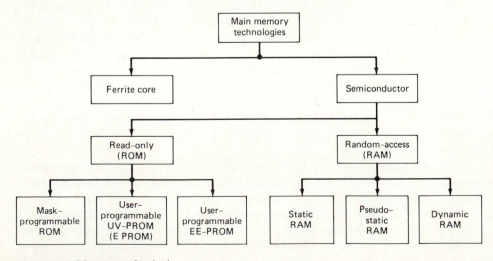

Figure 8.10 Memory technologies.

The underlying transistor and gate technologies confer their properties upon memory technologies. For example, CMOS memories will have very low power consumption compared with NMOS memories.

The memory technologies are divided into two types: random access memory (*RAM*) and read-only memory (*ROM*). RAM can be read or written to, and its contents are lost when power is shut off. *Battery backup* of RAM memory is often used to retain memory power during a power failure or short shut-down. ROM is reserved for a group of technologies in which it is not possible to modify the contents of memory rapidly or at all in some cases.

8.3.1 Random Access Memory (RAM)

There are two basic types of read/write memory, or RAM, at present:

- Static RAM
- Dynamic RAM

Static RAM uses a flip-flop to store each bit of data in memory. The logical state of the flip-flop will endure until it is changed or the power is shut off.

Dynamic RAM stores a charge on a small capacitor in each bit cell. Since this charge decays with time, the capacitor's charge must be refreshed periodically. Dynamic RAM requires fewer transistors per bit cell and thus can be more densely packed on a chip. However, the need to refresh the charges periodically (every several milliseconds) requires extra "refresh" circuitry. A simple READ operation will refresh a cell.

The *refresh circuits* are usually located off the chip. If this circuitry is included on the chip, the RAM is called "pseudostatic." In the Zilog Z8000 microprocessor, provision is made on the CPU chip for signals to refresh the RAM.

8.3.2 Read-Only Memory (ROM)

The second type of semiconductor memory, ROM, is either *user-programmable* or *mask-programmable*. Since this type of memory retains its contents without power, it is used for program storage of all types, both for systems software (operating systems, compilers, assemblers) and user software. It is also used for data storage for arithmetic look-up tables, code converters, and character generators.

Mask-programmable ROMs are produced by the semiconductor manufacturers with customer-specified contents. If a customer has a sufficient volume requirement for this type of ROM, then its use will be economical. Since programming and changes can be made only by the manufacturer, it is the least-flexible technology. All mask-programmable ROMs start out having the same pattern of transistors on the silicon substrate. All that remains is for the interconnections to be specified. When a customer gives the manufacturer a tape with the desired memory contents, the manufacturer makes up a photographic mask that will be used to deposit the final interconnection layer. In memory locations where the customer desires a binary 1, a connection is made; and where a 0 is desired, the connection is not made. For more details on the fabrication, the reader is referred to Sedra [2].

Another type of ROM is the user-programmable ROM. The most common type is the ultraviolet erasable, electrically programmable ROM called the UV-PROM, UV-EPROM, or simply EPROM. This device is usually programmed in a special device called a PROM-programmer connected to a microprocessor development system. The program is "down-loaded" into the EPROM. When program changes are made, the contents of the memory must be first erased by exposing the chip to ultraviolet light through a quartz window in the top of the chip package. The mechanism of data storage is that an electric charge is trapped between insulating layers. The charge can be released by the ultraviolet light energy. The erase-write process can be repeated many times if necessary.

The process is inconvenient, however, and a second type of electrically programmable memory is available which may replace it. This is called electrically erasable, electrically programmable ROM, or EE-PROM. These memories allow fast access reads but slow access writes. The technology has the advantage that the IC doesn't have to be removed from its socket to be electrically reprogrammed if programming provision is made on the board.

Table 8.2 compares some of the types of memory often encountered with regard to the basic operation of the chip.

8.3.3 Structure of a Memory Chip

When a binary-coded memory address is placed on the memory chip, the address must be *decoded* to select the correct memory cell (see Fig. 8.11). In addition the system must provide three control signals:

1. A *write select* signal that sets the chip in either read mode or write mode
2. A *chip-select* signal that when asserted, "enables" all chip functions except the output of data onto the data bus
3. An *output enable* signal that gates the data out of the chip onto the data bus during a read operation

Table 8.2

Type	Volatile?	Refresh?	Writable?
ROM	No	No	No
UV-PROM	No	No	Yes, after UV erase
EE-PROM	No	No	Yes, after electric erase
Static RAM	Yes	No	Yes
Dynamic RAM	Yes	Yes	Yes

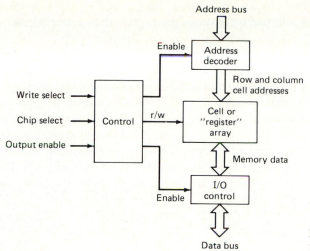

Figure 8.11 Internal memory chip functions.

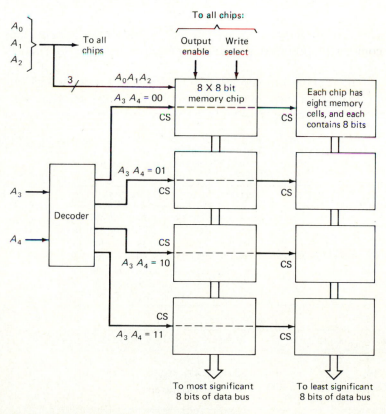

Figure 8.12 Addressing of a simple 32-word memory array.

The chip select is needed because there are many chips in a system and each chip has only a small number of address lines. The chip select signal is derived from higher address bits to select the correct memory chip.

The situation is shown in Fig. 8.12 for a simple system with RAM chips organized as eight cells, each containing 8 bits (64-bit RAMs). The data bus is assumed to be 16 bits. Thus two chips are enabled in parallel onto the data bus. The three least-significant address lines (A_0, A_1, A_2) are fed directly to all chips and provide access to the eight memory locations in each chip. The two most-significant address lines $(A_3$ and $A_4)$ are decoded to provide four control lines, one controlling each pair of RAM chips. Only one of these lines can be active at a time, corresponding to the four possible estates of the lines A_3, A_4. Thus the five address lines can uniquely address 32 memory locations consisting of four rows with eight locations per row.

8.3.4 Selection Factors

Aside from nontechnical factors such as availability, delivery, and cost (which very often dominate the selection), there are four important technical considerations:

1. *Memory density*. The density is expressed as the number of bits per chip and can range up to a megabit. Mask-programmed ROMs have the highest densities.
2. *Organization*. The contents of memory at any address in the chip consist of 1, 4, or 8 bits of data for most memory types. When 8 bits are addressed, the memory chip is called "byte-wide."

Table 8.3 Typical memory organizations and access time, ns

NMOS							
Dynamic RAM		Static RAM		ROM		EPROM	
$4K \times 1$	150–250	128×8	250–450	$1K \times 8$	250–350	$1K \times 8$	250–350
$16K \times 1$	100–200	$1K \times 1$	25–70	$2K \times 8$	350	$2K \times 8$	200–450
$32K \times 1$	150–200	$1K \times 4$	35–450	$4K \times 8$	350	$4K \times 8$	200–450
$64K \times 1$	150–200	$4K \times 1$	35–450	$8K \times 8$	200–350	$8K \times 8$	180–450
						$16K \times 8$	150–450
						$32K \times 8$	250
						$64K \times 8$	250

CMOS							
Dynamic RAM		Static RAM		ROM		EPROM	
$64K \times 1$	35–120	256×4	450–650	256×4	1200	$8K \times 8$	200
$256K \times 1$	120–200	$4K \times 1$	55–70	$2K \times 8$	430–550		
$64K \times 4$	150–200	$16K \times 1$	35–100				
		$2K \times 8$	150				

3. *Power requirements*. The power consumption (per bit) and the number of supply voltages required to operate the chip are important to many designs.
4. *Access time*. Access time is the maximum amount of time between the appearance of a valid address on the chip and the chip's response with valid data on its data lines.

Table 8.3 shows some standard organizations available from Intel and Motorola. With the 64K-bit RAM memory chips organized as 64K-bit by 1, for example, eight chips could be used in parallel to fill the full 64K-byte addressing range of a Zilog Z80 microprocessor. In such a main memory, all chips would be enabled for all addresses of memory. If byte-wide 64K-byte memories (8K × 8 bits) are used, each chip would represent 8K bytes and thus only one chip at a time is enabled.

8.4 MICROPROCESSOR TECHNOLOGY

As in the case of the memory technologies, the basic characteristics of the microprocessor are determined by the underlying semiconductor and gate technologies used in its construction. Thus a CMOS microproccesor will have a low power consumption that is related to its clock speed and an NMOS microprocessor will have a higher and relatively constant power consumption.

Aside from crucial overriding technical reasons for the selection of one microprocessor over another, the principal factors involved in the selection of a microprocessor in board-level design practice are

1. The familiarity of a design group with a particular microprocessor's hardware and support software
2. The in-house availability of development hardware and software for a particular microprocessor
3. The existence of a large already developed software base for a particular microprocessor (usually in assembly language, otherwise the code may be portable)
4. The multivendor availability of the microprocessor on cards that are compatible with a favored system bus
5. The availability of compatible subsystem boards such as arithmetic and input/output boards

The main technical reasons that a microprocessor type might not be acceptable for a certain board-level application can be classified as either

- Power consumption or
- Performance

Power consumption can be determined from the specification sheet, but performance is more difficult to assess.

8.4.1 Performance

There are a large number of ways to rate the performance of a microprocessor depending on what type of performance is important to you. The usual measures involve timing the execution of:

- Certain selected instructions
- Published benchmark programs
- User programs or code segments

The execution time of selected instructions on several microprocessors can be calculated from the data sheets. The important factors in the determination of the times are the clock rate, the operand size and the addressing modes for source, and destination operands. For example, the execution times for a register-to-register 16-bit word MOVE instruction for the six microprocessors in Table 8.4 have been listed [3].

The most relevant instructions to estimate are the ones that would occur in the most critical pieces of real-time code. One must be very careful to use the clock frequency of the microprocessor that is to be used in the application.

To get some idea of the performance of a program in a real-time task, one can gain some insight from execution time results on programs that have a similar instruction mix as the programs you will be developing. For example, if the real-time system you will be designing does a lot of digital filtering of signals, then the results of the published comparison [4] in Table 8.5 shows a comparison of five 16-bit microprocessors which might be of interest. The program's instruction mix was heavily loaded with multiply operations.

If a comparison of the preceding microprocessors had been carried out on the basis of the multiply instruction alone (the dominant instruction in the program used), the TMS9900 should have had about the same performance as the Z8000 [4]; but because of the use of registers in memory, rather than in the microprocessor, the TMS9900 fell to last place overall for the evaluation shown in Table 8.5. Thus it is important to assess a mix of instructions.

Table 8.4

Microprocessor	Clock rate, MHZ	Execution time, μs
Texas Instruments TI9900	3	4.6
Intel 8086	5	0.4
Zilog Z8000	5	0.75
Motorola MC68000	8	0.5
National NS16032	10	0.3
TI99110	6	0.5

Table 8.5

Microprocessor	Clock MHz	Program execution time, μs
Fairchild 9445	15.0	193.9
Motorola MC68000	8.0	327.0
Zilog Z8000	4.0	594.0
Intel 8086	5.0	855.6
Texas Instruments TMS9900	3.3	1000.0

In order to get a real appreciation of the performance of the microprocessor in a particular application, a user program should be written. Sometimes the result will be dominated by a few instructions and scrutiny of the instruction set may be all that is required.

8.4.2 Architectural Features

Aside from the performance, there may be a number of architectural features of the microprocessor that a designer might find essential to the application at hand.

For example, in an application that involves an operating system, the availability of a protected context such as the "supervisor state" in the MC68000 or "kernel mode" in the LSI-11 processors may be desirable.

Another feature that is important is the ability of the processor to detect and handle various types of error conditions, such as "nonexistent instruction."

The existence of vectored interrupts may be necessary for a real-time system. They allow program control to be transferred directly to an interrupt service routine. In some processors the interrupt vectoring is relegated to a separate chip, as in the case of the National NS16032 processor.

SUMMARY

- The two most common types of transistors used in the fabrication of integrated circuits are

  1. The bipolar junction transistor
  2. The metal oxide semiconductor field effect (MOS) transistor

- MOS transistors are the building blocks of the two predominant types of logic used in the manufacture of very large scale integrated circuits (VLSI) such as microprocessors:

  1. Complementary MOS (CMOS)—a low-power–high-noise-immunity process
  2. *n*-channel MOS (NMOS)—a process which is high density power consumption

- Board-level design is an intermediate design level that has the advantage that the development costs are relatively low and the design time is short. The approach has a definite advantage when only a few units are being manufactured (1 to 10).
- Component-level design leads to larger development costs and longer development times and has a definite advantage over 100 units.
- Random access memory (RAM) does not hold its data when the system is shut down unless the memory is *battery-backed*.
- Read-only memories (ROM) and programmable read-only memories (PROM) do retain their data when power is shut off, but their contents cannot be quickly modified (i.e., a ROM must be replaced, a UV-PROM must be removed and reprogrammed, and an EE-PROM cannot be programmed at the same speed as they can read).
- Memory chips vary in regard to the number of bits per chip (density), the organization of the bits (e.g., 8K bits × 8 vs. 64K bits × 1), the access time to receive data from the chip, and the power requirements.
- The selection of a microprocessor can be motivated by technical considerations (e.g., power consumption, interrupt facilities, performance) or by nontechnical factors (availability, familiarity, development facilities). The nontechnical factors are often the most important in practice.

EXERCISES

8.1 In an MOS transistor, the "source" is the source of majority charge carriers. (For an *n*-channel transistor, the majority carriers are electrons.) The "drain" must be biased by the right voltage polarity (with respect to the source) to "attract" the majority carriers. Meanwhile, the gate must be biased with the right voltage polarity (with respect to the source) to allow charge carriers to exist in the channel (see Sec. 8.1).

Draw symbols for the four types of field effect transistors, considering both *p*- and *n*-channel enhancement and depletion types and show with a + or − the polarity of the voltage on the gate and the drain (source grounded) required for the transistor to be on and conducting current.

8.2 Company XYZ is attempting to figure out whether to develop or buy a particular board for their system under development.

Task	Time, weeks	Cost, $
Board design	4	1500
Board layout	1	750
Layout photos	1	40
First pc card	1	200
Board debug (by designers)	1	400
Component ordering (by board assemblers)	4	300
Component cost	n.a.	50/card
pc cards	2	30/card
Board assembly	2	6/card

The commercially available STD bus boards cost $500 apiece, with a delivery time of 2 weeks in quantities of 1 to 24. In quantities of 25 to 49, a 5 percent discount and for 50 to 99, a 10 percent discount can be realized.

It was decided that if the company were to develop the boards, the job—including board design—would be contracted out. Estimates were obtained and the results were as indicated on p. 236.

Based on these figures, state how many boards would be required to make the component-level approach worthwhile. Compare the total times for the two approaches. Can you think of any likely tasks that might be omitted in the list on page 236?

8.3 Show how you would address the RAM chips in a memory system of a microprocessor that has an 8-bit data bus and 16 address lines. Use *byte-wide* 64K-bit RAM chips that have 8 data bits per cell. Fill the whole address space of the microprocessor with memory.

REFERENCES

1. Anon, "Custom Electronics," *Electron. Eng.*, July 1982, pp. 50–68.
2. Sedra, A. S., and K. C. Smith: *Microelectronic Circuits*, Holt, New York, 1982.
3. Gupta, A., and H-M. D. Toong: "Microprocessors—The First Twelve Years," *Proc. IEEE*, vol. 71, no. 11, November 1983, pp. 1236–1256.
4. Nagle, H. T. Jr., and V. P. Nelson: "Digital Filter Implementations on 16-bit Microcomputers," *IEEE Micro*, vol. 1, no. 1, February 1981, pp. 23–40.

BIBLIOGRAPHY

Mead, C., and L. Conway: *Introduction to VLSI Systems*, Addison-Wesley, Reading, MA, 1980. (This interesting book describes the mechanics of designing and fabricating integrated circuits based on a few simple design rules. The approach has been very successful and it is possible to design integrated circuits without an in-depth knowledge of the underlying semiconductor processes. However, the book does assume some digital electronics background.)

Sedra, A. S., and K. C. Smith: *Microelectronic Circuits*, Holt, New York, 1982. (Although intended as an undergraduate textbook for electrical engineering curricula, much of this book could be readily absorbed by students in other disciplines as well. The book describes how the various technologies of transistors and gates work and the basics of analog and digital circuits.)

Chirlian, P. M.: *Analysis and Design of Integrated Electronic Circuits*, Harper & Row, New York, 1981.

(This is an undergraduate electrical engineering textbook that gives the basics of transistor action and describes the conventional models of transistors in digital and analog circuits.)

Titus, C. A., J. A. Titus, A. Baldwin, and L. Scanlon: *16-Bit Microprocessors*, H. W. Sams and Co., Indianapolis, IN, 1981. (Six microprocessors are described and benchmarked: 8086, Z8001/2, LSI-11/23, TI9900, MC68000, and NS16032.)

Solomon, P. M.: "A Comparison of Semiconductor Devices for High Speed Logic," *Proc. IEEE*, vol. 70, no. 5, May 1982, pp. 489–509. (This paper compares the limitations of the various devices, including bipolar transistor and MOS transistors in the various logic configurations.)

NINE

INTERFACES TO EXTERNAL SIGNALS AND DEVICES

Signals on a system bus occur in a very orderly sequence. Each signal is initiated by a particular sequence of events and it is terminated either after a fixed time or by another sequence of events. Bus events are directly initiated by the microprocessor in a system containing only memory and a microprocessor.

Autonomous external devices and signals having no bus-compatible signals and no temporal relationship with the system bus signals cannot be connected to the system bus directly. Communication with the system bus is accomplished via an *input/output interface* (often called an *I/O port*).

Purposes of an interface   The main purposes of an input/output interface are the following:

- The interface must coordinate the transfer of data between the processor and independent outside devices. The *control lines* which carry out the coordination are sometimes called *handshaking* lines.
- The interface must convert data between a modality recognized by the processor (parallel binary usually) and a modality used in the external world (if different), such as serial data, analog data, or pulse-rate encoded data.

Other functions of the interface may be to provide electric isolation, amplification, noise rejection, temporary data storage, or data format conversion.

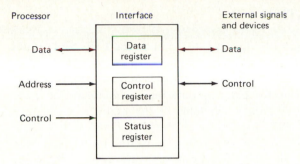

Figure 9.1 Input–output interface.

Interface registers   There are two sides to an interface—one viewed by the processor and the other viewed by the external world, as shown in Fig. 9.1. Inside the interface are a number of registers, usually including control, status, and data registers.

The bits in a *control register* are set (written into) by the processor and act as commands to the external device. For example, the command register bits may put the device into a read or write mode, reset the device, or start the execution of some device function.

The bits in the *status register* are set by the interface to indicate the status of the device. These bits are sometimes called *flags*. Error conditions, "task done" flag, and interrupt pending flag are examples of status data.

Interface registers accessible to the processor have an address that the processor puts on its address bus when it wishes to read from or write to the register. In the discussion to follow, it will be assumed that the registers are memory-mapped. That is, each register that can be accessed by the CPU has a unique address and can be read or written as if it is a memory location.

Types of interfaces   There is not a single standard interface that is suitable for all tasks. In general, the interfacing system gets more complex as the speed of input/output (I/O) transfers increases. Very often several kinds of interface capabilities can reside on the same printed circuit board or integrated circuit. For example, analog-to-digital and digital-to-analog channels are sometimes available on the same card. Thus the process of interface selection should include consideration of the total system requirements, to minimize the number of interface cards.

The interfaces to be considered here are

- Parallel input/output
- Digital-to-analog conversion
- Analog-to-digital conversion
- Real-time clock
- Direct memory access (DMA) interfaces

9.1 PARALLEL INPUT INTERFACES

The terms "input" and "output" usually refer to input to the CPU and output from the CPU. A simple standard form of interfacing is shown in Fig. 9.2. The input interface is controlled only from the processor. Whenever the processor addresses the input port, data on the external data bus is "enabled" onto the processor data bus and that this data is available at the moment the data is enabled onto the bus by the CPU.

In this interface there is no data register, only a "tristate" buffer (see App. B) which when it is enabled will force the processor bus to have the same binary value as the external data lines. When it is not enabled, the buffer output goes into a high impedance "third" state which effectively removes the data lines from the bus, allowing other data to use the system bus. The system control lines ensure that the data is enabled at the correct time for the processor to put it into one of its own internal registers.

The previous approach is not useful, however, when the external data is available for only short times or when the data could be in an unpredictable state at the time it is enabled onto the data bus. It is then necessary to use a control signal from the external device to store external digital signal levels into a register contained in the interface.

This can be done with the second form of input interface shown in Fig. 9.3. In this case, the input data is "latched" into an internal register in the interface by a *strobe* signal. A "strobe" is a pulse applied to the clock input of a *latch*

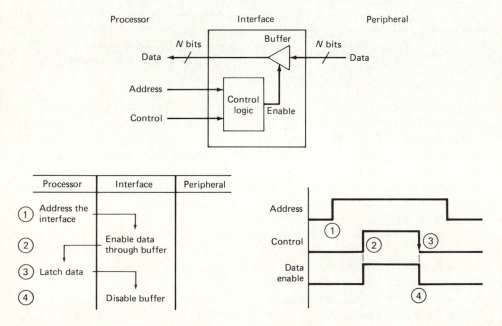

Figure 9.2 Simple parallel input interface.

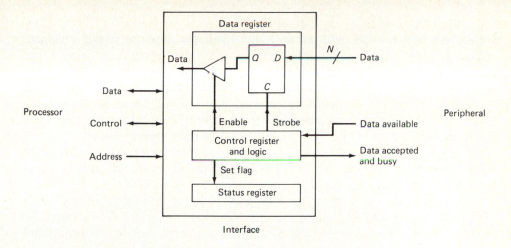

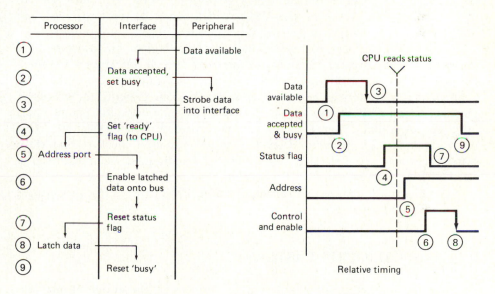

Figure 9.3 Parallel input interface with latch.

(*D*-type flip-flop). The transition of the pulse from one logic level to another will cause the latch to take on an output state corresponding to its input at the time of the strobe. In the case shown in Fig. 9.3, the strobing takes place on the falling edge of the data available pulse (marked with an arrow on the pulse edge). The storage of the data in the interface itself means that the external lines can take on new values while the processor gets around to reading the data out of the interface.

Handshaking (sequence used for request, permission, and transfer) on both sides of the interface is indicated in the figure. It starts out by the peripheral device asserting a line to inform the interface that data is available (step 1). The interface responds with a "data accepted and busy" signal (step 2), which will not fall until the processor reads the data out of the interface. Data is then strobed into the interface when the peripheral negates the data-available line (step 3). At about the same time, a "ready" flag (i.e., a bit equal to 1) is set in the status register to indicate to the CPU that data is available (step 4). In order to know this, the CPU must be continually "polling" the interface (reading the status register). The timing for this is not shown. When it finds the flag set, it reads the data register (steps 5 and 6), which resets the status flag (step 7) and finally resets the "data accepted and busy" line (step 9) after the CPU has latched the data (step 8). This informs the peripheral that the interface is now ready for more data.

There is an intermediate form of parallel input interface that has handshaking but no data storage in the interface. In such an interface, when the CPU reads the interface, the data (stored in a peripheral device register instead of an interface register) is passed directly through, as in Fig. 9.2. The presence of the handshaking enables the CPU to tell when to perform the read operation.

So far we have been discussing the use of polling, or *programmed I/O*. Many interfaces have the capability of *interrupt I/O*. In interrupt I/O, when the previously discussed data-available flag is set, the interrupt line to the CPU is also set. If the interrupt is *enabled* (this can be done either in the processor or sometimes in the interface), the processor will complete its current operation and then jump to an interrupt routine and read the interface data. This process relieves the processor from having to constantly check the interface status register to see if the flag is set. Interrupt handling has been discussed in Sec. 7.7.

The input task can also be carried out using *DMA* (*direct memory access*) *I/O*, discussed in Sec. 9.6.

9.2 PARALLEL OUTPUT INTERFACES

One difference between input and output interfaces is that an output interface *must* have a data register, since the processor data is constant for only a very short time on the CPU data bus; and while it is constant, it must be latched directly off the bus. Figure 9.4 shows a simple form of interface that accepts data whenever the CPU issues the correct address of the interface data register and performs a "write" operation. The data is constantly available to the external world after latching.

Figure 9.5 adds handshaking to the device side of the interface. As soon as data is written into the port register, the data-available line is raised. The peripheral device, when ready, strobes the data into itself on the rising edge of its data-accepted signal. Detecting this signal, the interface drops its data-available line. When the peripheral is ready to accept more data, it then drops its data-accepted line, which can be used by the interface to set a flag (and cause an interrupt if desired) to show the CPU it is ready for more data.

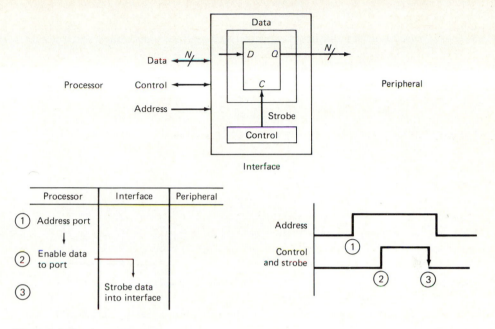

Figure 9.4 Simple output port.

There are many other possibilities for the organization of both input and output parallel interfaces. For example, Digital Equipment Corporation's DRV11 Q-bus interface provides two interrupt lines for external device use directly. A comprehensive coverage is not feasible here.

9.3 DIGITAL-TO-ANALOG CONVERSION INTERFACE

A digital-to-analog (D/A) converter is used to produce an analog voltage or current that is proportional to a digital binary number over a given "full-scale" range. For example, suppose that we wish to generate a voltage between the limits of 0 and 1 V using the contents of 1 byte of data. The digital values range from 00000000 to 11111111 and the corresponding analog values will range from 0.000 V to approximately 1.000 V. If 00000000 represents 0.000 exactly, and there are 255 more values that can be stored in 8 bits ($2^8 + 1$), then the maximum analog value that can be generated is $1 - (1/256)$V. This form of representation is called *straight binary*, or *binary* (see Fig. 9.6).

9.3.1 Digital Representation of Analog Voltage

Straight binary For a straight binary representation, the output voltage for any digital value can be found from

$$VOUT = VFS(b_7 2^{-1} + \ldots\ldots b_0 2^{-8})$$

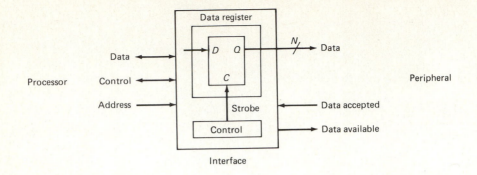

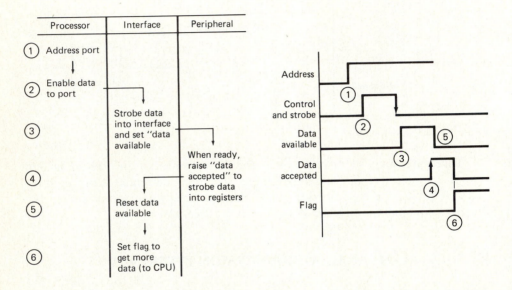

Figure 9.5 Output port with handshaking.

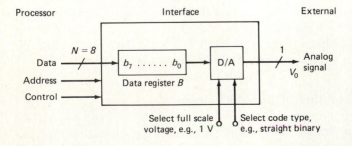

Figure 9.6 Digital-to-analog conversion.

Example For VFS = 1.000 V (the reference voltage), and the binary data = 11111111,

$$\texttt{VOUT} = \texttt{1.000} * (\texttt{1/2} + \texttt{1/4} + \ldots + \texttt{1/256})$$
$$= \texttt{1} - \texttt{1/256 volts}$$

Offset binary A bipolar output can be obtained for voltages between, for example, -1 and $+1$ V. If we use 00000000 to represent -1.000, then the maximum number that can be represented is $1 - (1/128)$V corresponding to 11111111. This type of representation is called *offset binary*.

Twos complement Another way of representing a bipolar output is to use *twos complement* numbers. If the most-significant bit is considered to be a sign, zero is 00000000, the most-positive number becomes 01111111, and the most-negative number becomes 10000000 (see App. A). Thus the positive binary numbers represent positive analog voltages and the negative binary numbers produce negative voltages.

Other possible representations are binary-coded decimal and sign-plus magnitude.

Selecting the representation On a D/A board there may be a choice of which binary representation is to be used. The choice can usually be made on the basis of programming considerations. The selection is achieved by a set of "jumper wires" (sometimes called *strapping options*) on the board that can be easily changed by the user.

9.3.2 Full-Scale Voltage

There may also be several *jumper-selectable* analog voltage ranges which the board can produce. The voltage ranges are determined by the D/A's full-scale reference voltage. The most common voltages are 5, 10, 5.12, or 10.24 V. The latter are for convenient voltage increments (see Prob. 9.7). Multiple D/A converters can often be accommodated on one board.

9.3.3 Other Specifications

In addition to the digital representation, and the full-scale voltages available, the specifications of a D/A converter should at least give its

- Resolution
- Slew rate
- Settling time
- Linearity
- Temperature coefficients of gain and offset

where

1. The *resolution* of an *N*-bit D/A converter is one part in 2^N. Typical resolutions for industrial applications are in the range of 1/256 for an 8-bit D/A to 1/4096 for 12 bits.
2. The *slew rate* gives the slope (in volts per microsecond) of the output voltage ramp when the digital input is suddenly changed from 0 to full scale.
3. The *settling time* measures how long the voltage will take to fall to within a prescribed fraction of the final value of voltage.
4. *Linearity* is the worst-case deviation between the actual output voltage for each binary increment, and the best straight line through the data.
5. The *gain* (or slope of the binary-input to voltage-output relation) and the *offset voltage* (the voltage when the binary input is zero) change with temperature. The *temperature coefficient* of this change may be important in the application.

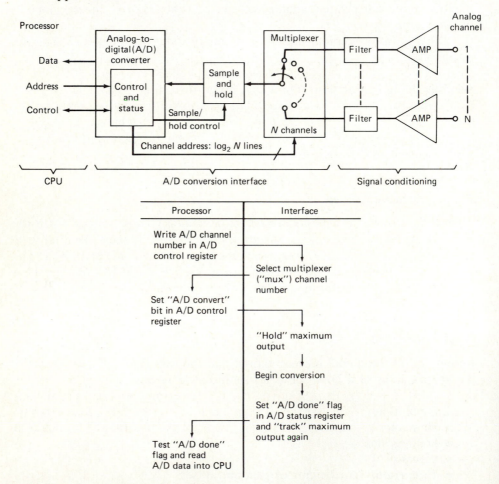

Figure 9.7 Analog signal acquisition system.

9.4 ANALOG-TO-DIGITAL CONVERSION INTERFACE

An analog signal acquisition system is shown in Fig. 9.7. Its purpose is to create a digital representation for the voltage on one of the N analog input channels at a particular instant of time, called the *sampling time*. As in the D/A converter, the desired digital representation is a design parameter that may be different for different available A/D interfaces (binary, offset binary, twos complement, binary-coded decimal, or sign-plus magnitude).

The analog voltage range at the input to the A/D gets mapped to a range of digital values. This voltage range is often jumper-selectable over either a high-voltage (e.g., 0 to 10.0 V) or low-voltage (<100-mV) range.

Digital specifications The A/D converter can be divided into an analog side and a digital side (Fig. 9.7). On the digital side, the designer must consider

- The integrated circuit technology (CMOS, TTL, ECL; see Chap. 8)
- Logic levels and tristate capability (see App. B)
- Resolution (number of bits converted)
- Conversion speed
- CPU handshaking (control, status)
- External controls

An external control such as a clock can be used to cause conversions rather than have processor-initiated conversion. Also useful is the ability to "short-cycle" the converter, which is possible with some types of converters to produce a lower resolution conversion at higher speed.

Analog specifications On the analog side of the interface the designer must be concerned about the

- Input voltage range (difference between the analog voltage that produces the maximum digital value and the analog voltage that produces the minimum digital value)
- The various error sources (quantization, linearity, and temperature errors)
- The equivalent input noise generated in the analog portions of the A/D

Several types of errors can occur. The quantization error in analog terms is equal to the analog voltage range divided by the total number of digital steps that are contained in the N-bit representation. The result is the size (in volts) of the least-significant bit (LSb). The quantization error is usually represented in digital terms as $\pm 1/2$ LSb.

Just as in the D/A converter, there is a possible gain and offset error (which can usually be adjusted to zero for a given temperature) and linearity errors. The temperature coefficients of these errors are also quite important, and their contribution to the total error should be calculated over the temperature range (see also Sec. 11.3).

9.4.1 Successive Approximation A/D

There is a wide variety of different techniques used in the construction of A/D converters. The most common is the successive approximation converter. It has the advantage of both moderate speed (up to approximately 100 Hz) and moderate resolution (up to 16 bits).

It functions by first comparing the input voltage with a test voltage set equal to one-half the full scale A/D reference voltage. The test voltage is obtained by using a calibrated D/A converter. If the input voltage is greater than one-half full scale after the first comparison is made, the most-significant output bit is set. If the input voltage is less than one-half full scale, the one-half full-scale voltage is removed from the test voltage; otherwise the test voltage remains the same.

Next, a one-fourth full-scale reference voltage is then added to the test voltage and the input voltage and test voltage are again compared. If the input voltage exceeds the test voltage (the test voltage is now either one-fourth full-scale or one-half full-scale plus one-fourth full-scale depending on the result of the first test), the next most-significant bit is set and the test voltage is held constant. If the input voltage does not exceed the test voltage, the second bit is reset (set to 0), and the one-fourth full-scale reference component is removed.

The process is repeated with successively smaller binary-weighted voltages until the least-significant bit has been tested. The method is procedurally similar to the determination of the weight of an object using a balance and a set of reference weights of 1/2 kg, 1/4 kg, 1/8 kg, etc.

Since the process is carried out from the most-significant to the least-significant bit, it can be stopped midway. The capability to short-cycle the conversion produces a lower resolution but a higher conversion speed, if required, from the same converter.

Because the converter is periodically comparing the internal references with the input voltage, the input voltage must remain constant during the full conversion time for a successive approximation converter. This is achieved by the use of a *sample-and-hold* circuit also controlled by the interface logic (see Sec. 9.4.4).

9.4.2 Dual Slope A/D

Toward the high-accuracy end of the performance spectrum lies the dual slope converter. In this converter, an unknown positive (constant) input voltage V is applied to an electronic integrator and is integrated from zero volts for a fixed time T, producing a positive-going output voltage ramp proportional to VT (the integral of a constant voltage over the time T). V is then removed and a known negative constant reference voltage U is integrated which produces a ramp down. This second ramp crosses zero at VT/U seconds from the time the reference was applied. The time is measured by a high-speed counter; and since T and U are constant, the counter holds a value that is proportional to the input voltage. If the input voltage, for example, is equal to the reference voltage U, the two integration times are equal and the counter would be set to reach its maximum. This type of converter is usually quite linear and converters up to 20 bits can be obtained, but the conversion time is relatively long.

9.4.3 Flash Converter

At the high-speed end of the performance spectrum, the parallel (flash) converter can provide conversion rates greater than 100 MHz. This is accomplished by providing internal voltage references for each of the $2^N - 1$ quanta into which the analog voltage range is divided. The voltage references are produced by a highly accurate resistive voltage divider. The analog signal is compared with all the reference voltages at once by a string of high-speed comparators whose outputs are used to generate the binary output. Because of the number and quality of the components required for this A/D, implementations are more common for 8 bits or less. Flash converters are usually expensive.

9.4.4 Sample-and-Hold Circuit

In front of the A/D converter in an analog conversion system is usually located a *sample-and-hold* circuit (see Fig. 9.7). A constant input is especially required in a successive approximation converter because the input is compared with a reference several times over the conversion period. The operation of this circuit is shown in Fig. 9.8.

Assume that a sample-and-hold circuit is holding a voltage V_0 from a previous conversion when it is commanded to sample (track) the signal again. Since the circuit cannot rise instantly to the new value, there is a time, called the "acquisition time" (usually between 0.35 and 25 μs), that is required until the output of the sample-and-hold circuit matches the input V_i.

A second type of delay, called the *aperture time*, takes place while the sample-and-hold control line switches to the "hold" mode. There can be an aperture delay of 30 to 100 ns before the switches to the input are opened. Thus the minimum time required before a stable output can be obtained is the sum of the acquisition time and the aperture time.

9.4.5 The Multiplexer

The final block of Fig. 9.7 to be discussed here is the *multiplexer*. It is conceptually like a rotary switch that can be rotated to "address" any input channel under control of the processor. For a single channel A/D, a multiplex-

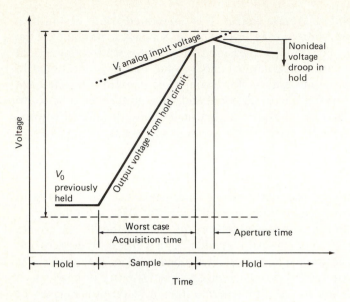

Figure 9.8 Operation of a sample-and-hold circuit.

er is not required and the analog signal is converted at the maximum A/D conversion rate. In a multiplexer with N channels, the effective per-channel conversion rate is reduced by a factor of N, since the A/D is being time-shared over N channels.

The multiplexer can be constructed from either mechanical switches (reed switches) or solid-state devices (such as CMOS switches). The multiplexing can be carried out in a variety of ways. Figure 9.9 shows both a "single-ended" and a "differential" connection. The single-ended connection is useful when a signal is referenced to ground. The differential multiplexer is useful when you are interested in the difference between two voltages, such as the two arms of a strain gauge bridge.

9.5 REAL-TIME CLOCK INTERFACES

There may be a number of functions for a real-time clock in a system. Often the main purpose is to time or count the occurrence of events rather than using the CPU for these tasks.

The real-time clock interface shown in Fig. 9.10 contains a crystal-controlled 10-MHz clock whose pulse rate can be divided by a programmable factor of 10. Output pulses from the divider are fed to a counter which counts down. The 16-bit counter can be preset to any value in its range by writing a data word to it via the interface with the processor. When the counter passes through zero, the processor is informed either by a flag or an interrupt. Alternatively, the zero state in the register can be "hard-wired" to another

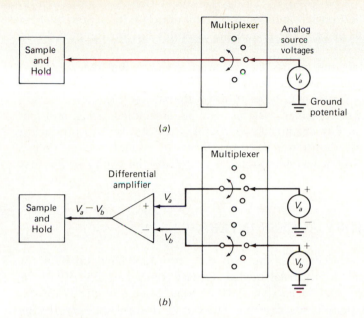

(a)

(b)

Figure 9.9 (a) Single-ended and (b) differential multiplexers.

system component (generating a "done-strobe," shown in Fig. 9.10). The done strobe can be wired, for example, to an A/D converter's "external convert" pin to initiate each conversion at a rate determined by the real-time clock. Upon completion of the conversion, the A/D can then cause an interrupt to the CPU.

An external event or clock can be used instead of the clock to decrement the counter. For example, if items on a conveyer belt each caused the counter to count down from an original setting of 12, then the CPU would be informed

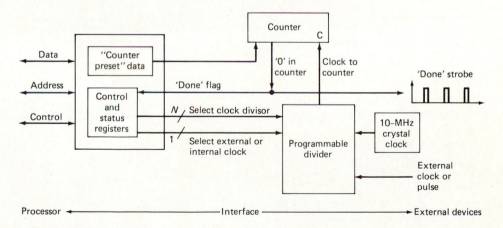

Figure 9.10 Typical real-time clock.

after a dozen items had passed and could take an appropriate action, such as move another bag into place at the end of the assembly line for the next dozen items.

For a final wrinkle, the control can be arranged such that the register is self-loading without processor intervention after every time it passes through zero. This produces a continuous pulse train on the strobe line.

More complex counter-timer systems can be assembled using multiple counters and registers. For example, two counters can be used to produce a square wave. The pulse duration is determined by one counter and interpulse interval is determined by a second counter triggered by the expiry of the first counter.

9.6 DIRECT MEMORY ACCESS INTERFACES

In all the interfaces described so far, the data transfers have taken place between the CPU and the external device, or vice versa (see Fig. 9.11). In many situations the external device data must be transferred too quickly for the CPU to process it. In these cases, external data can be transferred over the bus to and from memory under the control of the interface alone (see DMA in Fig. 9.11). The process is usually still initiated by the CPU and the CPU is informed of its completion.

DMA transfers occur in almost all high-speed disk drive interfaces, in graphics display interfaces, and in some high-speed analog-to-digital interfaces. We will be giving an example of a *floppy-disk* interface later in this section.

9.6.1 Interface Description

The characteristics of a DMA interface are

- The interface controls the direct transfer of data to and from the memory, without processor intervention.

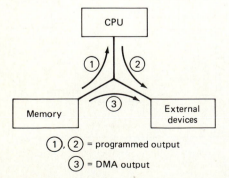

Figure 9.11 Data paths for programmed-vs.-DMA output.

- The time to transfer each byte of data is then limited only by the memory access time and the maximum bus transfer rate.
- There is generally more logic required to implement the DMA interface because the interface must now look after memory addressing and other bus control functions that the CPU normally looks after.

Registers The DMA interface must have at least the following registers in order to carry out all the DMA requirements:

1. A *starting address register* is needed to hold the first memory address to be accessed.
2. A *word count register* is needed to determine when the transfer should be stopped.
3. A *control register* is needed to hold the CPU commands to the interface when it is set up.
4. A flag or *status register* is required to display the current status of the interface to the CPU when the register is read.

DMA events Let's look at the sequence of events that would take place during the transfer of 1024 bytes of data into the memory from a peripheral device. The setup and initiation of the necessary registers can be carried out in a C program. The process is illustrated in Fig. 9.12.

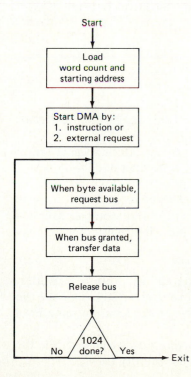

Figure 9.12 DMA transfer of 1024 bytes 1 byte at a time.

First the CPU sets up the interface by loading the starting address into the current address register and word count into the word count register. Then a bit in the control register is set which tells the interface to begin the transfer.

As each data byte becomes available from the peripheral device, the interface *requests* the bus from the processor by placing a logic level on a bus request line. The bus is *granted* by the processor (using another signal line) when the current bus operation is complete.

The interface then puts the current memory address on the address bus and makes the data available at the required time for latching into the memory.

The interface then *releases* the bus and waits for the next data to be available from the peripheral device.

The current address is then incremented and the word count is decremented. If the word count is equal to 0, the "done" flag is raised in the interface and an interrupt is delivered to the CPU to say that the transfer is complete. If the word count is not equal to 0, more peripheral data is awaited.

Variations As the data rate from a peripheral device increases, there can come a point at which the peripheral data being transferred to memory can "hog" the bus to the extent that the processor cannot get access to the bus. That is, there are no *bus cycles* (*machine cycles*; see Sec. 2.3) available for it. There are a number of ways to prevent this from happening.

The DMA channel in some interfaces can be set up to transfer in each DMA operation

1. One byte at a time
2. One block of N bytes at a time in bus "burst" mode
3. One byte every M bus cycles

In between DMA cycles the bus is released for the use of other devices or the CPU.

It is often necessary to start the DMA action by an external signal rather than by the CPU's setting the "go" bit. If there is more than one DMA device on the bus, then one must be concerned with the priority of DMA requests for the bus. The commercially available integrated circuits designed for the interfacing of DMA devices (e.g., Intel 8237A and 8257 and Motorola MC6844 and MC68450) all have four DMA channels per chip. Different priority mechanisms are available including:

- *Multilevel.* Each device is assigned a unique and fixed priority level.
- *Multilevel with round robin.* As above, but several devices can have the same priority; and if they do, the priority rotates on the same level so that each device gets a turn.
- *Round robin.* Each device has a turn at having priority until it gets the bus and then it falls to the bottom of the priority list.

It is also possible with some interfaces to move blocks of data around in memory by using the DMA interface to transfer from memory to memory and have no peripheral device involved (e.g., Intel 8237A DMA interface chip).

Another capability (Motorola MC68450) is to transfer into noncontiguous blocks of memory. The locations of the blocks are identified in several ways (e.g., using a table of addresses in memory).

9.6.2 DMA Interface to a Floppy Disk

One very common application of a DMA interface is to handle the large amount of relatively high-speed data that comes from magnetic disk storage (0.5 to 2 Mbytes/s). Magnetic disks are used for file storage of large amounts of data. *Removable disks* can be used for permanent storage, but *fixed disks* require some other form of backup storage such as tape or removable disks. One relatively inexpensive form of removable disk storage is the "floppy disk."

A floppy-disk drive requires a *disk controller* to perform all the low-level tasks necessary for reading and writing data onto the disk. The disk controller communicates with a *DMA controller*. Both controllers can be located on the same *interface card*.

Floppy disks A flexible, or *floppy*, *disk* is a digital magnetic storage medium in the form of a thin (e.g., 0.076-mm) flexible plastic disk coated with a magnetic material such as iron oxide and enclosed in a plastic protective jacket (Fig. 9.13). The disks are available in several diameters—called 8-in, 5-in, and 3-in disks.

The disk has a large hole in the middle through which a spindle in the *disk drive* projects. The flexible disk is clamped onto the spindle and rotated in its stationary jacket. A long radially oriented access slot in the disk allows a magnetic playback and recording head to be *loaded* onto the disk.

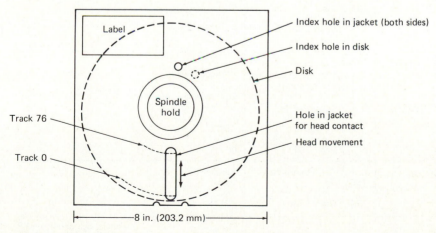

Figure 9.13 Floppy disk.

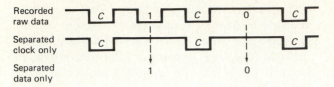

Recorded raw data

Separated clock only

Separated data only

Figure 9.14 Double-frequency (FM) recording pattern.

Information is recorded on the disk by causing a sequence of magnetic flux reversals to occur around a track under the head as the disk rotates (see Fig. 9.14). Data bits and clock.bits alternate. Since the data bits can be either 1 or 0, clock bits which are always 1 (in the data section of the sector) provide the flux reversals necessary to locate the data bits, even when the data is always 0. This recording method is called *double-frequency*, or *FM* (frequency-modulated), since the bit rate when all 1s are recorded is twice the bit rate when all 0s are recorded. A second method, called *modified frequency modulation* (*MFM*), is also used in *double-density* disk drives.

The head can be moved radially along the slot to an adjacent track. The drive for this disk is called a *moving-head disk drive*. The process of stepping to a new track is called *seeking* the track.

IBM format IBM originated the floppy disk and the *IBM format* [1] has been a major influence in the computer industry. The IBM format provides for 77 concentric tracks on an 8-in diskette. Each track is normally divided into 26 sectors with each sector containing a sector identifier and 128 bytes of data (see Fig. 9.15).

	Number of bytes	Field name	Contents
		Physical index	
	40	Gap	FF
	6	Sync field	00
	1	Track address mark	FC (clock = D_7)*
	26	Gap	FF
Sector ID	6	Sync field	00
	1	Sector address mark	FE (clock = C_7)*
	1	Track number	tr. #
	1	Side number	00 (one-sided diskette) < 00 side 0 / 01 side 1 } Two-sided
	1	Sector number	Sector#
	1	Sector length	Sect. length code (00, 01, 02)
	2	CRC code	CRC
	11	Gap	FF
	6	Sync field	00
Data	1	Data address mark	FB (clock = C_7)*
	128	Data	Data
	2	CRC	CRC
	27	Gap	FF

Repeated for each sector

*all other clock patterns = FF

Figure 9.15 IBM floppy-disk initialization format.

There is an additional hole in the magnetic disk and a hole in the jacket. Once per revolution, the disk hole and the jacket hole line up and a beam of light can shine right through the diskette. The detection of this alignment, called the *physical index mark*, is used by the disk controller and interface to define the "start" of the track that is currently under the head.

Prior to use, a diskette must be *formatted (initialized)*. This means that the information that defines the diskette format must be written on the diskette by the disk controller. Once this is done, the diskette can be used for data.

Figure 9.15 shows the arrangement of formatting information around one track starting with the physical index mark.

Each sector has a 7-byte sector identifier (sector ID) comprising a 1-byte *sector address mark* (a unique pattern of data and clock bits), a 1-byte track number, a 1-byte head number (for side identification on double-sided disk-ettes), a 1-byte sector number, a 1-byte sector data length code (128, 256, or 512 bytes), and a 2-byte error code, called a *cyclic redundancy check code (CRC code)*.

The CRC code is an error-detection code that consists of 2 bytes of data that have been calculated using the data in the rest of the sector ID. The CRC code is first calculated and stored on the track when the sector is formatted. When the sector ID is later read, the CRC code calculated by the disk controller should be the same as the one stored in the sector. There is a low probability that if an error occurs in reading the data in the sector ID, the CRC computed by the disk interface will be the same as the CRC stored in the sector ID—that is, that the error will not be detected.

The sector identifier is followed by a data section containing a *data address mark*, 128 bytes of data (or whatever other length has been specified), and a 2-byte data CRC code to verify the data section.

There are additional conditions defining the IBM format such as the location of the disk directory information and the definition of active and "spare" tracks used to replace a "bad track," but these go beyond the scope of this discussion (see Ref. 1).

A $5\frac{1}{4}$-in format (minifloppy) has been defined which is very similar to the one just discussed except that there are fewer tracks and sectors.

The type of disk format just described is called *soft-sectored* because each sector is identified by the information on the track alone. A *hard-sectored* disk is one in which each sector has a physical mark or hole that the controller can detect.

Disk-drive performance specifications There are a number of important parameters that can be used to assess the performance of a disk drive itself (independent of the controller):

1. The *data transfer rate* to disk gives the peak rate at which data bits can be written onto the disk. This rate occurs only in the data section of each sector. The average rate per sector depends on the amount of extra information being written (such as sector ID, data address mark, etc.).

2. The *track-to-track* time is the amount of time required to move the head over the next track.
3. The worst-case *latency* time is the longest delay experienced in getting to the data on a track. Thus it is the time for one complete revolution of the disk. The *average latency* is one-half the maximum latency.

Disk-drive functions   The interface must provide the proper signals to control the various drive functions. The drive has two major functions. The first involves the movement of the read/write head—from track to track and on/off the disk ("head load/unload"). The command lines to the disk for this purpose are

- "direction" of head movement (toward center or away)
- "step" one track in the selected direction
- "load" the head

The second major function involves the reading and writing activities. The lines associated with these activities are

- "write data": a line to the disk carrying the data to be written
- "above track 43": a line to the disk used for increasing the write current to the disk on tracks close to the center of the disk
- "read data": a line from the disk carrying the recorded data
- "read clock": a line from the disk carrying the recorded clock bits

A third set of lines are the disk drive status lines which sense various conditions in the drive and report them to the interface. These lines are

- Disk drive "ready" for operation
- "index" hole just passed
- head over "track 0"

The final line, called "select," enables all drive functions to be accessed and is useful when more that one drive is in a system.

Disk-drive controller   The *controller* for a floppy-disk drive is usually a single VLSI chip which takes a number of commands from the CPU and produces the necessary sequence of events to carry out the commands. The Intel 8271 and 8272A and Motorola MC6843 are three available integrated floppy-disk controller chips. Very often the controller is located in the computer interface itself. Such a system is shown in Fig. 9.16.

The system address lines can be decoded to select four unique addresses in the memory space of the system. These addresses correspond to the data

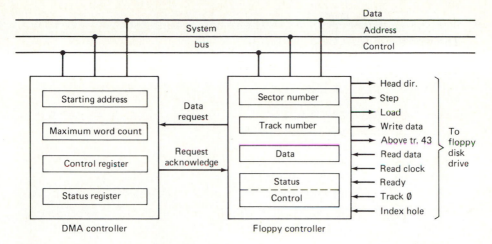

Figure 9.16 DMA controller and floppy-disk controller.

register, sector register, track register, and status register (when reading the "location") or the control register (when writing).

In addition to the usual read/write control lines, the control lines include a master reset line, an interrupt request line, and a DMA request line. The use of the last two lines provides a good example of a typical DMA situation.

In the normal operation of the controller, the interrupt is used to announce task completion or fault conditions, and the DMA request is used to pass each byte of data to memory on a READ and from memory on a WRITE.

A typical WRITE sequence consists of the sequence shown in Fig. 9.17. First the track and sector registers are loaded with the desired values. Then a SEEK command to find the desired track is invoked by loading its code into the command register. After executing the SEEK, the status register is checked for errors. Then a WRITE command is loaded into the command register. When the right track and sector have been detected and there are no error conditions in reading the sector ID, a DMA request is issued to the DMA controller. The DMA controller then requests the use of the system bus over which a data byte is delivered from memory to the disk controller. If no data is written into the data register, the disk controller will set an error flag and cause an interrupt. If the data is present, it will be written to disk. The DMA cycle repeats until all data in the block is written. The write sequence is finished by a CRC code and the completion is announced by an interrupt to the system.

Thus the CPU has only to look after the setup of the registers and start conversion. Other CPU tasks can be done while the DMA task is completing. It is also possible to interface the disk controller directly to the system bus, but this puts an undesirable load on the processor which then has to look after all device "ready" testing and memory transfers to the controller.

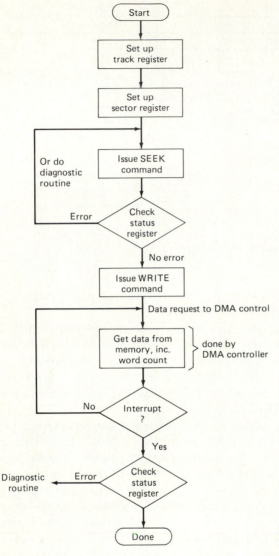

Figure 9.17 Typical WRITE_SECTOR function.

SUMMARY

An input/output interface in a computer system synchronizes the transfer of data between the system data bus and external signals or a peripheral device.

Interfaces usually have a number of registers, including data, status, and control registers.

There are three basic ways to control the data flow across the interface:

1. *Programmed I/O.* In this case the CPU continually tests the status flags in the interface to determine when data is ready to be transferred.

2. *Interrupt I/O.* In this technique, when the interface requires service, it informs the processor by causing a system interrupt. The CPU services this interrupt as described in Chap. 7.
3. *DMA I/O.* The fastest technique is direct memory access, although it has more hardware overhead. Multibyte transfers can occur without CPU involvement until the process is complete.

Analog data can be represented digitally in a variety of ways, including

- Straight binary
- Offset binary
- Twos complement
- Binary-coded decimal
- Sign-plus magnitude

Several methods for converting analog data to digital data are

- Successive approximation
- Dual slope conversion
- Flash conversion

Some real-time clock interfaces can be used to time or count external events as well as to generate flags, interrupts, or external signals at programmable rates.

A DMA interface contains a word count register and a current address register. These are used to control memory addressing.

A DMA controller can be used in conjunction with a floppy-disk controller in an interface to a floppy-disk magnetic storage device.

A floppy disk is formatted into multiple concentric tracks and multiple sectors on each track. A read/write head moves from track to track and when it is loaded onto the disk, it reads the desired sectors.

The worst delays in a floppy-disk system occur as a result of either head movement or rotational delays.

A sector contains a sector ID with its track and sector address, along with a fixed number of bytes of data (128, 256, or 512 bytes). An error-detection code, called a CRC code, is used to detect faulty data.

The disk controller looks after finding the desired sector and then reading and writing the data in it. The disk controller is also used to format the diskette.

EXERCISES

9.1 What are the purposes of an I/O interface?

9.2 Each interface usually contains several registers. What functions do the registers perform?

9.3 Why does a parallel output interface require a latch?

9.4 Describe the differences between programmed and interrupt I/O.

9.5 Define slew rate and settling time in a D/A converter.

9.6 For a given reference of 10.00 V, find the D/A output voltage corresponding to the following 8-bit digital number:

10010000

for (*a*) straight binary, (*b*) offset binary, and (*c*) twos complement representations.

9.7 If the reference voltage is changed to 10.24 V, what would the equivalent straight binary voltage be?

9.8 For an analog voltage of 7.500 V, a reference voltage of 10.00 V, and an 8-bit straight binary representation, what digital number would be produced by a successive approximation A/D converter?

9.9 What advantages does successive approximation have over dual slope conversion?

9.10 In a floppy-disk drive that has a rotational speed of 360 r/min and a track-to-track head movement time of 6 ms, find the worst-case amount of time it would take to read one sector of data on track 76 assuming that the head is presently over track 0. Assume that there are 26 sectors with 128 bytes per sector.

9.11 For the floppy-disk drive described in Prob. 9.10, what is the "average latency"?

REFERENCE

1. Intel Corporation, *Microprocessor and Peripherals Handbook*, Order No. 210844-001, Santa Clara, CA, 1983, pp. 6–211.

BIBLIOGRAPHY

Interface card descriptions

Digital Equipment Corporation, *PDP*-11 *Microcomputer Interfaces Handbook*, EB-23144-18, Maynard, MA, 1983.
Analog Devices Inc., *Data Acquisition Databook—Modules, Subsystems* 1982—*Volume* 2, Norwood, MA, 1982.
Intel Corporation, *OEM Systems Handbook*, Order No. 210941, Santa Clara, CA, 1983.

Integrated circuits for interfaces

Intel Corporation, *Microprocessor and Peripherals Handbook*, Order No. 210844-001, Santa Clara, CA, 1983.
Motorola Semiconductor Products Inc., *Motorola Microprocessors Data Manual*, Austin, TX, 1981.

Other

Analog Devices Inc., *Analog-Digital Conversion Notes*, Norwood, MA, 1980.
Jaeger, R. C.: "Tutorial: Analog Data Acquisition Technology. Part I. Digital-Analog Conversion," *IEEE Micro*, vol 2, no. 2, May 1982, pp. 20–37.
———: "Tutorial: Analog Data Acquisition Technology. Part II. Analog-to-Digital Conversion," *IEEE Micro*, vol. 2, no. 3, 1982, pp. 46–56.
———: "Tutorial: Analog Data Acquisition Technology. Part III. Sample-and-Holds, Instrumentation Amplifiers, and Analog Multiplexers," *IEEE Micro*, vol. 2, no. 4, 1982, pp. 20–34.

FOUR

CONNECTED SYSTEMS

SERIAL COMMUNICATIONS

In Chap. 2 we discussed the communication of data from card to card over the external processor bus (or "system bus"). The system bus is useful for high-speed internal system communication (e.g., 10 megawords/s) over distances of up to several meters.

Communication with external peripheral devices (e.g., line printers, terminals; see Fig. 10.1) and with other systems can be accomplished through standard serial interface cards attached to the system bus. The serial interface card is in turn connected by a set of lines to an external system.

One advantage of serial communications is that the number of lines required is less than would be required for full parallel communication of all the data lines on the system bus. In the 8-bit STD bus, for example, eight lines are required for data. The same 8 bits of data can be transmitted serially, 1 bit at a time, over only two lines (signal and ground) in a serial communications system. A further advantage of serial communications is that two systems with different internal system buses, timing, and handshake signals (e.g., Multibus

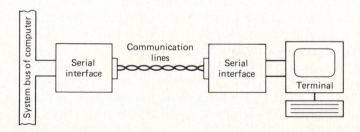

Figure 10.1 Serial communication between a computer and a terminal.

and STD bus) can communicate over a serial link without the necessity of designing a special interface between the two systems. All that is required is that each system have the same type of serial communications card available. A clear disadvantage of serial communications is that it takes longer to communicate serially.

Interface standards When an external system communicates with a serial interface in another system over an agreed set of lines and with an established *data protocol*, information can be exchanged in an orderly way.

National or international bodies are usually involved in the establishment of a widely accepted interface standard. Examples of such bodies referred to here are

- The Electronic Industries Association (EIA)
- The Institute of Electrical and Electronic Engineers (IEEE)
- The Comité Consultatif International Téléphonique et Télégraphique (CCITT)
- The International Standards Organization (ISO)

An interface standard should at least provide a clear definition of

1. Mechanical specifications of the lines and connectors along with the pin assignments to signals and mechanical connector designs for each end of the interconnecting lines
2. Functional specifications of the interface, including allowable subsets of the functions
3. Electrical specifications of the interconnecting lines including the electrical characteristics of the signal, the lines, the line driver and receiver impedances, and the short- and open-circuit properties.

There are two forms of serial interfaces in common use:

- The bit-serial interface
- The byte-serial interface

The bit-serial interface A *bit-serial interface* converts each data word taken from the computer bus data lines to a sequence of bits which are transmitted to an external device over a single pair of wires. The interface also receives data bit streams and converts them back to data words.

Suppose that the computer has the ASCII representation of the letter A (1000001) stored in a byte of its memory. In order to transmit this letter using the serial interface, the CPU would first fetch the byte from memory and store it in the interface's data register. The interface then converts the byte of data to a serial bit stream. The data is transmitted over a single pair of wires 1 bit at a time as a sequence of voltage levels.

Synchronous vs. asynchronous A *synchronous protocol* is one in which a clock signal must be transmitted across the serial interface. Each data bit must be synchronized with the transitions of the clock. An *asynchronous protocol* for data transmission is one in which the transmission of each data bit does not require a clock signal to be transmitted along with the data. The transmissions must be at a certain average bit rate, however.

The asynchronous protocol was developed for use in telecommunications using electromechanical teleprinters such as the KSR-33 manufactured by Teletype Corporation. Although the early asynchronous interfaces transmitted logical 1s and 0s by using two levels of current in the interconnecting lines (called a *current-loop interface*), later standards have evolved that use voltage levels.

Bit-serial protocol In the voltage-level protocols, one voltage represents a 1 and a second voltage a 0, as shown in Fig. 10.2.

Start bit The start of a character is signaled by a voltage change from the *marking* or logical 1 level to the *spacing* or logical 0 level. The line is held in the 0 state for one bit time to form a *start bit* that acts as a signal to the receiver that data bits are to follow.

Data bits The least-significant bit of the character code is transmitted next, followed by the remaining bits of the code. The data section usually contains 5 to 8 bits, depending on the code used. The normal ASCII code contains 7 bits, but there are also 5-bit, 6-bit, and 8-bit codes.

Parity bit *Even parity* is an error-checking technique that appends an extra bit, called the *parity bit*, to the transmitted data. If there is an odd number of bits set to 1 in the data to be transmitted, the parity bit will be set to 1 to make an even number of "set" bits in the transmitted frame. If there is already an even number of bits in the character, then the parity bit will be set equal to 0. Start and stop bits (to be described next) are not included in the calculation of the number of data bits. *Odd parity* works the same way to force an overall odd number of "set" bits in the frame between the start and stop bits.

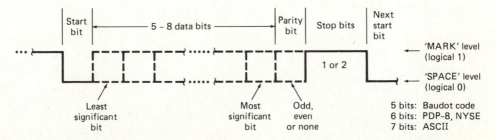

Figure 10.2 Bit-serial communications protocol.

At the receiving end of the communication link, the receiver can then check each frame for the correct parity. If an odd number of bits is received when even parity is expected, then an error has occurred in transmitting the data. Of course, undetected errors can occur, such as a 2-bit error that leaves the frame parity unchanged. Another form of transmission error checking is the *checksum*. The checksum, which operates on multibyte data, is discussed in Sec. 15.5.1.

Stop bits   The data bits are followed by either one or two *stop bits* that signal the end of the character. Upon reception of the stop bits by the interface, the data bits are repacked into a digital byte of data, and the byte is gated onto the system data bus at the proper time.

It is necessary to set up the external device to match the serial protocol of the computer interface (number of data bits, number of stop bits, number of bits per second, etc.).

Types of bit-serial interfaces   The most common bit-serial interface standard arose out of the need to agree upon the connection between computing devices and devices used for the communication of data over the switched telephone network. This standard, called RS-232-C, will be discussed in Sec. 10.1.

Other bit-serial interfaces are

1. 20-*MA current loop*. Current loop uses two levels of current to encode the data. This type of interface is still available as an alternative to RS-232-C on some computer interfaces. There are usually no control lines with this interface.
2. *Hewlett-Packard interface loop* (*HPIL*). This is a two-wire transformer-coupled interface that was developed for use with battery-powered processors, peripherals, and measuring instruments. Its protocol is like the IEEE-488 (see Sec. 10.2) but on two wires only.
3. *Local area networks* (*LAN*). The number and type of local area networks are still expanding owing to the need for information sharing between computers (especially microcomputers). A brief outline of some major terminology follows.

Local area networks   The ISO has defined an *open-system reference model* for the levels of hardware and software needed to communicate data.

At the lowest level—the *physical layer*—there is a cable (coaxial, twisted pair, or optical fiber) connecting the systems. Data bits are carried on this cable in *broadband* or *baseband* mode. Broadband has the capability of carrying voice and TV as well as multiple data channels and requires a coaxial cable or fiber-optic cable, whereas baseband can operate on a twisted pair of wires and carries only the serial data.

Information bits to be transmitted are preceded and followed by framing bits generated by the *data link layer*. This layer must be able to recognize an incoming *frame* consisting of data and framing bits.

The next-higher layer is called the *network layer*. Its job is to control the flow of information packets in the network. A *packet* consists of data, source and destination addresses, and control bits. The two contending methods of scheduling packet transmissions are called *carrier sense multiple access/collision detection (CSMA/CD)* and *token passing*. In CSMA/CD, if two nodes on the network begin to talk at the same time, each will wait a random delay before restarting. In token passing, only the node that has the "token" is allowed to talk. The token is then passed on.

The Ethernet LAN developed by Xerox has found fairly wide industrial acceptance and is available in the form of interface boards for microcomputers and industrial controllers. Ethernet is a CSMA/CD protocol transmitted in baseband over a coaxial cable.

The IBM PC network is another network using CSMA/CD. It is based on a broadband physical link. IBM has also announced the availability of a token-passing protocol.

A complete discussion of local area networks goes beyond the scope of this book. For further information on LANs, see the bibliography at the end of this chapter.

The byte-serial interface A *byte-serial interface* sends data bytes serially over eight lines to a receiving interface. All 8 bits in a byte are transmitted simultaneously.

One particular byte-serial interface was developed by the Hewlett-Packard Company for the connection of measuring instruments to computing devices. This interface became an Institute of Electrical and Electronics Engineers (IEEE) standard in 1975—designated as IEEE-488 (1975). In 1978, the standard was further clarified and became designated as IEEE-488 (1978).

Since 8 bits are transmitted in parallel on IEEE-488 interfaces, a higher data throughput can be achieved in this protocol compared with a bit-serial protocol with the same bit rate over each data line (1 to 2 Mbytes/s compared with 1 to 2 Mbits/s).

Other standard interfaces are

- CAMAC is an interface standard that started in the nuclear electronics industry for connecting modular measurement instruments together within the same enclosure, or "crate." This bus structure was specified in the IEEE-593 standard. This standard could be classed as a parallel interface standard, since up to 24 bits can be transferred in parallel over the bus at one time. IEEE-596 defined a high-speed parallel interconnection between up to 7 crates and IEEE-595 defined a low-speed serial communication between up to 62 crates.
- A common byte-oriented interface is the "Centronix parallel interface." This proprietary interface has become a de facto standard as an interface for connecting line printers to computers. The interface consists of data lines (eight), control lines, and ground lines.

Probably the most common interfaces between computers and instruments are the RS-232-C and IEEE-488 interfaces. In the next two sections we will have a look at each of these standards.

10.1   THE EIA RS-232-C INTERFACE STANDARD

Because of its origins in the data communications industry, the RS-232-C standard describes the connections to data communications equipment so that two computer systems can send data over the switched telephone network (see Fig. 10.3).

Each computer in the figure has a *serial interface* connected to its internal bus. For the purposes of describing the RS-232-C standard, the combination of the computer and the serial interface is called the *data terminal equipment* (*DTE*). The serial interface carries out the conversion of parallel bus data to a sequence of bits which can be seen as a sequence of dc voltage levels, as shown in Fig. 10.2.

Since the switched telephone network cannot reproduce dc voltage levels, each logical state must be converted to a tone (i.e., modulated) at one of two *standard frequencies* within the audible frequency range that can be transmitted by the network. This conversion is carried out by the *data communication equipment* (*DCE*). The RS-232-C interface standard specifies the signals that pass between the DTE and the DCE.

The DCE is normally a *modem* (for *mo*dulator-*dem*odulator). A modem converts the interface voltage corresponding to a logical 0 to one tone in the audible frequency range transmitted by the telephone network. Each logical 1 is converted to a second tone. Thus a series of voltage levels at the input to the modem become translated to a series of tones by the modem. These tones are transmitted over the telephone lines and are picked up by the modem at the other end of the line, which translates the tones back into voltage levels.

Most modems can simultaneously transmit data on one pair of frequencies and receive data on another pair of frequencies. This is known as a *full-duplex* link. A *half-duplex* link is one in which transmission must cease while reception

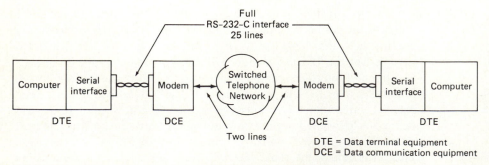

Figure 10.3 Data communications over a switched telephone network.

takes place. This, of course, increases the time it takes to complete a data exchange compared with the full-duplex link. In a *simplex* link, transmission occurs in only one direction. We will assume in the following discussion that a full-duplex link is implemented.

The RS-232-C standard was established by the Electronic Industries Association in 1969. (The complete standard is available from the EIA; see Bibliography.) Although intended to describe the DTE-DCE interface only, it has been widely used for the interconnection of all sorts of digital equipment. This can result in confusion when both pieces of equipment are DTEs (e.g., two computers).

10.1.1 Mechanical Specifications

The mechanical interface consists of a cable containing 25 wires. The cable is terminated at the DCE end by a female connector and at the DTE end by a male connector. The cable is recommended to be less than 15 m, but regardless of its length, no signal wire can have a capacitance to signal ground exceeding 2500 pF.

The names of the signal lines, the connector pin number each line is connected to at each end of the cable, and the direction of data flow in the signal line with respect to the DTE is shown in Table 10.1. The lines can be divided into six functional groups of lines:

Table 10.1

Group	Label	Line	To/from DTE	Pin
Ground	AA	Protective ground	n.a.	1
	AB	Signal ground	n.a.	7
Signal	BA	Transmitted data (TD)	From	2
	BB	Received data (RD)	To	3
Major control	CA	Request to send (RTS)	From	4
	CB	Clear to send (CTS)	To	5
	CC	Data set ready (DSR)	To	6
	CD	Data terminal ready (DTR)	From	20
Ancillary control	CE	Ring indicator	To	22
	CF	Received line signal	To	8
	CG	Signal quality detector	To	21
	CH/CI	Data signal rate selector	From/to	23
Timing	DA	Transmitter signal timing	From	24
	DB	Transmitter signal timing	To	15
	DD	Receiver signal timing	To	17
Secondary	SBA	Secondary transmitted data	From	14
	SBB	Secondary received data	To	16
	SCA	Secondary request to send	From	19
	SCB	Secondary clear to send	To	13
	SCF	Secondary received line signal detector	To	12

Note: Pins 9 and 10 reserved for testing; pins 18, 11, and 25 unassigned.

1. Two ground lines
2. Two signal lines, which carry the data
3. Four major control lines, which initiate and control data flow
4. Four ancillary control lines, which provide telephone line status and select data rate
5. Three timing lines, which provide a clock for synchronous transmission
6. Five secondary lines which allow a second channel to share the interface

The remaining 5 lines of the 24 allowed by the standard are either unassigned or used for testing.

Unfortunately, the standard does not specify a required interface connector. The industry has generally standardized on a 25-pin connector, but the pin numbering on the connectors may differ, and the sex of the connector may be incorrect (DTE should be male; DCE, female). This confusion can result in the need for a user to assemble a special cable to interconnect two RS-232-C-compatible devices.

10.1.2 Functional Specifications

The signal ground line is a common ground return path for all other signal lines. In the DCE this line is taken from a single common ground point for all circuits. The protective ground connects the frame of the DTE with the frame of the DCE. The frame may be connected to the signal ground point.

The only lines that are absolutely necessary for communication are the two data lines and the signal ground. The names of the lines are given with respect to the DTE. Thus the transmitted data (TD) line carries data from the DTE to the DCE. The transmitted data line is held in the marking state when no data are being transmitted. The received data (RD) line carries data from the DCE to the DTE. Most computer interfaces to peripherals make provision for the four data control lines as well as the data lines in order to control the flow of data.

In order for the interface to transfer data, all four data control lines must be asserted (on). The data terminal ready (DTR) signal from the DTE commands the DCE to prepare a communication link. When the switched telephone network rather than a dedicated line is being used, this signal is necessary. The data set ready (DSR) line from the DCE informs the DTE that all local functions required to establish a communication link are complete.

Once a link is established, the DTE then issues a request to send (RTS) to the DCE, commanding it to prepare the communications channel for transmission and reception. When ready, the DCE responds by raising the clear to send (CTS) line, which it keeps asserted while data is interchanged between the DTE and the DCE.

When a modem is present and operating over the switched telephone network, the line status signals can be used to provide ring indication and to indicate the presence of a received line signal. A line is provided to indicate line signal quality to the DTE. The DTE can also select one of two possible signaling rates available in the DCE using the data signaling rate selector line.

For synchronous transmission, the transmitter and receiver signal element timing lines can be used to carry a clock signal across the interface.

A secondary channel can share the same interface by using the five secondary lines (two data lines and three control lines).

Since not all interface lines are required for all applications, the RS-232-C standard allows 12 defined interface subsets (named A, B, . . . , L) of the complete set (named M) and a manufacturer-specified circuit (named Z).

DTE-DTE connection (no modem)   In connecting serial interfaces it is important to establish whether the serial interface card and the external device interface you may be dealing with are configured as DTE or DCE. One problem that often occurs when nonmodem equipment such as printers, terminals, and other computers are connected to a computer is that both sides of the interface may be defined by their manufacturers as DTEs. Thus the "transmitted data" line, for example, would be driven from both ends if connected. The solution is to create a special device called a *null modem*. The null modem is often simply a cable that is wired such that the TD line at one end is connected to the RD line at the other, and vice versa. If the control lines are used in the interface, they must also be appropriately connected.

10.1.3 Electrical Specifications

All signals are transferred across the interface over a circuit that has the electrical equivalent circuit shown in Fig. 10.4.

For each signal line in a RS-232-C interface the equivalent driver output voltage V_0 will not exceed a magnitude of 25 V. The combination of the driver voltage and the driver resistance R_0 will not allow a short-circuit current greater than 0.5 A to flow.

The load resistance R_L will be between 3 and 7 kΩ, and the driver voltage, driver resistance, and load resistance will produce a dc voltage V_1, of 5 to 15 V at the interface point when the open-circuit terminator voltage E_L is zero. The cable and terminator shunt capacitance C_L will not exceed 2500 pF.

On the data lines, a logical 1, or marking state, is represented as a voltage less than −3 V. A logical 0, or a spacing state, occurs when the line voltage exceeds +3 V.

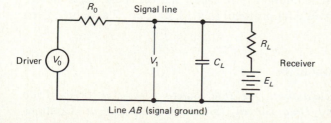

Line *AB* (signal ground)

Figure 10.4 Specified components of RS-232-C equivalent circuit.

On the control and timing circuits, a "high-true" convention is used. This difference between the data and control and timing conventions can cause some confusion unless it is noted. When a control or timing circuit is true, the line voltage is greater than $+3$ V, and when the circuit is false, the voltage is less than -3 V.

The standard allows a wide range of voltages to be used. The nominal supply voltages to the voltage source electronics is usually ±12 V. Integrated circuits are commercially available for driving and receiving RS-232-C signals. The use of ±5 V can also be found, and in some cases, manufacturers are using 0 to $+5$ V in contravention of the standard.

The line drivers must be able to withstand both open circuit and short circuits to any other line or to ground.

The maximum bit rate for transmission allowed by the standard is 20,000 bits/s. The maximum standard data rate for modem transmission that is less than 20,000 bits/s is 19,200 baud. The *baud rate* is the number of unit time intervals per second in the transmission. For example, if a 7-bit ASCII character is transmitted along with one start bit and two stop bits, there are 10 bits associated with each character. Thus in a 300-baud transmission system, 30 characters (300 divided by 10 bits) are transmitted each second.

Procedure for connecting RS-232-C interfaces   When two interfaces are to be connected together for the first time, there are several operations that must be carried out:

1. Determine whether the interface at each end is DTE or DCE.
2. Determine the sex of the connector on both the computer end and the peripheral end. According to the standard, the DTE should have a male connector and the DCE, a female connector.
3. Check the pin numbering on each connector. Even if the connector is a 25-pin connector as required, the second row's pin numbers may start at one end or the other.
4. Purchase or make up a cable that has the correct matching connectors on each end. The wiring should connect pins of the same number at each end if the cable is for the standard DTE-DCE connection. You may not require all 25 wires in the cable. If the cable is to connect DTE-DTE devices, use a null-modem connection with control lines connected as needed by the interfaces.
5. Set any switches in the computer interface and the peripheral device interface to give the desired data protocol (e.g., 1 start bit, 7 data bits, even parity, 1 stop bit).

10.1.4 Other Related Standards

1. CCITT V.24: *List of Definitions for Interchange Circuits between Data Terminal Equipment and Data Circuit Terminating Equipment*. Contains line connection definitions that correspond to the RS-232-C definitions.

2. CCITT V.28: *Electrical Characteristics for Unbalanced Double-Current Interchange Circuits*. Corresponds to the equivalent RS-232-C specifications. Thus a manufacturer of a serial interface might designate the interface as "compatible with RS-232-C/CCITT V.28."

3. RS-422: *Electrical Characteristics of Balanced Voltage Digital Interface Circuits*. Provides for both differential line drivers and differential line receivers to be used instead of ground-referenced signals. Thus there is common-mode voltage rejection (discussed in Sec. 11.4). The standard also allows for a higher data rate that varies with distance from 10 Mbaud at 12 m to 100 kbaud at 1200 m. The standard allows more than one receiver on the lines. This is known as a *multidrop* capability.

4. RS-423: *Electrical Characteristics of Unbalanced Voltage Digital Interface Circuits*. Provides for the common mode source's signal ground to be brought to one input of a differential line receiver while the signal line is brought to the other. Thus common mode voltage differences between source ground and receiver ground are rejected. Data can be transmitted at 100 kbaud at at 12 m and 1 kbaud at 1200 m.

5. RS-449: *General Purpose 37-Position and 9-Position Interface for Data Terminal Equipment and Data Circuit-Terminating Equipment Employing Serial Binary Data Interchange*. The RS-449 standard has renamed the RS-232-C lines and provides a new 37-pin connector for the primary lines and an optional 9-pin connector for the secondary lines. A particular interface card might state that the serial interface is "RS422/449-compatible," meaning that the RS-422 standard (differential drivers and receivers) is met and that the mechanical connection is a 37-pin connector.

10.2 THE IEEE-488 INTERFACE STANDARD

The IEEE-488 interface standard describes a multiwire communication system that allows both data and control information to pass between devices. When one device is the "bus controller," the comunication lines, or *bus* lines, have some functional similarities to a microprocessor system bus. For example, the controller can "talk" or "listen" to each device attached to the bus and devices can be equipped to request service and the controller polls the devices to find out who requested service.

A major difference between the IEEE-488 bus and an intrasystem bus is that the IEEE-488 bus interface protocol in each device is more complex. With the IEEE-488 bus there is more handshaking required to start and stop communication with the controller and other devices. On a microprocessor bus, most communication is between the CPU and each device. On the IEEE-488 bus many different devices produced by different manufacturers can talk to each other under the controller's supervision. Even the job of controller can be passed on to another suitably equipped device. Since it would not be appropriate for all devices to act as a controller, for example, not all the capabilities are required for each instrument on the bus. An example of a particularly simple device that would never need the full bus control capabilities is a digital multimeter. The multimeter would simply report its readings (talk) to some other device on the bus.

The development of the IEEE-488 interface was originated by Hewlett-Packard Company for connecting multiple measuring instruments to computers and other instruments. (The full standard is available from the IEEE; see Bibliography.)

In contrast to the unidirectional RS-232-C interface lines, the IEEE-488 lines are bidirectional. Up to 15 IEEE-488-compatible device interfaces can be attached to the multiwire IEEE-488 bus. A system diagram is shown in Fig. 10.5.

A device interface attached to the bus can be activated as a *talker* to one or more other interfaces activated as *listener* interfaces. Messages can be passed between devices (via the interfaces) as well as between interfaces (for the transfer of bus control information). Only one of the attached devices may be active as a bus *controller* at any time. Bus control can be passed to another device, however.

The system provides very great flexibility in its complete implementation. There are now integrated circuits available from several manufacturers that incorporate all the logic necessary for the state transitions of the IEEE-488 interface functions (e.g., Intel 8291A talker/listener, 8292 controller, and 8293 GPIB transceiver). The interface can also be designed around a general-purpose I/O integrated circuit such as Motorola's MC6821 peripheral interface adapter. The interface functions are then carried out by software in the CPU.

The total length of the GPIB bus must not exceed 20 m (compared with 15 m for RS-232-C). The maximum data rate on each line is limited to 1 Mbit/s. Since 8 bits are transmitted in parallel, this is equivalent to 8 Mbits/s on a single line, which is 400 times the maximum allowable speed of the RS-232-C standard.

The following description of the IEEE-488 interface is intended to enable a system designer to select the necessary interface functions for a particular application. The designer can then select a commercially available interface card with the proper set of implemented interface functions.

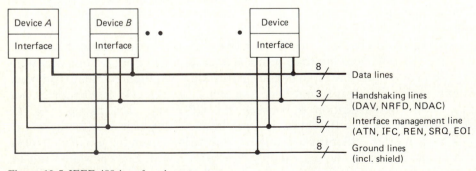

Figure 10.5 IEEE-488 interface bus structure.

10.2.1 Mechanical Specifications

The bus consists of 24 conductors organized into

- Eight data lines (DIO1-DIO8)
- Three data control (handshaking) lines (DAV, NRFD, NDAC)
- Five interface management lines (IFC, ATN, SRQ, REN, and EOI)
- Seven ground lines
- One cable shield

Attached to each device containing an IEEE-488 interface is a female 24-pin connector with two rows of 12 pins. Pin 13 is located in the second row under pin 1. In order to allow several cables to be connected to the same device connector, each end of the cable has both a male and a female connector. For a linear bus, only two connectors are stacked on the device connector, but the arrangement also allows for star and other bus topologies (see Fig. 10.6).

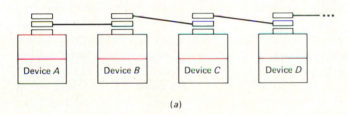

(a)

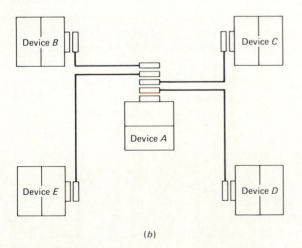

(b)

Figure 10.6 IEEE-488 bus topologies: (a) linear; (b) star.

10.2.2 Functional Specifications

Within each IEEE-488-compatible apparatus such as a computer or measurement instrument one can identify an application-dependent *device* (e.g., voltage measurement electronics) with certain *device functions*, and an *interface* that can have one or more IEEE-488 *interface functions*. One side of the interface comunicates with the device logic and the other side with the IEEE-488 bus. Some messages on the bus are directed from one interface to another interface. These are called *interface messages* (e.g., service request). Other messages are passed from one device through both the local and remote interfaces to the remove device. These are called *device-dependent messages* (e.g., device data and device status).

The basic functions that the interface can carry out are listener, talker, and controller. An interface that has a listener function in its repertoire must be *addressed by* the IEEE-488 bus controller as a listener before it can receive messages over the bus. A talker function must also be activated by being addressed by the controller. If the capability is present in both interfaces, a controller in one interface can pass control to another interface. A device can have more than one active interface function (e.g., controller and talker).

Although there are only 15 devices allowed to be connected to the bus, there are 31 possible addresses available. Thus within each interface there can be more than one interface function address. A listener can therefore have one address, and a separate listener function in the same interface can have another address. A talker and a listener in the same device may have the same address but two talker functions must have different addresses. An address of 31 (address bits all set $= 1$) is not allowed, since it is used as part of the UNT (untalk) message that resets all addressable functions to the idle state.

There is a second level of addressing possible for talkers and listeners called the extended listener and extended talker functions. These functions require a 5-bit secondary address message after the primary 5-bit address has been passed, before the talker or listener function is enabled. This allows the IEEE-488 bus system to have up to 961 (31×31) allowable function addresses, since a secondary address of 31 is also not allowed. Another use for the secondary address is to select some internal subdevice or register. There is a speed penalty in the use of a secondary address, since two addresses must be sent over the bus as opposed to the use of primary addresses only.

Listener function (L)   A listener is an interface function that allows a device to accept data over the bus from another device or interface. An example of a device needing a listener function is a line printer.

A device interface can be activated as a listener by the bus controller. To become a listener, a particular interface must be addressed by the bus controller by issuing an MLA (my listen address) message containing the interface's 5-bit interface address in the message. The interface can optionally be locally designated (by a switch) as a listen only interface.

The listener function requires a local service function called the acceptor handshake (AH) function in order to be able to accept data.

The extended listener function (LE) allows a secondary 5-bit address [via my secondary address (MSA) message] to be sent after the primary 5-bit address has been sent to the interface (via the MLA message). If an instrument has an extended listener function, both primary and secondary addresses must be received before the interface is enabled to listen. An interface can have both L and LE functions provided that they have different MLA addresses.

The allowable subsets to the listener and extended listener functions are given in Table 10.2.

Talker function (T) A talker is an interface function that sends data to one or more listeners. An example of a device that requires a talker interface function is a voltage-measuring the device.

The talker can be activated by being addressed over the bus by the controller. Alternatively, the talker function can also be locally designated as a talk-only interface by a switch.

If the controller places an address on the bus that corresponds to the talker function address, the address is referred to in the IEEE-488 standard as my talk address (MTA).

As with the LE function, an extended talker function (TE) can be used instead of the T function. The secondary address message (MSA) is issued by the controller after the MTA message to convey the secondary address to the device.

Once an interface is addressed to talk, if another talker is addressed by the controller (other talk address, OTA, or other secondary address, OSA) the initially addressed talker will revert to its unaddressed initial state to prevent bus contention.

The talker requires at least the presence of both the acceptor handshake (AH), in order to receive its address, and a service function, called the source handshake (SH) function, in order to transmit its data to a listener.

Table 10.2

	Capability			
Subset	Basic listener	Listen only*	Unaddress if talker[†]	Other functions needed
L_0, LE_0	No	No	No	None
L_1, LE_1	Yes	Yes	No	AH_1
L_2, LE_2	Yes	No	No	AH_1
L_3, LE_3	Yes	Yes	Yes	AH_1, T_1 to T_8, or TE_1 to TE_8
L_4, LE_4	Yes	No	Yes	AH_1, T_1 to T_8, or TE_1 to TE_8

* The interface must be able to be set to the "listen only" mode by a switch.

[†] This condition would occur when the same interface is addressed as a talker while already addressed as a listener. The interface would respond by unaddressing itself as a listener if this capability is selected.

Table 10.3

| Subset | Capability | | | | |
	Basic talker	Serial poll	Talk only	Unaddress if made L	Other functions needed
T_0, TE_0	No	No	No	No	None
T_1, TE_1	Yes	Yes	Yes	No	SH_1, AH_1
T_2, TE_2	Yes	Yes	No	No	SH_1, AH_1
T_3, TE_3	Yes	No	Yes	No	SH_1, AH_1
T_4, TE_4	Yes	No	No	No	SH_1, AH_1
T_5, TE_5	Yes	Yes	Yes	Yes	SH_1, L_1 to L_4 or LE_1 to LE_4
T_6, TE_6	Yes	Yes	No	Yes	SH_1, L_1 to L_4 or LE_1 to LE_4
T_7, TE_7	Yes	No	Yes	Yes	SH_1, L_1 to L_4 or LE_1 to LE_4
T_8, TE_8	Yes	No	No	Yes	SH_1, L_1 to L_4 or LE_1 to LE_4

When an interface requests service, it places a service request (SRQ) onto the bus (see the next subsection). The controller will respond with a serial poll enable (SPE) request for a status byte from each interface. A talker function in each interface polled must have the capability to respond to the SPE by placing the status byte on the data lines. The talker's response is described along with the service request function.

The allowable subsets to the talker function are given in Table 10.3.

Service request function (SR)   The SR function allows a device to request service via its IEEE-488 interface by issuing a service request(SRQ) message on the control lines. The normal response to this event is for the controller to conduct a serial poll to find the device requesting service.

When an SRQ is detected, the controller issues a serial poll enable (SPE) command and addresses each interface in turn as a talker in order to find out which device requested service.

When the interface that initiated the request is SPE-enabled and addressed as a talker, the interface issues a request for service (RFQ) message to the controller and simultaneously places its device status byte (STB) on the data lines, using a source handshake function. The status byte is then read by the controller. The SR function requires only the talker and source handshake functions to operate.

There are only two subsets of the SR function:

SR_0 = no capability
SR_1 = complete capabilty and requires one of T_1, T_2, T_5, T_6 or one of TE_1, TE_2, TE_5, TE_6

Table 10.4

Subset	Capability	Other functions needed
PP_0	No capability	None
PP_1	Remote configuration	L_1 to L_4 or LE_1 to LE_4
PP_2	Local configuration	None

Parallel poll function (PP)   Unlike the serial poll that is initiated by any interface asserting the SRQ line on the bus, a parallel poll is initiated by the controller itself.

With a PP function, up to eight device interfaces can simultaneously present a 1-bit status (using the PPR message) on a unique preassigned data bus line in response to a command (IDY for "identify") from the controller. The controller thus receives the status of up to eight devices (that need not be addressed as talkers) in response to a single command.

The device interfaces can initially be configured by switches either in the local device (function subset PP_2) or over the bus (function subset PP_1). When configured over the bus, a sequence of parallel poll configure (PPC) and enable (PPE) commands from the controller are required. These commands require the acceptor and listener functions.

The allowable parallel poll subsets are given in Table 10.4.

Remote-local function (RL)   In a device attached to the IEEE-bus, there may be some functions that are under local control only, some functions that are under remote (via IEEE-bus) control only, and some functions that can be either remotely or locally controlled (e.g., voltage range, sampling rate).

The device can be placed under remote control (for device functions that can have either remote or local control) by a message from the controller on the remote enable line (REN), while the remote device is addressed as a listener. The local controls can then be prevented (i.e., locked out) from being locally returned to local control by issuing an LLO message from the controller.

The allowed remote-local subsets are listed in Table 10.5.

Device clear function (DC)   The DC function within a device interface can be commanded to clear or initialize device functions. It can be activated in all interfaces by a device clear (DCL) message from the controller. A DC function in a particular interface can also be activated (in DC_1-type interfaces) by

Table 10.5

Subset	Capability	Other functions required
RL_0	None	None
RL_1	Complete	L_1 to L_4 or LE_1 to LE_4
RL_2	No local lockout	L_1 to L_4 or LE_1 to LE_4

Table 10.6

Subset	Capability	Other functions required
DC_0	None	None
DC_1	Complete	L_1 to L_4 or LE_1 to LE_4
DC_2	No selected device clear	AH_1

issuing a device clear (DCL) message while a selected interface has been addressed as a listener.

The allowed device clear function subsets are given in Table 10.6.

Device trigger function (DT)   An interface with the DT capability can provide a trigger signal to its device to start its operation. One or more devices can be addressed as listeners and then the controller issues a "group execute trigger" command which the device interface recognizes.

There are only two allowed subsets of the DT function:

DT_0: No capability
DT_1: Complete capability requiring L_1 to L_4 or LE_1 to LE_4 functions.

Source handshake function (SH)   The SH function is a service function required by both the talker and controller functions. Of the three handshaking lines used on the IEEE-bus, the SH function controls the data available (DAV) line.

The source handshake function is invoked by multiline messages from

- The talker interface function for the exchange of data between devices and for the transmission of device status information during a serial poll (see SR function)
- The controller interface function for the transmission of multiline messages

There are only two subsets allowed:

SH_0: No capability.
SH_1: Complete capability. The function subsets that require the source handshake function are T_1 to T_8, TE_1 to TE_8, and C_5 to C_{28}.

Acceptor handshake function (AH)   The AH function is another service function that provides each interface with the capability to receive multiline messages from other interfaces. It controls two of the three handshake lines connected to the interface: the ready for data (RFD) and data accepted (DAC) lines.

The AH function is invoked

- By the attention line (ATN message) becoming true, as would occur when the controller sends any message to the interface
- When the interface's listener function is addressed or active

There are only two allowed subsets of the AH function:

AH$_0$: With no capability
AH$_1$: With complete capability

No other function subsets are required.

Controller function (C) The controller can address or unaddress device interfaces as listeners or talkers, respond to a service request from a device, conduct polls to determine device status, and remotely control device parameter settings or trigger a device. Means also exist for one controller to pass the control of the bus to another controller.

The controller has the exclusive control of two of the five interface management lines:

- Attention (ATN line), which the controller asserts whenever it issues a multiline message
- Interface clear (IFC), which it asserts to reset all listener, talker, and inactive controllers to their initial (power-on) states

The commands that can be issued by the controller include

1. *Activating commands.* A group of commands exist that allow the controller to activate an interface as a talker, listener, or controller:

   - Address a listener
   - Address a talker
   - Pass control (TCT): In order to pass control, the present controller addresses the next controller as a talker and then issues a take control (TCT) message.
   - Receive control: When an interface has been addressed as a talker and the TCT command has been issued, the interface will become the new controller.

2. *Deactivating commands.* A second set of commands can be used to deactivate all listeners and talkers:

   - Unlisten (UNL): Unaddresses (returns to idle state) all addressed listener functions.

- Untalk (UNT): Initializes all talker interfaces to idle state. Used as a system function only, since each active interface returns from addressed to idle state when a new talker is addressed or the interface is cleared.
- Device clear (DCL): Clears (returns to initial state) all devices. Has no effect on interfaces.
- Selected device clear (SDC): Clears only addressed devices, not interfaces.
- Interface clear (IFC): Clears all interfaces only (returns them to idle state). No effect on devices.

3. *Service request*
   Service request (SRQ): The service request can be asserted by any interface whose device requires the controller's attention.

4. *Serial service response*
   - Serial poll enable (SPE): The controller's response to a SRQ request is to unlisten all interfaces and issue an SPE message that places the talker function of any service-requesting interface(s) into serial poll mode. The controller then addresses each interface in the system to talk. The SPE-enabled talker interface places its status byte (STB) on the data lines and issues a request service message (RQS). If the addressed interface is not requesting service, the RQS, message is not sent and the controller polls the next interface.
   - Serial poll disable (SPD): Returns all talker functions to the idle mode from the serial poll enabled mode.

5. *Parallel service response.* In a controller-initiated parallel poll, the listener function of an interface to be polled must be first addressed. The PPC and PPE messages are then used in turn to configure the interface to provide its 1-bit status on a particular data line. The sequence just described is repeated for each interface to be assigned a response line. The controller's IDY message is then used to simultaneously poll the configured interfaces.

   - Parallel poll configure (PPC): the PPC is first issued to prepare the interface to accept a data line number given in the PPE command.
   - Parallel poll enable (PPE): The PPE message from the controller tells the addressed interface which data line it is to return its 1-bit status on. The configuration sequence is terminated by an unlisten message.
   - Identity (IDY): When an IDY message is issued by the controller, the parallel poll function of each configured interface becomes active and places its 1-bit status (using PPR1-PPR8 message) on the data lines.

- Parallel poll disable (PPD): The controller must issue a PPD message to return the parallel poll function in all interfaces to the configured but inactive state.
- Parallel poll unconfigure (PPU): This message commands the parallel poll function of each interface to return to the idle (unconfigured) state.

6. *Device remote or local control.* The following commands control the source of device control for device functions where either remote or local control is possible.

- Remote enable (REN): The transmission of a REN on the bus after a device has been listener-addressed will cause the addressed device to come under remote control. In this state the device can still be returned by a user to local control by a local "return to local" (rtl) switch.
- Local lock out (LLO): To prevent the device from being returned to local control by the rtl switch, the controller can issue an LLO command to the RL function in all listener-addressed interfaces, which will lock out the rtl switch.
- Go to local (GTL): The remote-enabled and locked-out interface can be returned to local control by the GTL message.

7. *Device function triggering.* A device function can be triggered via its interface by the controller using the group execute trigger (GET) command. The controller first addresses to "listener" each interface to be triggered. Once all interfaces are addressed, the controller issues a GET message which simultaneously triggers all addressed devices.

The allowed controller function subsets are given in Table 10.7.

10.2.3 Electrical Specifications

The IEEE-488 bus lines operate on TTL levels for which 5.25 V is the maximum and 0 V is the minimum. All lines can be driven by open-collector TTL drivers which must be capable of sinking 48 mA continuously.

A typical circuit that fulfils the requirements is shown in Fig. 10.7.

Each signal line attached to the bus must be terminated with a resistive load. The load is divided into a pull-up resistor $R_P = 3$ K and a resistor to ground $R_G = 6.2$ K. The line-to-ground capacitance must not exceed 100 pF for each line in a device interface. The negative voltage excursions must be limited. In Fig. 10.7, this is done with a diode to ground.

The bus cable must contain 24 conductors (16 signal and the remainder grounds). Each signal line must have less than 150 pF/m capacitance to ground and must have all other lines connected ground. The cable must be shielded.

Table 10.7

Subset	Capabilities*										Other functions required				
	1	2	3	4	5	6	7	8	9	10	C_1	C_2	LG†	SH₁	TG‡
C_0	N§	N	N	N	N	N	N	N	N	N					
C_1	Y§														
C_2		Y									R§				
C_3			Y								R				
C_4				Y											
C_5					Y	Y	Y	Y	Y	Y			R	R	R
C_6					Y	Y	Y	Y	Y	N				R	R
C_7					Y	Y	Y	Y	N	Y			R	R	R
C_8					Y	Y	Y	Y	N	N				R	R
C_9					Y	Y	Y	N	Y	Y			R	R	R
C_{10}					Y	Y	Y	N	Y	N				R	R
C_{11}					Y	Y	Y	N	N	Y			R	R	R
C_{12}					Y	Y	Y	N	N	N				R	R
C_{13}					Y	Y	N	N	Y	Y	R	R	R		
C_{14}					Y	Y	N	N	Y	N	R		R		
C_{15}					Y	Y	N	N	N	Y	R	R	R		
C_{16}					Y	Y	N	N	N	N	R		R		
C_{17}					Y	N	Y	Y	Y	Y			R	R	R
C_{18}					Y	N	Y	Y	Y	N				R	R
C_{19}					Y	N	Y	Y	N	Y			R	R	R
C_{20}					Y	N	Y	Y	N	N				R	R
C_{21}					Y	N	Y	N	Y	Y			R	R	R
C_{22}					Y	N	Y	N	Y	N				R	R
C_{23}					Y	N	Y	N	N	Y			R	R	R
C_{24}					Y	N	Y	N	N	N				R	R
C_{25}					Y	N	N	N	Y	Y			R	R	
C_{26}					Y	N	N	N	Y	N				R	
C_{27}					Y	N	N	N	N	Y			R	R	
C_{28}					Y	N	N	N	N	N				R	

* 1 = System controller
   2 = Send IFC
   3 = Send REN
   4 = Respond to SRQ
   5 = Send interface messages
   6 = Receive control
   7 = Pass control
   8 = Pass control to self
   9 = Parallel poll
  10 = Take control synchronously
† LG = AH_1, L_1 to L_4 or LE_1 to LE_4
‡ TG = T_1 to T_8 or TE_1 to TE_8
§ R = required, Y = yes, N = no; one or more of C_1 to C_4 may be selected; only one from C_5 to C_{28} may be selected.

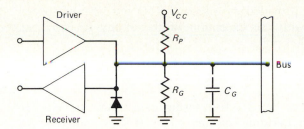

Figure 10.7 Typical bus I/O circuit.

The construction of the cable is further constrained by the IEEE-488 standard, which the reader should see for additional details.

The timing of certain state transitions within various interface functions is also constrained by the standard.

Minimum configurations The minimum set of signal lines that are required by a system with no controller are

- Eight data lines
- Three handshaking lines

When a system has a controller, the interface clear line (IFC) and the attention line are required.

Each system must contain at least one talker (at least a T_3 containing a talk only) and one listener function (at least an L_1 with listen only).

Examples

1. *Digital Equipment Corp. Q-Bus compatible IBV-11A* The specifications for this board show that it is compatible with the following IEEE-488 subsets: SH_1, AH_1, T_5, TE_5, LE_3, SR_1, RL_1, PP_2, DC_1, C_1, C_2, C_3, C_4.

These can be interpreted from the previous tables to mean that the board (with its software driver) has complete capability for source handshake, acceptor handshake, talker, extended talker, extended listener, service request, remote or local, and device clear; and has some controller functions.

The board cannot be remotely configured nor can it issue a device trigger.

2. *Ziatech ZT 7488/28 STD-bus compatible GPIB (IEEE-488) controller interface* The specifications for this board show that Ziatech's interface conforms to the following IEEE-488 subsets: SH_1, AH_1, T_3, L_1, SR_0, RL_0, PP_0, DC_0, DT_0, C_1, C_2, C_3, C_4, C_9.

The interpretation of this information is that the interface has complete capability for source handshake, acceptor handshake, and all controller capabilities except the ability to pass control to itself.

It also has the basic talker and talk-only functions as well as the basic listener and listen-only functions. It has no other capabilities.

The other capabilities are provided on a second STD bus board (Ziatech ZT 7488/08 talker/listener) with the exception of the ability to carry out remote configuration for parallel poll.

SUMMARY

- Serial communications can be used to transmit data from one sysem to another. The communicating systems may be two computers, a computer and a modem, a computer and a peripheral device, or a computer and some instrumentation or data-acquisition equipment.
- Various standards committees study the needs of the industry and generate standards.
- Two very common communication standards are RS-232-C and IEEE-488 (1978).
- The RS-232-C standard describes the communication between one data terminal device such as a display terminal or computer and one data communication device such as a telecommunications modem. The data bits within a byte are sent serially over a single data line. The maximum data rate allowed by the standard is 20K bits/s.
- Since most computers and terminals have the RS-232-C connectors, other equipment manufacturers (of line printers, plotters, measuring instruments, digital recorders, etc.) began to produce RS-232-C-compatible equipment for interconnection to other devices. The result is that the compatibility between systems must be investigated carefully, since the equipment may have features that lie outside the scope (and intent) of the standard.
- The IEEE-488 (1978) standard allows up to 15 devices with IEEE-488 compatible interfaces to communicate over a 24 conductor bus. All the bits of each message byte are sent simultaneously over the 8 data lines in the bus. The maximum transfer rate is 1 Mbyte/s. The IEEE-488 bus is more versatile, faster, and more complex than RS-232-C communication. The complexity is offset somewhat by the existence of integrated circuits that can carry out the complete bus protocol.
- The defined subsets of the standard allow a designer to choose which of the many IEEE-488 capabilities are appropriate for the application at hand.

EXERCISES

10.1 Compare the RS-232-C interface and the IEEE-488 (1978) bus with respect to the following points:

- Purpose for which the standard was developed
- Number of interfaces attached to the interconnecting lines defined in the two standards

- Number and type of interconnecting lines
- Restrictions on the length of the interconnection system
- Types of information transmitted over the data lines
- Maximum data rate, in bytes per second
- Method of synchronizing data between sender and receiver.

10.2 List the sequence in which the four major control lines become true in the normal setup for data transmission over an RS-232-C interface.

10.3 List the exact sequence of logical transitions that the transmitted data line would undergo in the transmission of the three letters in the word BUS. Assume that the data protocol is 1 start bit, 7 data bits, odd parity, and 1 stop bit. What would the transmission time be for a 1024-byte data buffer at a baud rate of 4096?

10.4 For the IEEE-488 bus, give the sequence of commands that the controller would have to issue to send 4 bytes of data from a new talker (T) to three new listeners (L, L, and LE).

10.5 If a device on the bus requests a serial poll and there are three valid bus addresses, list the steps that the controller must take to find the device that requested service.

10.6 Suppose that you wish to select suitable IEEE-488 bus interfaces to periodically pass the voltage read from a sensor connected to CPU_1 to a nearby CPU_2. Assume that the times that the information is to be passed is under control of CPU_1 (since it is determining the sampling rate).

What are the minimum interface functions required for the IEEE-488 bus interfaces connected to CPU_1 and CPU_2?

10.7 Suppose that you are developing a line printer that has an internal CPU and bus architecture. You wish to connect the line printer to a computer (which, for the purposes of this problem, you can assume has a complete IEEE-488 bus available). The line printer requires that the following messages be conveyed over the IEEE-488 bus to or from the computer:

- Data bytes to be printed.
- A printer reset caused by the computer.
- A printer error condition that should cause the computer to be notified immediately. The particular error will set a bit in a status register. The status register should be able to be passed to the computer for error identification.
- A line feed command from the computer.

Select the interface functions required to implement the printer's IEEE-488 bus interface. Assume that the bus controller will always be the computer's interface.

BIBLIOGRAPHY

Standards

Electronic Industries Association: *EIA-RS-232-C: Interface between Data Terminal Equipment and Data Communication Equipment Employing Serial Binary Data Interchange*, Washington, D.C., August 1969. (Material used here with permission of EIA.)
Institute of Electrical and Electronics Engineers: *IEEE-488: Standard Digital Interface for Programmable Instrumentation*, New York, 1978. (Material used here with permission of IEEE.)

Telecommunication interfaces

McNamara, John. E. (ed.): *Technical Aspects of Data Communication*, Digital Press, Bedford, MA, 1977.

Local area networks

Tanenbaum, Andrew S.: *Computer Networks*, Prentice-Hall, Englewood Cliffs, NJ, 1981.
Shapiro, S. F. (ed.): "Special Report on Data Communications Systems Design," *Computer Design*, vol. 22, no. 3, 1983, pp. 157–216.

Product data

ISB-3700 Dual Independent Channel, Synchronous/Asynchronous Communications Card— Product Specification, Intersil Inc., Systems Division, Sunnyvale, CA.
IBV11-A IEEE-488 Interface, *PDP-11 Microcomputer Interfaces Handbook*, Order code EB-23144-18, Digital Equipment Corporation, Maynard, MA, 1983–1984.
ZT-7488/28 IEEE-488 (GPIB) Interface Specification Sheet, Ziatech Corp., San Luis Obispo, CA.

ELEVEN

INPUT SYSTEMS

The *input system* is defined here as the *sensor* plus the *input conditioning circuitry* (see Fig. 1.6). The term "sensor" designates a package which contains

1. A *transducer* that produces an electrical signal as a function of a measured variable such as pressure. A pressure transducer itself is a particular physical implementation (e.g., strain gauge mounted on a thin metallic diaphragm) of a transduction phenomenon (resistive change with mechanical strain). The same phenomenon can be used in different transducers (e.g., load cells also use strain gauges).
2. The electronics required to produce a signal that can be easily "conditioned" (amplified, level-shifted, filtered, or isolated; see Sec. 1.3) to match the requirements of a standard computer interface. The aforementioned pressure transducer consists of a strain gauge bridge and an amplifier.

The sensor is often a single physical package. It connects to the *input conditioning circuitry* consisting of any electronics between the sensor and the *computer interface*. In other situations the designer will add electronics to the basic transducer to form a sensor, as shown in some of the examples to follow. The dividing line between the "sensor electronics" and the "signal conditioning electronics" is sometimes open to individual interpretation.

The subject of transducers and sensors is very broad, so it would not be appropriate to attempt to cover it fully in this text. A number of sensors will be discussed, however, to illustrate some common transduction mechanisms and the characteristics of sensor outputs that would be of importance in microprocessor system design.

Sensor selection criteria For any input variable to be monitored, there are usually several different types of sensors available, based on different physical transduction phenomena. In the example given in Chap. 1, the water level in a tank can be measured with an ultrasound sensor, pressure sensor, capacitance sensor, conductivity sensor, float and potentiometer, optical sensor, as well as in many other ways. A selection must be made based on the measurement environment and other requirements of the design (see Chaps. 3 and 4).

There are many considerations that should be taken into account in the selection of a sensing system, some of which are

- Performance
- Sensor and signal conditioning electronics
- Packaging
- Cost and delivery time

We will be discussing only performance and sensor and signal conditioning requirements in the sections to follow.

From the standpoint of the microprocessor system designer, the signal waveform characteristics at the computer input interface determine the nature of the computer hardware and software components of the system. We will use these characteristics to categorize sensors into three basic groups:

1. Sensors that produce binary digital levels such as an on-off switch
2. Sensors such as the anemometer in Sec. 11.2 that produce a binary pulse-coded signal (pulse train) from a sensed physical variable (e.g., wind velocity for the anemometer), where the physical variable has been converted to a pulse rate (or in some sensors, a pulse duration or pulse interval)
3. Sensors that generate continuous analog signals

11.1 SENSORS WITH BINARY-STATE OUTPUTS

Mechanical switches The simplest type of sensor consists of a single-pole–single-throw (SPST) mechanical switch. This device is shown in Fig. 11.1 connected to a "pull-up" resistor. When the switch is open, the resistor pulls

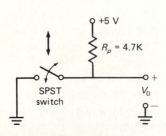
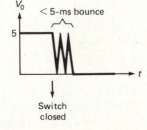

Figure 11.1 Simple switch.

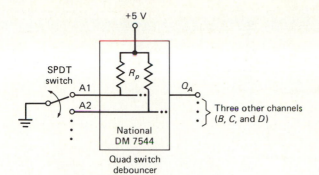

Figure **11.2** Integrated switch debouncer.

the output to 5 V, which is interpreted by a TTL-level compatible gate in the computer interface as one logical state. When the switch is closed, the output is pulled to ground potential, which is interpreted as the other logical state.

One problem with using a mechanical switch is that when it is closed, it "bounces" for a few milliseconds (approximately 5 ms). When it is only important that the first closure be detected, such as in a limit switch or in a "panic button," the subsequent opening and closing bounces after the first closure is detected need not be monitored. When the opening of the switch must be detected after a closing, the switch must not be interrogated until after the switch "settling time" has expired. The use of a programmed delay is one means of overcoming the effects of switch bounce. The software approach to *debouncing* the switch is described in Chap. 6.

A hardware approach to debouncing using a single-pole–double-throw (SPDT) switch plus an integrated debouncing circuit is shown in Fig. 11.2. When the grounded moving contact touches either input, the input is pulled low and the circuit is designed to latch the logic state corresponding to the first contact closure and to ignore the subsequent bounces.

Digital integrated level sensor An electronic switch can be activated by the detection of a physical event. This happens in the sensor shown in Fig. 11.3 as a result of the detection of a lack of conduction between two electrodes in water.

Conduction is sensed by generating an ac current waveform through the two electrodes and the intervening water. When the voltage across the electrodes is relatively low (i.e., when the electrodes are "shorted" by the water), the output transistor is off (the "switch" is open) and the sensor output voltage is pulled to the supply voltage by R_p. When the electrodes are out of the water, the electrode voltage rises, the output transistor turns on, and the output is pulled to ground.

The ac frequency is set by the capacitor C_2 and the device sensitivity is set by the potentiometer VR_1. The values depend on the electrode and fluid properties and can be empirically determined.

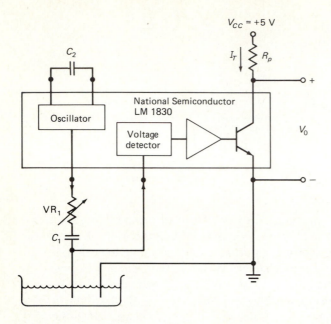

Figure 11.3 Liquid-level sensor.

The pullup resistor R_p should have a value

$$R_p \gg \frac{V_{cc}}{I_{T,\max}}$$

where V_{cc} = logic power supply for interface
$I_{T,\max}$ = maximum transducer circuit on current

in order to prevent damage to the output transistor by excessive current.

Digital integrated metal sensors Sensors are available to detect the presence of magnetic fields. These sensors rely on a transduction phenomenon called the *Hall effect*. The sensor shown in Fig. 11.4 (e.g., Micro Switch AV type) is one configuration of these sensors which is used to detect the presence of a ferrous metal vane in the slot.

The sensor shown has a built-in magnet which generates a field that crosses the gap and is sensed by the Hall device. The output transistor is normally on in this state ($V_{on} = 0.4$ V, $I_{on} < 10$ mA). When a ferrous-metal vane is inserted into the slot, the magnetic field is shunted back to the magnet, bypassing the Hall device, and the output transistor turns off. As a result, the output voltage will be pulled up by the resistor shown.

Hall-effect sensors are also available in other configurations for sensing magnetic fields. For example, a "slotless" version of the sensor just described can be used for sensing the presence of ferrous metal (counting gear teeth, for example).

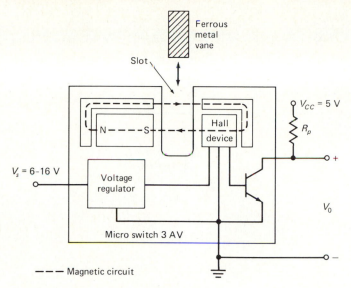

Figure 11.4 Hall-effect ferrous-metal sensor.

Other integrated sensors embodying a different principle, called *eddy current* sensors, can be used to sense either ferrous or nonferrous metals (e.g., Micro Switch FYCB16A1) at ranges of up to several inches.

11.2 SENSORS THAT PRODUCE BINARY PULSE TRAINS

Some sensors produce binary pulse trains encoding the sensed variable as a function of the pulse rate. The pulses can be counted in the processor or the interface over a fixed time to produce an estimate of the amplitude of the sensed variable.

For a variable that changes very slowly, a transduction mechanism that produces a pulse rate as a function of the sensed variable is an attractive one. The pulse rate can be averaged over very long periods of time, yielding high resulting accuracy, limited only by the stability of the transduction mechanism.

Another advantage of this form of output is that the output is relatively insensitive to noise pickup in the transmission of the pulses from sensor to interface, since arbitrarily high voltage (or current) pulses can be used, overriding the noise. In Sec. 11.5 we will discuss circuit modules which transform analog variables to digital pulse rates for signal transmission over long distances.

Thus by comparison with an analog voltage carried over the same pair of conductors, pulse rate can be used to encode a signal with equivalent accuracy and less noise sensitivity. However, it takes longer to extract an estimate of the analog variable because of averaging of the pulse rate. It is interesting to note that information carried along the nerve fibers of biological systems is encoded as a pulse rate.

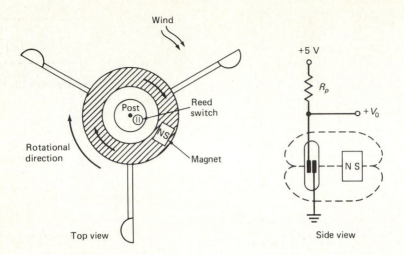

Figure 11.5 Anemometer.

Anemometer   A simple example of a pulse-generating sensor is shown in Fig. 11.5.

The device embodies a simple magnetically operated switch in a wind-driven rotor assembly to sense wind velocity. The rotating part contains a magnet that closes the contacts on a glass-encapsulated reed switch (e.g., Gordos, Hamlin). A pull-up resistor is used in the circuit as before to limit the switch current. Provision must be made to ignore the switch bounces.

Other wind-velocity sensors are available that use a *dc tachometer* (essentially a dc generator) to generate an analog voltage as a function of rotor speed.

Flow meter   Vehicles can be equipped with a fuel-flow meter that measures the instantaneous fuel consumption. One of these devices is illustrated in Fig. 11.6.

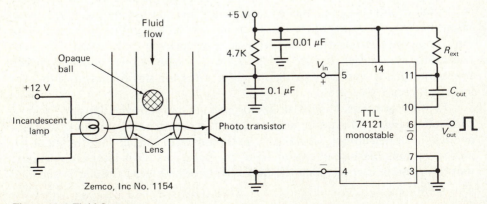

Figure 11.6 Fluid-flow sensor.

In the flow sensor shown (Zemco, Inc. No. 1154; see also FloScan Instrument Co. Series 100 and 200) the fluid passes once around a toroidal channel. A small opaque ball circulates around the channel carried by the fluid. A small incandescent lamp shines a beam of light across the channel (through two small lens-equipped "portholes"). The beam is focused on the active area of a phototransistor. As long as light is focused on its active area, the phototransistor will be on, with a low resistance between its collector and emitter. When the ball passes between the lamp and the phototransistor, the beam of light is interrupted and the transistor goes into its off state. The output is then pulled high (to 5.0 V) by the 4.7K resistor. The frequency at which the ball passes between the lamp and the phototransistor is proportional to the fluid-flow rate.

Because the pulses out of the phototransistor are rather variable in length, and somewhat irregular in shape, some signal conditioning is required in the form of a *monostable pulse generator*. This device produces the pulse shown in the diagram whenever its input goes high. The length of the pulse can be varied by adjusting R_{ext} and C_{out} (see manufacturer's specifications sheet). The output is TTL-compatible and can be connected to the computer interface directly or via isolation circuitry.

Pressure transducer Another example of a pulse-generating transduction mechanism occurs in some pressure sensors. Pressure sensors can be manufactured using a crystal "tuning fork" that is deformed by the ambient pressure, thereby changing its natural frequency of vibration. With attached electrodes, the crystal is used in a tuned electric circuit and produces a pulse frequency that is proportional to pressure.

This particular type of transducer has found application in measuring ocean pressures at various depths with very high accuracy.

11.3 SENSORS THAT PRODUCE CONTINUOUS ANALOG SIGNALS

Many sensors produce a continuous output signal whose amplitude is a monotonic function of a primary input variable. If the input/output function is linear, changes in the output can be related to changes in the input by a gain factor called the sensor *sensitivity* and a bias called the sensor *offset*. Using these parameters, the data-collection system can then easily calculate the value of the input variable given the sensor output.

Calibration curves When the nonlinearity is significant, a manufacturer will normally provide a calibration curve showing the input-output function for the sensor. The microprocessor system can then be used to "look up" in the calibration data the input variable corresponding to the present sensor output value —possibly interpolating between data points.

Figure 11.7 shows the input/output relationship for a typical "linear" sensor.

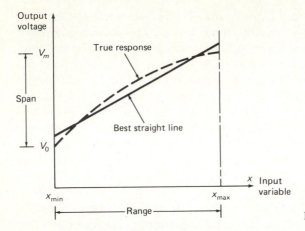

Figure 11.7 Sensor transfer function.

The allowed input variable x has a minimum value x_{min} and a maximum value x_{max}. The total input difference is called the input *range*. These two inputs produce output voltages V_o and V_m. The total output difference is called the output *span*. V_o is called the *offset voltage*.

Thus the sensitivity (S) can be defined as

$$S = \frac{\text{span}}{\text{range}}$$

and the output voltage V for a given input x becomes

$$V = V_0 + S(x - x_{min}) \tag{11.1}$$

In analog sensors, one of the major problems is that the sensors can exhibit many different kinds of errors due to nonlinearity, sensitivity to temperature, and other effects. We will discuss some of these errors in the next paragraphs to enable us to use the terminology while discussing the various analog sensors.

11.3.1 Error Sources

1. Linearity error   Most transducers do not have exactly a straight-line transfer function. The best straight line through a measured set of data is usually taken. That is, the values for offset and sensitivity are determined using a least-squares fitting procedure. The *linearity error* can be defined as either the maximum deviation from the best straight line or the deviation at some point such as the midrange point.

2. Response to minor variables   In most transduction systems, the measured output is often a function of both the desired primary input variable and other undesired, or minor, secondary variables. As an example, a pressure transducer will usually be sensitive to *temperature* and sensor power-supply voltage, which is called *excitation voltage*. One would expect that if the sensitivities to these variables were known and were repeatable, they could, in principle, be measured and compensated. In general, however, this is not done, and these variables become error sources.

3. Repeatability errors   Repeatability errors depend on the input history of the sensor. One form of repeatability error is called *hysteresis* error. Hysteresis error refers to the maximum measurement difference ΔV in sensor output achieved when the input variable is increased to a final value $x = a$ compared with the final value achieved when the input variable is decreased to $x = a$. The term *repeatability* incorporates not only hysteresis but other less-predictable forms of input-history-dependent output errors.

4. Stability   Stability refers to the variation in the output with time that could be expected while both primary and secondary inputs are held constant.

5. Calibration error   When a manufacturer provides calibration information for a sensor series, the individual sensors in the series will differ from the calibration data by the amount of the calibration error. This error can be reduced by carrying out an individual calibration on each device.

11.3.2 Error Analysis

In a measurement system it is necessary to estimate the worst-case error that could be encountered. The contribution of the sensor itself to the calculation of overall error must first be determined.

  If we assume that the transducer will take on two different transfer functions (see Fig. 11.8) corresponding to the two extreme values of one error source, then the transducer function can be written as

$$V_1 = V_{01} + S_1(x - x_{min})$$

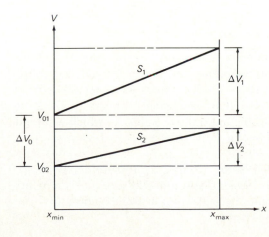

Figure 11.8 Offset and span errors.

for one extreme of the error source (e.g., low temperature) and

$$V_2 = V_{02} + S_2(x - x_{min})$$

Taking the difference between these equations, we have

$$\Delta V = \Delta V_0 + (\Delta V_1 - \Delta V_2) \frac{(x - x_{min})}{(x_{max} - x_{min})} \tag{11.2}$$

where
$$\Delta V = V_1 - V_2$$
$$\Delta V_0 = V_{01} - V_{02} = \text{offset error}$$
$$\Delta V_1 - \Delta V_2 = \text{span error}$$

$$\frac{(\Delta V_1 - \Delta V_2)}{(x_{max} - x_{min})} = S_1 - S_2 \tag{11.3}$$

The total error at any value of x is thus a function of the offset error (ΔV_0) and the span error $(\Delta V_1 - \Delta V_2)$. Offset and span errors are shown in Fig. 11.8.

To calculate the error over the full range, set $x = x_{max}$ in the error equation [Eq. (11.2)] and sum the offset and span errors to get the total voltage error. Repeat for each of the error sources and sum the totals to get the worst-case error ΔV.

To calculate the error over only a portion of the range, set x equal to the upper limit of the range expected and use Eq. (11.2).

The calculated worst-case error can be referred to the output (in volts) as it is above or the input (in input units) or it can be calculated as a percentage of full scale, in which case it can apply to either input or output.

Since the same type of error analysis should be carried out for the A/D converter in the interface, it is convenient either to use percent full scale or to refer all the sensor errors to the sensor output and the A/D errors to the same point in order to sum them and achieve a total system measurement error (assuming no intervening signal conditioning and a high-resolution digital arithmetic representation in the CPU).

11.3.3 Integrated Semiconductor Pressure Sensors

All the types of error described above can be seen in integrated pressure sensors in which the pressure sensing elements and the electronics are fabricated using integrated circuit technology. These sensors have been relatively expensive until recently. It is expected that increasing numbers of these or similar devices will find their way into the automotive industry, which will tend to further decrease costs through economies of scale.

Example: Sensym model LX1440A Let us consider a pressure sensor (e.g., Sensym model LX1440A) suitable for making absolute pressure measurements over the range 0 to 1000 psia. Estimate its maximum error, assuming that the input pressure swings over a full 1000 psi.

We will calculate errors in output volts. The specification sheets show a sensitivity of 10 mV/psi. Thus over a 1000 psi the output will span 10 V.

The manufacturer considers the following types of errors:

1. Offset calibration (OC) error. These errors include the unit-to-unit variations and also any stability variations that could occur over 1 year.

$$\text{OC error} = 0.25 \text{ V (2.5 percent)}$$

2. Offset temperature coefficient (OTC) error. The change in offset voltage as a function of temperature is given as ± 1.65 percent full scale over a temperature range of 0 to 80°C. For this temperature range,

$$\text{OTC error} = \frac{1.65 \text{ percent full scale} \times 10 \text{ V}}{100 \text{ percent}} = 0.165 \text{ V (1.65 percent)}$$

3. Offset repeatability (OR) error. The maximum deviation (at 25°C) in offset voltage as the unit is cycled is

$$\text{OR error} = \frac{0.4 \text{ percent full scale} \times 10 \text{ V}}{100 \text{ percent}} = 0.04 \text{ V (0.4 percent)}$$

4. Span sensitivity calibration (SSC) error. The span error resulting from errors in span sensitivity [Eq (11.3)] include both unit-to unit variation and long-term stability.

$$\text{SSC error} = 0.2 \text{ mV/psi} \times 1000 \text{ psi} = 0.2 \text{ V (2.0 percent)}$$

5. Span temperature coefficient (STC) error. This error in span voltage due to changes in temperature coefficient is given as ± 2.75 percent full scale over the 80°C temperature range.

$$\text{STC error} = \frac{2.75 \text{ percent full scale} \times 10 \text{ V}}{100 \text{ percent}} = 0.275 \text{ V (2.75 percent)}$$

6. Span repeatability (SR) error. This error includes all trial-to-trial differences resulting from hysteresis and any other repeatability error sources. It also includes linearity errors. It is given as ± 1.0 percent full scale

$$\text{SR error} = \frac{1.0 \text{ percent full scale} \times 10 \text{ V}}{100 \text{ percent}} = 0.1 \text{ V (1.0 percent)}$$

Total error

$$\text{Total error} = (0.25 + 0.165 + 0.04 + 0.2 + 0.275 + 0.1) = 1.03 \text{ V}$$

Expressed as a percentage of full-scale voltage:

$$\text{Total error} = \frac{1.03 \text{ V}}{10.0 \text{ V} \times 100 \text{ percent}} = 10.3 \text{ percent}$$

If sensor excitation voltage is expected to vary, the amount of additional associated error is listed as 0.5 percent full scale, which could increase the total error to 10.8 percent.

Temperature compensation   The largest single contributor to the total error is temperature. If either the temperature range (through the control of environmental temperature where possible) is reduced or the effects of temperature are compensated by measuring the exact temperature at the sensor, and subtracting its effect, then the total eror could be reduced to less than 7 percent.

Individual calibration   The calibration error can also be reduced by individually calibrating each transducer—thus eliminating the device-to-device variation in the calibration error (2.0 percent in span and 2.5 percent in offset error for a total of 4.5 percent).

However, the *long-term stability error* (defined for 1 year) must then be taken into account. The specification sheets show a stability offset error of 1 percent and stability span error of 0.4 percent, adding back 1.4 percent to the full-scale error for a net error reduction of 4.5 percent − 1.4 percent = 3.1 percent in total full-scale error when individual calibration is performed.

Offset compensation   In principle, offset errors can be neutralized on-line by periodically applying a reference pressure (x_{min}) to the sensor which will produce the present offset voltage. A similar technique is used in Chap. 14 to remove a torque sensor offset. Any differences from the reference offset voltage are then considered to be errors and are subtracted from all measurements until the next reference pressure check. This does require some additional valving and a means to generate a reference pressure.

Errors in other types of pressure sensors   Each sensor technology will have its characteristic error problems. For example, in mechanical bellows-type pressure sensors there is a relatively large amount of diaphragm movement required to produce an output signal. When there has been no pressure variation, the diaphragm is stationary. A change in pressure at the input must be sufficiently large to overcome the static friction of the system before the sensor output will change. As with the familiar weather barometer, a gentle tap will overcome the friction and produce a closer reading. This has led to the definition of "static" and "dynamic" error. The static error (typically 0.6 percent) exceeds the dynamic error (typically 0.4 percent).

11.3.4 Potentiometers

One of the most common transducers is the potentiometer (Fig. 11.9) consisting of a wiper, moved by the input variable (a rotary or linear motion), that taps off a voltage distributed monotonically on a resistive material along the

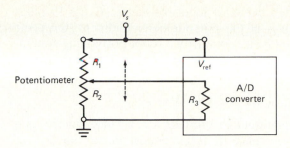

Figure 11.9 Potentiometer transducer with ratiometric sampling.

path of the wiper. The voltage variation is usually a logarithmic, linear, or sinusoidal function of the wiper displacement. Potentiometers are used in several types of industrial sensors. For example, the previously mentioned bellows-type pressure sensor uses a potentiometer to monitor the metallic "bellows" movement. Potentiometers are also used directly to measure shaft position in some industrial robots.

Although the principle is very simple, a potentiometer with very high linearity (<0.1 percent), low temperature error (<0.01 percent/°C), and long mechanical life (50 million turns) can be relatively expensive.

The normal resistive materials are wire, carbon, conductive plastic, and cermet. Each material has its application niche.

A major advantage of potentiometers is that there is no amplification or other signal treatment required to produce a computer-compatible signal.

Ratiometric connection   Although signal conditioning will be discussed in the next section, one important error-reducing technique worth mentioning here is to excite the potentiometer with the A/D reference voltage (see Fig. 11.9). This connection is called *ratiometric*. If the excitation voltage and reference voltage are separate, then a drift of one with respect to the other will cause span errors in the total measurement. With a ratiometric connection, no specific voltage regulation is required for the common reference and excitation voltage.

Linearity errors   When the potentiometer is connected to the subsequent circuitry, a linearity error is introduced by the load resistance (e.g., the A/D input resistance). The transducer function can be shown to be

$$\frac{V}{V_s} \approx x \left[1 - \frac{x(1-x)}{r} \right] \tag{11.4}$$

where $V =$ output voltage (across R_2)
   $V_s =$ supply voltage
   $x = R_2/(R_1 + R_2) =$ (percent full scale input)/100
   $r = R_3/(R_1 + R_2)$ assumed $\gg 1$

Thus it can be seen that when the A/D input resistance (R_3) is 10 times ($R_2 + R_1$) and the input position is at the potentiometer midpoint, the error will be 2.5 percent. This would not be acceptable in a robot-positioning system, for example, unless the processing system subsequently corrected for the error.

11.3.5 Resistance Temperature Detector (RTD)

An RTD uses a metal such as platinum to produce a resistance change as a function of temperature. The resistance R is an almost linear function of temperature T (°C) from a reference temperature $T_0 = 0°C$ (see Fig. 11.10). Expressing the ratio of the resistance at temperature T to the resistance at the reference temperature, we have

$$R/R_0 = 1 + aT + bT^2 + \cdots$$

where a = *resistance temperature coefficient* ($0.004°C^{-1}$ approximately for platinum)
b = a positive or negative constant ($-0.59 \times 10^{-6} °C^{-2}$ approximately for platinum)

The sensitivity is very small for this sensor, and any current used to sense the resistance change will heat the sensor and thus change the measurement by an amount proportional to the square of the current.

The resistance output can be measured with a differential amplifier and bridge configuration. These will be considered in Sec. 11.4.

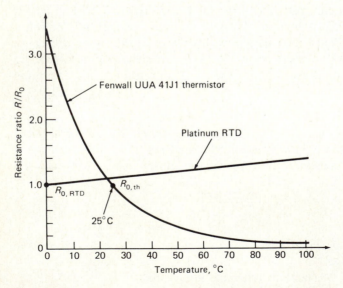

Figure 11.10 Temperature characteristics for thermistor and RTD.

11.3.6 Thermistor

A thermistor ("thermally sensitive resistor") uses a semiconducting material as the basic transduction process. Although more nonlinear than the RTD, and producing a characteristic with a negative slope (see Fig. 11.10), the thermistor has a temperature coefficient that is typically 10 times the temperature coefficient of the RTD. These devices are also available in very small beam-shaped packages (e.g., 0.35 mm diameter).

The ratio of the resistance R for a thermistor at temperature T (in °K) to its resistance R_0 at a reference temperature T_0 (usually $T_0 = 298$°K) is approximately

$$\frac{R}{R_0} = \exp\left[b\left(\frac{1}{T} - \frac{1}{T_0}\right)\right] \tag{11.5}$$

where b is a constant (typically 3000 to 5000°K).

The local slope of this curve, normalized by R/R_0, is called the *temperature coefficient* and is a function of temperature:

$$a = \frac{1}{(R/R_0)} \frac{d(R/R_0)}{dT} = \frac{-b}{T^2}$$

The value of the temperature coefficient a is typically between -0.03 and -0.06°C^{-1} or -3.0 to -6.0 percent/°C at 25°C.

Example: Fenwal UUA41J1 thermistor   For example, the Fenwal UUA41J1 thermistor has a resistance of 32.650 K at 0°C, 10.000 K at 25°C, and 1.752 K at 70°C, which is a nonlinear function. A simple circuit for reading out the temperature over 0 to 70°C is shown in Fig. 11.11.

The Zener diode maintains a constant voltage of 9.1 V across the thermistor and output resistor. It is assumed that the Zener diode and resistor are in an approximately constant temperature environment to prevent temperature drift. When the resistance of the thermistor changes as a function of temperature, the voltage drop across the 10-K resistor will change from a maximum

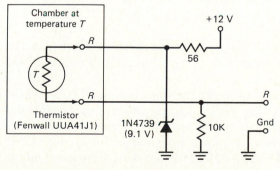

Figure 11.11 Thermistor sensing circuit.

value at 70°C of 9.1 $[10000/(10000 + 1752)] = 7.74$ V to a minimum value at 0°C of 9.1 $[10000/(10000 + 32650)] = 2.13$ V.

Self heating Because a thermistor has resistance, current passing through it will generate heat and raise its temperature above its surroundings. At 25°C the power dissipated in the thermistor is $RI^2 = 10000 (9.1/20000)^2 = 0.002$ mW. For this thermistor with a *dissipation constant* = 1 mW/°C, the device temperature (when the thermistor is in air) will rise by 1°C for each milliwatt of power dissipated.

Thus the amount of temperature rise in air is 0.002°C. In water, the dissipation constant is about five times that in air.

Linearizing the sensor With the circuit shown, the voltage out must be related to a temperature. Since the output voltage is not a linear function of temperature, either the nonlinearities must be compensated in analog electronics (called "linearizing" the sensor) or a software alternative must be taken.

One way to do this in software is to use a calibration table in the computer memory. The table can be either one supplied by the manufacturer of the thermistor or one experimentally determined by comparing thermistor readings against a known temperature standard. The table can also be approximated by a function such as the inverse of Eq. (11.5). Other approximations can also be devised.

11.3.7 Integrated Semiconductor Temperature Sensor

Integrated temperature sensors are available with close to linear responses to temperature (e.g., National Semiconductor LM335A, Analog Devices AD 590). These devices exhibit the various types of errors described earlier, including calibration error, nonlinearity error, and stability error. Individual calibration eliminates the calibration error component.

Example: National LM335A This device produces an output voltage that is proportional to absolute temperature with a slope of 10 mV/°K. Thus at 0°C the output should be 2.73 V. The device is shown in a circuit (Fig. 11.12) that has provision for calibration of the sensor, and offset and gain adjustment so that the overall circuit provides an output voltage $V_0 = T$(in °C)/10. Thus 0 V is produced at 0°C and 10 V at 100°C.

In the circuit, the differential amplifier takes the difference between the sensor voltage V_B and a constant offset voltage V_A. This amplifier is described in the next section.

VR_1 (a potentiometer) is used to calibrate the transducer gain as described in the manufacturer's data sheets. The Zener ZD_1 (at room temperature), the potentiometer VR_2, and the unity-gain buffer are used to produce a constant offset voltage of 2.73 V which when subtracted from the sensor output at 0°C produces an output voltage of 0 V. The difference between the sensor output

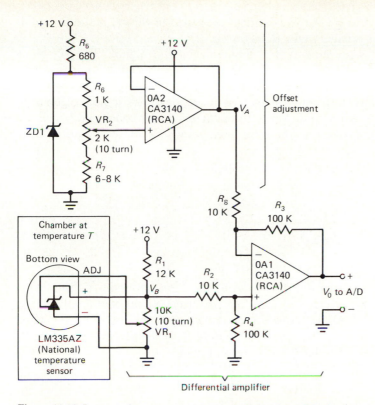

Figure 11.12 Integrated temperature sensor and signal conditioning.

and the offset voltage is then amplified by the differential amplifier (OA_1, R_2, R_3, R_4, and R_8) by a factor of 10 to give a 10-V output for 100°C. This output can be sampled by the A/D conversion interface.

11.4 SIGNAL CONDITIONING CIRCUITS

The remaining elements of the input system, as shown in Fig. 1.6, constitute the *transmission circuits* which are used when the sensors are physically remote from the computer, and the *signal conditioning circuits*, which are used for signal filtering, level and gain conversions, isolation, noise rejection, and protection. We will first look at the signal conditioning circuitry which is usually present in all systems where analog data is acquired.

11.4.1 Differential Amplifier

One of the most common problems with measuring voltages between two points, *A* and *B*, in the real world is that there is often a *common mode* voltage between analog ground in the A/D converter and each of the two points in the environment. The voltage that we want to measure, however, is the potential difference between points A and B.

Consider the measurement of a biomedical EEG signal. The signal of interest is the very small potential difference (generated by cortical and subcortical neurons) between each electrode pair on the scalp. Other voltages generated by more remote sources, such as room electrostatic noise, generate approximately the same potential at both of the scalp electrodes and hence the noise voltages do not appear in the difference voltage taken by the differential amplifier.

We can define the common mode voltage as

$$V_C = \frac{V_A + V_B}{2}$$

and the *differential voltage* as

$$V_D = V_B - V_A$$

The amplifier shown in Fig. 11.13 can be used for amplifying a differential signal. It uses an *operational amplifier*. An idealized operational amplifier has infinite gain and when in the configuration shown (with voltage sources having zero resistance, that is, $R_A = R_B = 0$, two equal R_2 resistors and two equal R_1 resistors) can be shown to have a differential gain:

$$\frac{V_{\text{OD}}}{V_D} = \frac{R_2}{R_1}$$

The common mode gain V_{OC}/V_C for the differential amplifier connection depends on the rated intrinsic *common mode rejection ratio* (*CMRR*) of the operational amplifier, where

$$\text{CMRR} = \frac{\text{differential gain}}{\text{common mode gain}}$$

Thus the common mode output voltage is

$$V_{\text{OC}} = V_C \frac{\text{differential gain}}{\text{CMRR}}$$

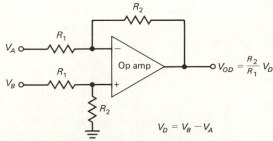

$$V_D = V_B - V_A$$

Figure 11.13 Differential amplifier.

The expressions above apply only when there is a perfect match within each of the R_1 and R_2 resistor pairs. Any imbalance leads to a reduction of the achievable CMRR for the differential amplifier to a value below that of the operational amplifier itself. It is difficult to match the resistors perfectly in practice.

Also, when source resistances R_A and R_B are present, any imbalance in the source resistances can lead to the generation of a common mode voltage. This situation can be quite severe in biomedical applications where the sources are biomedical electrodes. The electrode-skin interface resistance can exceed 10K with a 50 percent imbalance between electrodes. The solution is to use an instrumentation amplifier connection.

11.4.2 Instrumentation Amplifier

The *instrumentation amplifier* connection shown in Fig. 11.14 avoids the problems of the differential amplifier connection at the expense of two more operational amplifiers. There is a single resistance (R_3) used in setting the gain. Lack of exact equality in the R_4 pair produces only a differential gain error and does not affect the common mode gain. Also, the operational amplifier itself has a very high input resistance, $(10^{10}\ \Omega$ input impedances are available) and thus source resistance imbalance has only a small effect upon CMRR.

The second differential amplifier stage normally has a low gain, which keeps the common mode output very small.

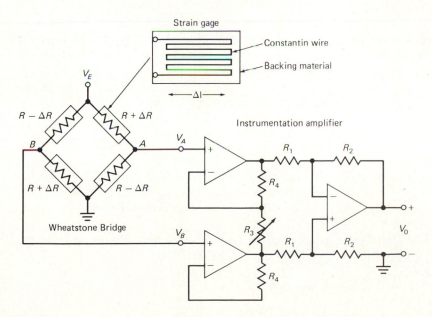

Figure 11.14 Instrumentation amplifier and strain gauge bridge.

The total differential gain for the instrumentation amplifier is

$$\frac{V_0}{V_B - V_A} = \frac{R_2}{R_1}\left(1 + \frac{2R_4}{R_3}\right)$$

Example: Strain gauges (BLH Electronics Model FAE-37-12-S6 gauge) Both common mode and differential voltages are generated when, for example, a *Wheatstone bridge* is used to measure the small resistance changes that are caused by strain in *strain gauge* elements.

A strain gauge bridge is shown in Fig. 11.14 connected to a dc excitation voltage source V_E. When all four arms of the bridge contain the same resistance, the bridge is balanced and the voltage between points A and B is 0, although the voltage to ground from either point is $V_E/2$. Thus

Common mode voltage $V_C = \dfrac{V_E}{2}$ and   differential voltage $V_D = 0$

As temperature increases, the bridge is still balanced, provided that the temperature coefficient of all arms is the same.

If strain occurs in all the arms in such a way that the resistances in opposite arms vary in the same direction, the bridge becomes unbalanced—a condition which results in a differential voltage between A and B. The common mode voltage, however, stays the same.

$$V_C = \frac{V_E}{2} \quad \text{and} \quad V_D = \frac{\Delta R}{R}V_E$$

This condition would occur, for example, if two strain gauges on opposite arms were placed on the side in tension of a beam in bending and the other two gauges were placed on the side in compression.

The percent change in resistance as a function of strain is defined as the *gauge factor* (GF) and is given by the gauge manufacturer:

$$\text{GF} = \frac{\Delta R/R}{\Delta l/l} \qquad \frac{\Omega/\Omega}{\text{in}/\text{in}}$$

The gauge used in this example is made of a Constantin (copper-nickel alloy) conductor with a temperature coefficient of expansion that matches stainless steel. The gauge factor $\text{GF} = 2.1$ and the nominal resistance of the gauge is $120\,\Omega$.

For a 0.1 percent maximum strain and an excitation voltage of 10 V, the expected differential voltage from the bridge is

$$V_D = \text{GF}\,\frac{\Delta l}{l}\,V_E$$
$$= 2.1 \times 0.001 \times 10$$
$$= 21.0\,\text{mV full scale}$$

Because this voltage rides on a common mode voltage of 5 V, the use of an instrumentation amplifier would be appropriate.

For the example, if the CMRR is 10,000 (often expressed in decibels as CMRR [dB] = 20 log CMRR − 80 dB), the amplifier differential gain is 100, and the common mode voltage is 5 V, the common mode component of the output voltage will be 50 mV. This voltage represents an effective offset voltage for the sensing system.

11.4.3 Analog Isolation Amplifiers

When the common mode voltage exceeds the power-supply voltages of the instrumentation amplifier, the source ground must be isolated from the amplifier ground. *Isolation amplifiers* are used to achieve this. As can be seen in Fig. 11.15, the *floating input state* has a local ground that is isolated from the system-ground-referenced portion of the device.

Isolated power is supplied to the floating input stage. This is achieved by the transformer coupling of ac power from the nonisolated part to the isolated part, with no conductive connection between the parts. The amplified input signal is transmitted across the isolation barrier by one of two means:

1. Using the amplified signal to modulate an ac waveform, transformer coupling the modulated waveform across the isolation barrier, and then demodulating the waveform to recover the signal
2. Converting the electric signal to an optical signal which is transmitted across the isolation barrier and reconstructed as an electric signal

Transformer-coupled devices are available, for example, from Analog Devices Inc., and both transformer and optically coupled devices are available from Burr-Brown.

Although there is a very high common-mode isolation (>1000 V), the

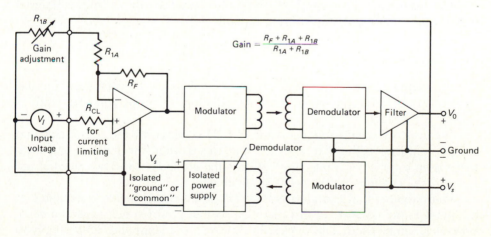

Figure 11.15 Isolation amplifier.

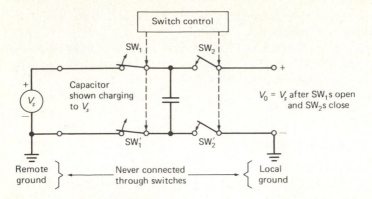

Figure 11.16 Flying capacitor isolation.

specifications such as frequency response, nonlinearity, and noise vary depending on the technology used by the manufacturer for a particular model.

There can be a significant nonlinearity in some of these devices. Because of the presence of the nonlinearity, there will be an effect on the overall system accuracy. Thus it is insufficient to simply calculate the error contributions of the sensor. Error introduced in all the preprocessor stages must be added into a total *error budget* for the input system.

Another technique exists for isolation which is called *flying capacitor* isolation (see Fig. 11.16). It is based on the fact that a capacitor can be momentarily connected through a pair of reed switches across the two lines carrying the signal to be measured. It charges up to the voltage across the lines very quickly. Then the switches are opened and a second pair of switches is closed that connects the capacitor across the A/D input lines, thus transferring the voltage to the A/D converter. Thus the input devices are never electrically connected directly to the computer and only the difference voltage is applied to the A/D converter. The capacitor can be considered to "fly" from the external lines to the A/D input, transferring the voltage.

11.4.4 Digital Isolation

A good method for isolating digital source signals from the computer is to use optical isolators. The optical isolator is usually located at the computer end, as shown in Fig. 11.17.

It consists of a *light-emitting diode* (*LED*) powered by the sensor side and a phototransistor acting like a switch that closes when the LED is on. Since there is no conductive connection between the grounds of the two sides of the isolator, the common mode signal is eliminated.

One problem that can arise is that only a low-power signal is available at the input to the isolator. In this case, an isolated power supply, called a *dc-dc converter*, can be used. This device first converts the computer +5-V supply to ac and is passed through a transformer which produces the isolation, and the

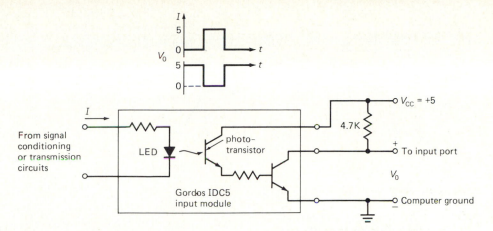

Figure 11.17 Optoisolated dc input.

power is then converted back to +5 V dc. The isolated side of the converter can then be used to power the LED.

Optical isolators are available as modules for both input and output signals. Since there may be a large number of isolators used in a system, bus-compatible driver boards are available that contain memory-mapped I/O registers in which each bit in a data word communicates with an optoisolator. The isolator modules are generally available as plug-in units with screw connections on a rack-mountable panel. Such a system is manufactured by Gordos Corporation, for example (see also Chap. 14).

11.4.5 Low-Pass Filtering

One type of analog prefiltering that is usually required is the low-pass filter—a signal filter that passes signal frequencies that are below a usually defined "cutoff frequency" and rejects frequencies above the cutoff frequency. A low-pass filter is often used to eliminate high-frequency noise.

In order to reconstruct a sampled signal perfectly, it is known that the signal must be sampled at a rate equal to or exceeding the *Nyquist rate*—a frequency that is twice the highest significant frequency component in the incoming analog signal (including signal and noise). If the signal is sampled at a frequency lower than that of the Nyquist frequency, the samples do not contain sufficient information to reconstruct the original signal without a distortion known as *aliasing*.

To ensure that the Nyquist criterion is met, an analog low-pass filter, called an *antialiasing filter*, is often applied to the input just before the sampler. In some cases, the analog signals are naturally low-pass in nature and the filter is not necessary. In other cases, a simple single-pole RC filter can be used (see Fig. 11.18), often in the form of a capacitor (C) across the feedback resistor(R) in an operational amplifier such as a differential amplifier that is

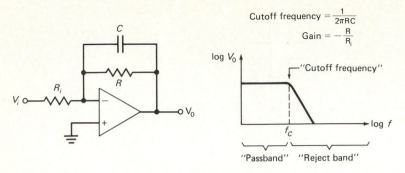

Figure 11.18 Simple low-pass filter with amplification.

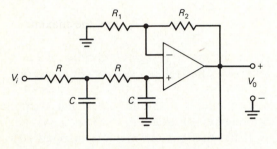

Cutoff frequency $= \dfrac{1}{2\pi RC}$

Gain $= \dfrac{R_1 + R_2}{R_1}$ = 1.586 for Butterworth frequency response

Figure 11.19 Second-order Butterworth low-pass filter.

required to be in the circuit for other purposes. The attenuation of this filter above the *cutoff frequency*, $f_C = 1/(2\pi RC)$ [Hz] is a factor of 10 for every decade increase in frequency above the cutoff frequency.

A family of low-pass filters (called *Butterworth filters*) with maximally flat low-pass characteristics are often used for antialiasing purposes. One circuit for this filter is given in Fig. 11.19. The filter shown has an attenuation factor of 100 for every decade above the cutoff frequency.

In practice, a sampling rate that is 5 to 10 times the Nyquist rate may be chosen to allow for the finite slope of the filter's *transition band* (the portion of the frequency spectrum between the passband and the reject band). Also, the higher sampling rate allows the trend of the digital data (when plotted raw point by point as recorded) to be more easily seen.

11.4.6 Gain Conversions and Level Shifting

The final job that can be carried out by input conditioning circuitry is the alignment of the dynamic amplitude range of the signal with the dynamic range of the A/D conversion interface. This consists of two processes:

- Offset adjustment
- Gain adjustment

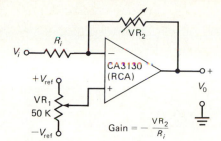

Figure 11.20 Gain and bias adjustment.

Both types of adjustment can be carried out with a circuit such as the one in Fig. 11.20. The offset adjustment is set when the input signal is minimum by adjusting the VR_1 such that the dc level is the same as that required by the A/D converter. The gain adjustment VR_2 is set so as to allow the maximum input signal to drive the A/D input to its maximum converted value.

For example, if a 2- to 7-V sensor output is to be connected to an A/D converter that converts over the range of 0 to 10 V, the 2-V bias must be level-shifted to 0 V, and the 5-V sensor span must be amplified by a factor of 2 to cover the 10-V A/D range.

11.5 TRANSMISSION CIRCUITRY

Very often it is not convenient to have the sensors close to the computer system. As the distance increases, however, the amount of noise on the input lines also increases and the signal levels become attenuated. There may be common mode voltages generated. The common mode signal can be depicted as a voltage source between the ground at the remote sensor and the ground at the computer. This voltage could be generated, for example, by a large 60-Hz ground current caused by some unrelated device that happens to be connected to the same ground circuit. The methods described can be used for either transmitting sensor signals to the input interface or transmitting output signals from the output interface to the actuator systems.

The conductors used to carry the signals are

1. Coaxial cable (one or two conductors surrounded by a shield)
2. Twisted pairs of wires side by side
3. Untwisted pairs side by side in a flat cable

A pair of conductors forming a transmission line has a characteristic impedance, expressed in ohms, when viewed from either end. The propagation velocity down a cable is at the speed of light (approximately 1 ns/ft or 3.3 ns/m). Unless the cable is matched at each end with the characteristic impedance, a signal will reflect back up the cable, bouncing off each end, causing the output voltage to appear to "ring."

In Fig. 11.21, the 50-Ω resistor matches the cable impedance at the output

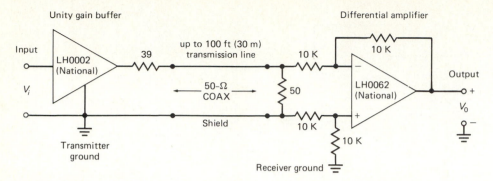

Figure 11.21 Short-range analog transmission.

end. The 39-Ω resistor added to the output impedance of the buffer amplifier also comes close to 50 Ω. For amplifiers that are limited to low frequencies, this matching is not so important as it is in the application shown in Fig. 11.21.

11.5.1 Short-Range Analog Signal Transmission.

A circuit that will pass signals up to about 1 MHz is shown in Fig. 11.21. The National LH0002 is a unity-gain amplifier that has a low output impedance and high speed. The circuit shown can drive about 100 ft (20 m) of coaxial cable. The coaxial cable shield is grounded at the driver end and is connected to one side of the differential amplifier at the receiver end. Thus any common mode signals generated by ground differences are rejected.

11.5.2 Medium-Range Analog Signal Transmission

Longer transmission distances can be achieved at lower frequencies (up to 10 Hz) using *current-loop transmission*. In this technique, the sensor output voltage is converted to a current in the range of 4 to 20 mA. Since a sensor output voltage of zero maps to 4 mA, if the loop becomes broken, the 0 current condition can be detected as an error condition. The current is converted back to a voltage at the receiver by a 250-Ω resistor in the loop (see Fig. 11.22).

The current loop is a standard transmission technique (Instrument Society of America, Standard S50.1).

The advantage of the current loop is that most noise sources are incapable of generating a significant voltage drop across the 250-Ω load, since their source impedances tend to be high.

The input and output of a converter can also be isolated for protection from high common mode voltages. Both isolated (Analog Devices Inc. Model 2B22) and nonisolated versions are available (Analog Devices Inc. Model 2B20, and Burr-Brown XTR-100).

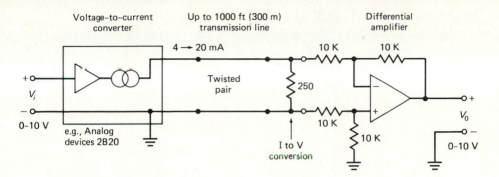

Figure 11.22 Medium-range analog transmission.

11.5.3 Long-Range Analog Signal Transmission

A third technique (see Fig. 11.23) can be used that provides the same benefits as digital transmission—the capability of driving long transmission lines (several thousand feet) with very high noise immunity. However, the analog frequency range is lower than the previous methods (several hundred hertz).

More recently, data transmission over optical fibers has become practical. The digital electrical signals are converted to light pulses which are conducted over thin glass fibers with very high internal reflectance for low light losses. At the receiver, the optical signals are converted back to electrical signals. Aside from the total isolation, the optical signals are immune from electric and magnetic field disturbances, making this form of transmission suitable for distances of up to 1 km or more. In addition, the data rate over the fibers can be higher than any of the previously described techniques (in excess of 10 Mbits, and experimental systems of 1000 Mbits/s).

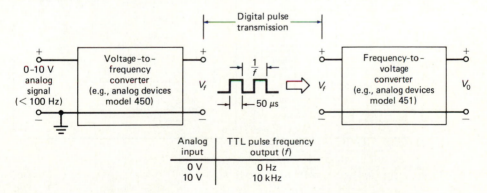

Figure 11.23 Long-range analog signal transmission by conversion to digital pulses.

11.6 BUS-COMPATIBLE INPUT SYSTEMS

Several manufacturers supply bus-compatible boards that can acquire data directly from certain common sensors—4 to 20 mA current loops or voltage sources in standard ranges. One such system from Analog Devices Inc. is shown in Fig. 11.24.

The two-card STD-bus system has the capability of acquiring data from

- Thermocouples
- 100-Ω platinum RTDs
- Strain gauges
- Current loops
- AD590, AD2626 temperature sensors
- Voltage sources in five ranges from ±25 mV to ±25 mV to ±10 V

The input card provides protection, low-pass filtering, 1000-V isolation, and necessary offset and gain adjustments.

A second card is a control card containing its own local microprocessor and 13-bit A/D converter. The processor linearizes the sensor data and stores the result in a 1K buffer for reference by the bus processor. The system has only two sampling rates—11 and 30 Hz.

Other system configurations are also available that, for example, separate

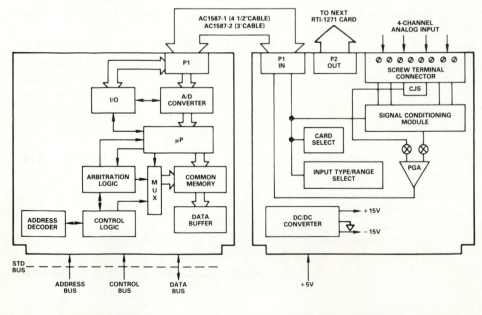

RTI-1270
A/D-CPU BASE CARD

RTI-1271
SIGNAL CONDITIONING CARD

Figure 11.24 STD bus-compatible analog signal acquisition system. (*Courtesy Analog Devices, Inc.*)

the data-acquisition function from the system bus entirely. An example of such a system is the Burr-Brown PCI-3000 and Analog Devices microMAC. Both systems can acquire and linearize data using a local microprocessor. The data can then be accessed by a host processor over a industry standard data link such as RS232, IEEE-488 (see Chap. 10).

SUMMARY

- Prior to analog/digital conversion, a transduced signal may pass through the following stages:

Transduction→ sensor   amplification→ transmission→ isolation→ filtering → gain and offset conversion→ sampling→ analog/digital conversion

- An *error budget* should be computed that includes all sources of error in all the stages shown in the first paragraph, including transduction and A/D conversion. This will allow the designer to detect errors that could drive the design out of specification.
- When the voltage between two terminals on a device in the environment is to be measured, there is often a significant voltage between either terminal and computer ground, called the *common mode* voltage. The common mode voltage can exist even when one of the remote terminals is the remote "ground."
- Common mode voltages may be generated either by the measurement system (as in the strain gauge bridge) or by noise sources. An instrumentation amplifier is used to reject the common mode voltage and amplify the difference voltage.
- When signal transmission is present in a system, noise, signal distortion, and attenuation can occur. Common mode noise can be reduced by the use of a differential line receiver for both analog and digital signals. In digital signal transmission, electrical and magnetically coupled noise can be eliminated by the use of optical fibers.
- For the protection of computer input electronics and the amplification of signals with common mode voltages that exceed the power-supply voltages in the interface, isolation devices can be used. These include transformer-coupling and optical isolation and the use of "flying capacitor" couplers.
- In order for information about an analog input not to be lost in the process of sampling, the analog input must be sampled at a rate that is at least twice that of its highest significant frequency component. For any desired sampling rate, this can be achieved by sharply low-pass filtering the signal plus noise prior to the A/D converter.

EXERCISES

11.1 For the Sensym pressure sensor (model LX1440A) used as an example in this chapter, suppose that you wish to reduce the error by restricting the pressure range to 50 percent of full scale, individually calibrating each device, measuring ambient temperature, compensating for temperature changes, and, finally, applying periodically the minimum pressure to the transducer from some reference pressure source. Find the worst-case pressure error as a percent of full scale.

11.2 Show that the approximate equation [Eq. (11.4)] is valid. Find the point of worst-case nonlinearity error for a 10K potentiometer and a 100K load resistance.

11.3 Given Eq. (11.5) as an approximate expression for the thermistor, find a constant b to fit the following data for the Fenwal UUA41J1thermistor over the temperature range given.

Temperature, °C	Resistance, Ω
25	10000
30	8056.8
35	6531.4
40	5326.4
45	4368.4
50	3602.3
55	2986.1
60	2487.8
65	2082.7
70	1751.6

If you were going to use the expression to find the temperature, given the resistance reading of the thermistor, and you could know the thermistor resistance to within 0.01 percent, what would your worst-case error in temperature be? What suggestions can you make to improve the temperature estimate?

11.4 Write a program function to implement the technique you finally select. The function that you create should take the resistance reading as an argument and deliver the temperature to the calling routine. You should also have an error flag that indicates resistance out of range.

11.5 Write an expression for the total output voltage from a differential amplifier (including common mode and differential voltages) as a function of the input voltages V_A, V_B, and V_C.

11.6 Suppose that a transformer-coupled isolator were used in conjunction with the sensor of Prob. 11.1. The isolator has a gain nonlinearity of 0.25 percent, a gain stability of 0.009 percent/year, a gain temperature coefficient of 0.0075 percent/°C, a gain calibration error of 3 percent, an offset calibration error of $(5 + 50/\text{gain})$ mV, and an offset temperature coefficient of $(0.008 + 0.250/\text{gain})$ mV/°C. Assuming an ambient temperature range for the isolator of 30°C, a gain of 1, and a peak voltage of 10 V in and out, find the contribution to total error due to the isolator.

11.7 If the output of the isolator in Prob. 11.6 is fed to an 8-bit ± 10 V A/D converter, with total errors of 0.3 percent (including linearity error but not including quantization error), a sensor total error of 10 percent, and the errors found in Prob. 11.6, find the total error for the data-acquisition system.

11.8 If a Wheatstone bridge (see Fig. 11.14) contains two strain gauges, in opposite arms of a strain gauge bridge (i.e., the two strain gauges have no common connection to each other), find an expression for the output voltage across the bridge when the excitation voltage is held constant. Assume that both strain gauges respond in the same way (both increasing or both decreasing) and that the remaining two resistances in the bridge are fixed and match the strain gauge resistance prior to straining.

11.9 Repeat Prob. 11.8 for the case of a single strain gauge in the bridge. What are the disadvantages of the two arrangements by comparison with the four-gauge bridge discussed in this chapter?

BIBLIOGRAPHY

Sensors and sensing systems

Cobbold, R. S. C.: *Transducers for Biomedical Measurements*, Wiley, New York, 1974.
Doebelin, E. O.: *Measurement Systems: Application and Design*, McGraw-Hill, New York, 1983.
Garrett, P. H.: *Analog I/O Design, Acquisition: Conversion: Recovery*, Reston, VA, 1981.
Graeme, J. G., G. E. Tobey, and L. P. Huelsman (eds.): *Operational Amplifiers: Design and Application*, McGraw-Hill, New York, 1971.
Jones, B. E.: *Instrumentation, Measurement and Feedback*, McGraw-Hill, New York, 1977.
Wobschall, D.: *Circuit Design for Electronic Instrumentation: Analog and Digital Devices From Sensor to Display*, McGraw-Hill, New York, 1979.

Noise

Morrison, R.: *Grounding and Shielding Techniques in Instrumentation*, 2d ed., Wiley, New York, 1977.
Ott, H. W.: *Noise Reduction Techniques in Electronic Systems*, Wiley, New York, 1976.

Transducer manufacturer application books

Sheingold, D. H. (ed.): *Transducer Interfacing Handbook*, Analog Devices, Norwood, MA, 1980.
Micro Switch (Honeywell): *Hall Effect Transducers*, Freeport, IL, 1982.

Manufacturer data books

Analog Devices Inc.: *Data Acquisition Handbook*, vol. II, Norwood, MA, 1984.
Burr-Brown Corp.: *Product Data Book*, Tucson, AZ, 1984.
SenSym Inc.: *Pressure Transducer Handbook*, Sunnyvale, CA, 1983.

TWELVE

OUTPUT SYSTEMS

Real-time microcomputer systems have two types of output devices. Information display devices, such as printers, plotters, and video displays, are used to present information to human operators. *Actuators*, on the other hand, are output devices used to control a system or process. In this chapter we concentrate on the design of output systems involving actuators, since they distinguish the output systems of real-time computer systems from those of data-processing computer systems.

An actuator system can often be divided into two elements, a *transducer* and an *amplifier*. The transducer in an actuator converts electric energy to the required control output. For instance, an electric motor is a transducer that converts electrical energy into mechanical motion. The amplifier portion of an actuator system amplifies the low-level control signal from the computer-output interface to a level sufficient to drive the transducer. In some cases, both the transducer and the amplifier portion of the actuator system are available in a single unit or at least can be purchased as a complete system from a single vendor. However, there are still many instances where the amplifier and transducer portions must be selected or designed individually.

12.1 OUTPUT SYSTEMS INVOLVING TWO-STATE ACTUATORS

In many situations sufficient control of a system can be achieved if the actuator has only two states: one with electrical energy applied (on) and the other with no energy applied (off). This simplifies the design of the output system

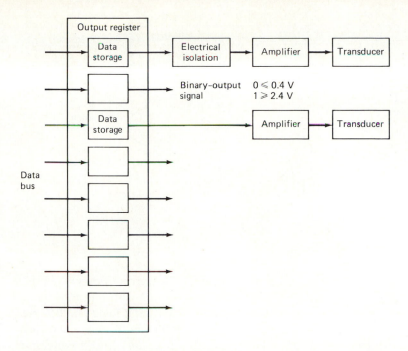

Figure 12.1 Output system with two-state actuators.

considerably, since no digital-to-analog (D/A) converter is required and amplification can be performed by a simple switching device rather than by a linear amplifier.

An output system with several two-state actuators is shown in block diagram form in Fig. 12.1. The output register is a computer output port. The computer can address this port and write data to it. The data is stored in the register and is available as a binary signal on the register output lines. The signal is used to control a switch which applies power to the transducer. In some cases, the signal is first passed through an isolation element which electrically isolates the switch and transducer from the computer system.

12.1.1 Output-Port Current and Voltage Ratings

The output lines from a computer output port can supply only small amounts of power. Typically a high-level output signal has a voltage between 2 and 5 V and a low-level output signal has a voltage less than 1 V. The current capacity of the output lines often depends on how the load is connected. In the configuration shown in Fig. 12.2a, the output port is said to be *sourcing* current. Current flows out of the output port into the load when the output signal is at a high level. No current flows when the output signal is at a low level. In the alternative configuration shown in Fig. 12.2b, the output port is said to be *sinking* current. Current flows through the load and then into the

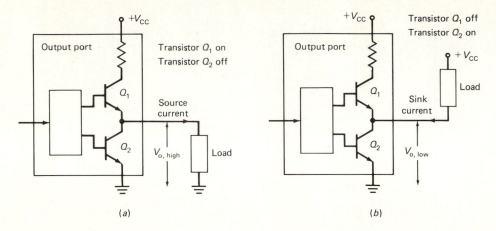

Figure 12.2 Sourcing and sinking current.

output port when the output signal is at a low level. No current flows when the signal is at a high level. Many digital circuits can sink substantially more current than they can source. Ratings for some typical digital logic families are given in Table 12.1. For most types of digital logic circuits both source and sink current ratings are less than 20 mA. Thus an output line can directly switch no more than 100 mW of power.

12.1.2 Low- and Medium-Power Switches

Most actuators require considerably more than 100 mW of power. Therefore, the signal from the output port of the computer must be used to control a

Table 12.1 Output-current ratings for digital logic circuits
(Supply voltage (V_{cc}) is 5 V in all cases.)

a. Standard TTL logic
   Guaranteed source current with V_{out} at 2.4 V: 400 μA
   Guaranteed sink current with V_{out} at 0.4 V: 16 mA
b. Low-power Schottky (LS) TTL logic
   Guaranteed source current with V_{out} at 2.4 V: 400 μA
   Guaranteed sink current with V_{out} at 0.4 V: 8 mA
c. Low-power Schottky (LS) TTL high-current buffers (e.g., 74LS241)
   Guaranteed source current with V_{out} at 2.0 V: 12 mA
   Guaranteed sink current with V_{out} at 0.5 V: 24 mA
d. Standard CMOS logic
   Guaranteed source current with V_{out} at 2.5 V: 1.6 mA
   Guaranteed sink current with V_{out} at 0.4 V: 500 μA
e. High-speed (HC) CMOS logic
   Guaranteed source current with V_{out} at 4.2 V: 5 mA
   Guaranteed sink current with V_{out} at 0.4 V: 5 mA

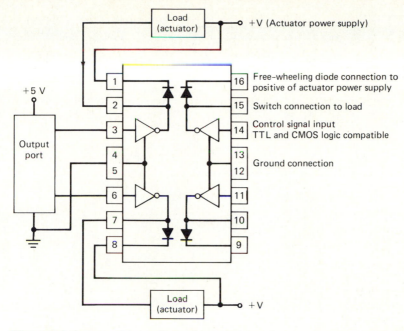

Figure 12.3 Actuator driver using an integrated circuit.

switch that can handle the power levels required by the actuator. A number of switching devices are available to perform this function. We have space to discuss only a few representative samples of these devices.

Integrated circuits Integrated circuits containing transistor switches can be used when the actuator voltage is under 80 V and the actuator current is under 1.5 A. The quad driver circuit shown in Fig. 12.3, a Sprague ULN2069B, contains four transistor switches that can be driven directly by TTL or CMOS logic circuits. When the output line is high (i.e., $V_{out} \geq 2.4$ V), the transistor it controls is on and current flows through the actuator. When the output line is low (i.e., $V_{out} \leq 0.4$ V), the transistor it controls is off and no current flows through the actuator.

Discrete solid-state switches For actuators requiring higher currents, switches based on discrete power transistors or power MOSFETs (metal oxide field effect transistors) can be designed. Figure 12.4 shows two such switches. The circuit in part *a* can carry about 5 A of current and block up to 100 V. The switch in part *b* can carry about 10 A of current and block up to 120 V.

The circuit in part *a* of the figure uses a power *Darlington* transistor (Q_2) as the switch. The transistor requires approximately 20 mA of base current to turn on completely when it conducts its maximum rated current of 5 A. Most digital logic cannot source this much current. Therefore, an additional stage of

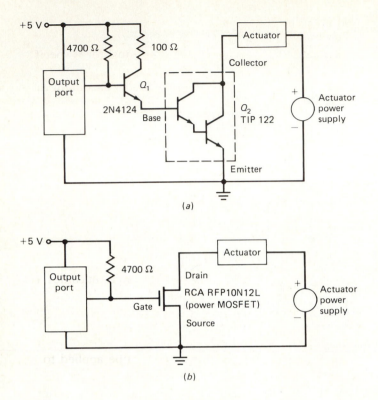

Figure 12.4 Medium-power switches. (*a*) Transistor switch. (*b*) Power MOSFET switch.

amplification is required in order to allow the output line to control the power switch. When the output line is high, a total of about 0.5 mA of current flows into the base of transistor Q_1, turning the transistor on. When Q_1 is on, approximately 25 mA of current flows through the 100-Ω resistor and through Q_1 into the base terminal of the power transistor, Q_2, causing it to switch on as well. With Q_2 on, current flows through the actuator. When the output line is low, it sinks the current flowing through the 4700-Ω resistor (now about 1 mA), diverting it away from the base terminal of transistor Q_1. As a result, both transistors stay off and no current flows through the actuator.

The circuit in part *b* of Fig. 12.4 uses a power MOSFET. This is a voltage-activated device; its control terminal draws no current in the steady state. When the output line is high, the control terminal on the MOSFET (the *gate* terminal) is pulled up to 5 V by the 4700-Ω resistor and the MOSFET is switched on. Current then flows through the actuator. When the output line is low, the voltage at the MOSFET's gate terminal is less than 0.4 V and the MOSFET is in the nonconducting state. As a result, no current flows through the actuator.

Both the power transistor and the power MOSFET still have some internal resistance when they are switched on. They therefore dissipate some power when carrying current and will overheat if they are not mounted to an adequate *heat sink*. For example, the TIP 122 Darlington transistor is specified to have a maximum voltage drop of 4 V across its collector-emitter terminals when it is fully turned on and is conducting 5 A of current (the maximum rated current). Therefore, the TIP 122 can be expected to dissipate up to

$$4 \text{ V} \times 5 \text{ A} = 20 \text{ W}$$

The circuits described above are illustrative of how low- and medium-power switches are designed. However, in many applications it is possible to avoid the time and expense of designing these circuits, since output boards containing these types of switches are commercially available for most microcomputer bus systems.

12.1.3 Electrically Isolated Switches

As the power-handling capability of the switch increases, it becomes less desirable to have a direct electrical connection between the control signal from the computer output port and the switch. When the switch and the computer share a common electric circuit, the high curents flowing through the switch may generate electrical noise which affects the operation of the computer. Also, if the switch fails, it is possible that high voltages will be applied to the computer output port, causing damage to circuits within the computer.

Electromechanical relays The most common electrically isolated switch used in control applications has been the electromechanical relay (see Fig. 12.5). When a current flows through the relay coil, the armature is pulled away from the normally closed contact and pulled on to the normally open contact. A typical relay coil may be rated at 12 V and 0.5 A. It cannot therefore be driven directly by the computer output port. The current through the coil must instead be switched by one of the devices discussed above. Figure 12.6 shows a typical relay driver circuit.

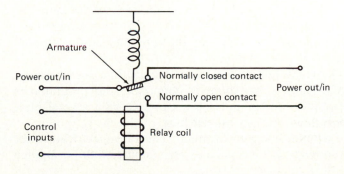

Figure 12.5 Electromechanical relay.

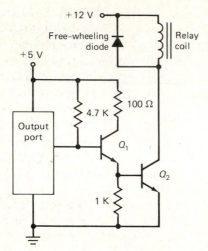

Figure 12.6 Relay driver circuit.

Electromechanical relays that handle any substantial amount of power are too big for boards that must be plugged into a microcomputer motherboard. They are therefore normally installed outside the microcomputer. However, boards containing low power *reed relays* are available for several microcomputer bus systems. These boards can be used for the isolated switching of signals and small amounts of power.

Electromechanical relays have several advantages:

- A variety of contact configurations are available.
- They can switch direct or alternating current.
- They are available in a wide range from small reed relays used to switch millivolt level signals to large contactors used to switch hundreds of kilowatts of power.
- They can handle momentary overloads without damage.
- They are well understood by maintenance personnel.

On the other hand, they have some disadvantages when used with computers:

- They usually cannot be driven directly by the computer output lines so they require relay driver circuits.
- They switch in milliseconds rather than microseconds. In cases where the timing or sequencing of switching is important, the programmer must remember that the computer can easily execute several hundred instructions while a relay is switching.
- They suffer from contact bounce which can generate electrical noise. This noise, if inductively or capacitatively coupled to the computer circuits, can affect computer operation.

Solid-state relays   *Solid-state relays* have been introduced in an attempt to rectify some of the deficiencies of electromagnetic relays. These modules have a control input which is coupled optically or inductively to a solid-state power switching device. If the relay is to switch direct current, the switching device is a power transistor or a power MOSFET. If the relay is to switch alternating current, the switching device is a *triac* or possibly two *SCRs* (silicon-controlled rectifiers) connected in an antiparallel configuration.

Triacs and SCRs are members of a class of semiconductor power switches called *thyristors*. Thyristors exhibit a latching action when they are switched on. Once a thyristor is *fired* (i.e., switched on), it will remain on as long as current flows through the switch. Removing the firing signal will not cause the thyristor to switch off as in the case of power transistor or power MOSFET switches. The thyristor switches off only if current ceases flowing through it. As a result, thyristor switches are most commonly used to switch alternating current, since the current passes through zero at regular intervals, allowing the thyristor to shut off if no firing signal is present. Thyristors have the advantage that for the same cost, they can handle considerably more power than power transistors or power MOSFETs. Readers interested in learning more about these devices are referred to the books listed at the end of this chapter.

Most solid-state relays used to switch alternating current are designed to switch on at a zero crossing of the alternating voltage as well as switch off at zero crossings. This *zero-voltage switching* eliminates the electrical noise caused by the rapid rise of current if the switch is closed while the voltage is nonzero (see Fig. 12.7).

The control inputs on solid-state relays are commonly designed to be driven directly from digital logic circuits. As a result, they are often easier to use with computers than are their electromechanical counterparts. Figure 12.8 shows the schematic for an output interface using a high-current ac solid-state

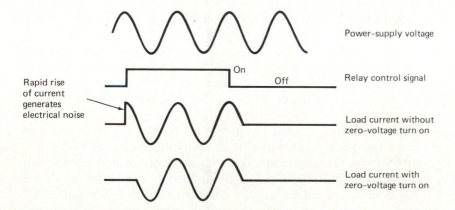

Figure 12.7 Effect of zero-voltage switching in solid-state relays.

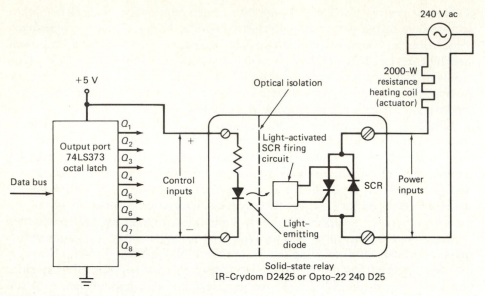

Figure 12.8 Output interface using a solid-state relay.

relay to switch power to a heating coil. The control input to the solid-state relay has the following specifications:

Control signal range	3 to 32 V dc
Must operate voltage	3 V dc maximum
Must release voltage	1 V dc minimum
Input current	4 mA maximum at 5 V dc

When the output port line is low, current can flow through the control inputs of the solid-state relay, turning the relay on. Since the output port consists of low-power Schottky (LS) TTL components, it has no trouble sinking the 3 or 4 mA flowing through the relay's control inputs. Also, the output low voltage of the LS TTL components is guaranteed to be less than 0.4 V, so the voltage across the relay control inputs will be greater than the 3 V required by the specifications. When the output port line is high, no current flows through the control inputs of the relay and the relay is off.

Computer-output modules Among the easiest solid-state relays to use with computers are those packaged specifically as computer-output modules. The modules are packaged in a standard plug-in enclosure which allows easy interchanging without disturbing wiring. The dc module can switch up to 3 A and block up to 60 V. Ac modules are available for both 120-V and 240-V circuits and can carry up to 3 A RMS. The control inputs are optically isolated from the switching device and are designed to be driven by digital logic circuits.

These modules are available from a number of companies, including

Gordos, IR-Crydom, Opto-22, and Potter & Brumfield. The modules plug into standard racks that contain screw terminals for connecting the loads and a connector for a ribbon cable that connects the rack to an interface card plugged into the microcomputer backplane. Interface cards are available for most common microcomputer bus systems. It is possible to put together an isolated output system capable of switching both ac and dc current simply by purchasing the appropriate interface board and I/O module rack and then plugging in the output modules required for the application. We will examine a system in which these modules are used in Chap. 14.

Solid-state relays are replacing electromechanical relays in an increasing number of applications. There are still some applications where solid-state relays are not suitable. For instance, solid-state relays are not suited to switching analog signals, since their transmission characteristics are nonlinear at low voltages. Solid-state switches with linear transmission characteristics are available. Called *analog switches*, they are used when low-level analog signals must be switched.

Solid-state relays also have limitations at higher power levels, since solid-state relays are not presently available in current ratings above about 20 A dc and 45 A ac. Electromechanical contactors or specially designed solid-state switches must be used for higher-current applications.

12.1.4 Influence of the Actuator on Switch Ratings

The characteristics of the actuator must be considered when determining the ratings of the switch in an output system. If the actuator can be modeled as a constant resistive load, then the switch rating can be determined quite easily. The switch must be rated to carry the current that will pass through the actuator when it is switched on and must be rated to block the actuator supply voltage when the actuator is switched off. In practice, it is wise to choose a switch that is conservatively rated, since the reliability of the switch increases if it is not operated at its maximum rating.

Surge currents Unfortunately many actuators do not act as constant resistive loads. Some actuators draw a *surge current* when they are switched on which is several times greater than the current they draw during normal operation (see Fig. 12.9). Incandescent lamps have this characteristic, since a cold filament has a resistance that is much lower than a hot filament. Some electric motors also draw high surge currents until they reach their rated speeds. Motors can also draw higher-than-rated currents if they are stalled due to excessive loading. When these types of actuators are to be controlled, the switch must be able to handle the expected surge current.

Electromechanical relays can normally handle short-term overloads well in excess of their steady-state current rating. Solid-state relays using thyristors have a somewhat more limited overload capability. Typically they can handle surges up to 10 times the rated current for 10 to 20 ms. For a 1-s surge, the

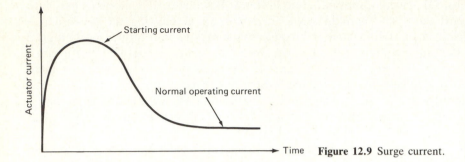

Time   **Figure 12.9** Surge current.

capability drops down to two or three times the rated current. Switches using power transistors or power MOSFETs have almost no overcurrent capability. Even very short pulses of current above the current rating can damage the switch. Therefore, if these switches are to be used, they must have a steady-state current rating sufficient to handle the surge current.

Inductive loads   Inductive loads, such as relay and solenoid coils and electric motors, present another problem. When the switch is opened to turn the actuator off, the voltage across the actuator is

$$V = L \frac{di}{dt}$$

where L = load inductance
di/dt = rate of change of load current

Since the current through the actuator is switched off rapidly, the voltage across the actuator can become very high while the current is being switched off (Fig. 12.10). This transient voltage spike can permanently damage solid-

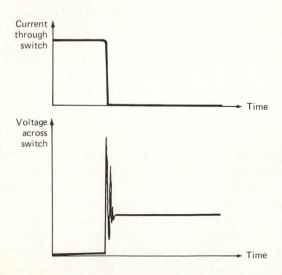

Figure 12.10 Example of a voltage spike that occurs in an inductive circuit when a switch is opened.

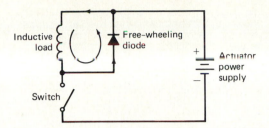

Figure 12.11 Use of a free-wheeling diode to suppress switching transient voltage.

state switches. In electromechanical switches it can cause arcing which reduces the life of the switch. When inductive loads are switched, it is normally necessary to use some means to eliminate or at least partially suppress the voltage spike.

A common technique is to place a *free-wheeling diode* across the load (Fig. 12.11). When the switch is opened, the load current starts to flow through the free-wheeling diode. The voltage across the actuator is clamped to about 0.8 V (the voltage drop across the diode) and therefore the switch must block only the actuator supply voltage. In fact, there will always be some stray inductance in the free-wheeling diode circuit, so a small voltage spike is still possible. The voltage rating of the switch should have some safety margin to account for this.

The load current flowing through the free-wheeling diode decays exponentially:

$$i(t) = I_o \exp \frac{-tR}{L}$$

where I_o = load current when switch is opened
L = load inductance
R = load resistance

In some cases the decay rate (determined by the *time constant L/R*) may be too slow, resulting in an actuator that takes too long to turn completely off. A solution is to insert a resistor, R_1, in series with the free-wheeling diode (Fig. 12.12). The peak voltage across the switch will now be

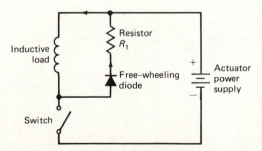

Figure 12.12 A resistor in series with a free-wheeling diode increases the rate of decay of the current in the load.

$$V = R_1 I_o + 0.8 + V_{source}$$

The current will now decay as follows:

$$i(t) = I_o \exp \frac{-t(R + R_1)}{L}$$

A faster rate of decay is gained at the expense of a higher peak voltage across the switch.

12.2 OUTPUT SYSTEMS WITH CONTINUOUS ACTUATORS

Not all systems can be controlled with two-state actuators; sometimes the output of an actuator must be continuously variable over a range of values. Figure 12.13 shows a block diagram of an output system involving a continuous actuator. The computer sends binary data to the output port where it is applied to a digital-to-analog (D/A) converter to produce a voltage proportional to the value of the binary data. This voltage signal is applied to the control input of a linear amplifier which controls the power applied to the actuator.

12.2.1 Low- to Medium-Power Amplifiers

Amplifiers are required since the output power capabilities of D/A converters are limited to milliwatts. At power levels up to several hundred watts, it is possible to use amplifiers that are essentially similar to the power amplifiers used in audio equipment. A representative example of such amplifiers is the Intersil ICH8530 (Fig. 12.14). This hybrid integrated circuit is capable of sourcing or sinking up to 2.7 A over a range of ±24 V. Similar amplifiers, often called *power op amps*, are available from a number of vendors in power ratings up to a few hundred watts. Basic ratings for several of these devices are presented in Table 12.2

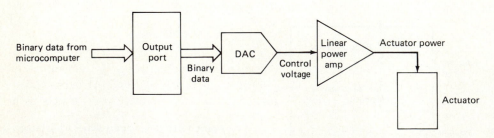

Figure 12.13 Output system with a continuous actuator.

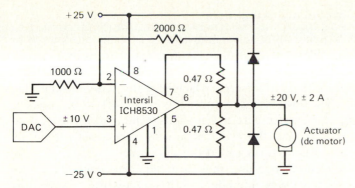

Figure 12.14 Power operational amplifier used to amplify control signal from digital-to-analog converter.

Table 12.2 Representative power operational amplifiers

Manufacturer	Model	Maximum output current, A	Maximum output voltage, V	Maximum rate of change of output voltage (slew rate), V/μs
Apex Microtechnology	PAO3	±30	±72	50
Burr-Brown	OPA501	±10	±20	1.5
National	LM675	± 1.25	±20	8
Silicon General	SG1173	± 3.5	±20	0.8

At higher power ratings, suitable amplifiers are often listed as *servo amplifiers* or *programmable power supplies*. Servo amplifiers are normally designed to supply power to electromechanical transducers such as motors or electrically controlled valves. Their characteristics, such as frequency response and output current and voltage ratings, are optimized for these types of applications. Programmable power supplies are variable-output power supplies whose output can be controlled by an analog or digital signal. They are designed for high accuracy and stability but usually do not respond as quickly to a change in the control signal as do servo amplifiers. The maximum rate of change of output voltage for a programmable power supply is typically on the order of 1 V/ms.

Amplifiers are normally designed to be *voltage sources*. That is, the control signal controls the output voltage of the amplifier. The output of some transducers is a linear function of the current flowing through the transducer rather than the voltage across the transducer. An amplifier which acts as a *current source* is better suited to such transducers. Servo amplifiers and programmable power supplies are available which can be operated in a current source mode as well as a voltage source mode.

12.2.2 Pulse-Width-Modulated Amplifiers

At higher power levels many amplifiers make use of a technique known as *pulse-width modulation* (PWM) to control the power supplied to the load. A PWM amplifier's power-control section consists of solid-state switches (i.e., transistors, power MOSFETs, or thyristors) which turn the power to the load fully on or fully off. The switches are turned on and off at a constant frequency. The average voltage (or current) applied to the load can be varied by varying the amount of time that the switch is on with respect to the time the switch is off (Fig. 12.15). This ratio is called the *duty cycle*.

PWM amplifiers are used because they are very efficient. Since the power-handling portion of a PWM amplifier consists of switches, the power dissipated by the amplifier is low. Another advantage of PWM amplifiers is that they can be very easy to implement in a microprocessor-based system. Since the amplifier consists of switches which are either on or off, the switches can be controlled directly by a digital output port. Software in the microcomputer, perhaps operating in conjunction with a counter-timer IC, can be used to generate the signals to control the switching. In this way, microcomputer control of a continuous actuator can be achieved without the need for a D/A converter and a separate amplifier.

PWM amplifier switching frequency A PWM amplifier can be used successfully only if its switching frequency is so high that the actuator or the system driven by the actuator responds to the average output rather than to each output pulse. Consider the output interface shown in Fig. 12.8, where a solid-state relay is used to control a heating coil. If the control signal to the solid-state switch is pulse-width-modulated, the heating coil can be used to control the temperature of a water bath. As long as the switching interval is a matter of a few seconds, the water-bath temperature remains almost constant, once a steady state is reached, since a single pulse from the heating coil cannot significantly alter the temperature (see Fig. 12.16). On the other hand, if the switching interval is increased to several minutes, the water temperature will rise and fall significantly during each switching interval.

Similar arguments apply to electromechanical actuators such as motors or solenoids except that the required switching intervals are on the order of a

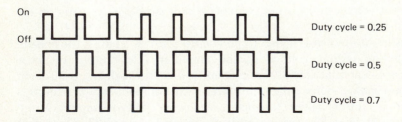

Duty cycle = 0.25

Duty cycle = 0.5

Duty cycle = 0.7

Figure 12.15 Voltage applied to actuator by PWM amplifier.

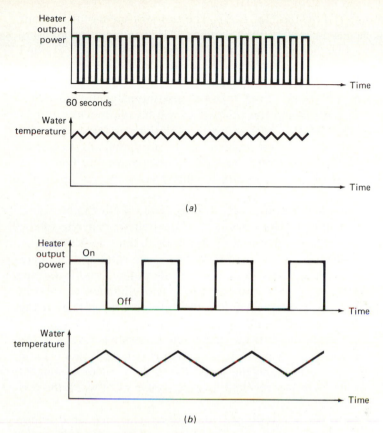

Figure 12.16 Effect of switching interval in pulse-width-modulated control. (*a*) Short switching interval. (*b*) Long switching interval.

millisecond. The current in an electromechanical actuator will circulate through the free-wheeling diode when the amplifier switches are off. Therefore, the current will decay only slightly if the switching interval is a fraction of the time constant (equal to L/R) of the load.

12.3 EXAMPLES OF ACTUATOR SYSTEMS

In order to illustrate some of the interfacing concepts we have discussed in this chapter, we will examine a few representative actuators and their interfaces to computers.

12.3.1 Stepping Motors

Basic operation A stepping motor converts electric pulses applied to its windings into incremental motion. A simplified diagram of a *bifilar permanent* *magnet* stepping motor is shown in Fig. 12.17. The stator of the motor has four poles, each of which has a center-tapped winding. The windings on opposing poles are connected together so that only five wires—*A*, *B*, *C*, *D*, and *V* + —leave the motor. A winding is excited by sending a current into the *V* + wire and out one of the other wires. When a winding is excited, the two opposing poles on which it is wound will take on opposite magnetic polarities. That is, one pole will be north and the other pole will be south. The rotor of the motor has six teeth which have magnetic polarities determined by a permanent magnet in the rotor. The three teeth at the north end are skewed 60 degrees with respect to the teeth at the south end. The operation of the motor relies on the simple principle that opposite magnetic poles attract while like poles repel.

Consider the situation in position 1: pole 1 is energized as north and pole 2 as south by exciting winding *B*. The rotor will align itself so that one of the south teeth lines up with pole 1 and one of the north teeth lines up with pole 2. If we now remove the excitation from winding *B* and excite winding *C* instead, pole 4 is energized as south and pole 3 as north while poles 1 and 2 are not activated. As a result the rotor will rotate in a clockwise direction (a *step*) in order to align a north tooth with pole 4 and a south tooth with pole 3. If the remaining windings are excited in the sequence shown in the figure, the clockwise rotation can be continued. If the excitation sequence is reversed, the direction of motion will also be reversed. If excitation is removed, there is still some attraction between the poles and the teeth due to the permanent magnet in the rotor. As a result, there is a residual *holding torque* even when there is no power applied to the motor.

This simple motor takes 12 steps per revolution. The angular distance covered by a single step (the *stepping angle*) is 30 degrees. In order to increase the number of steps per revolution, many stepper motors are designed with additional teeth on the rotor and have teeth machined into the stator poles. The stepping motion now aligns rotor teeth with stator teeth. With this kind of construction, a typical stepping angle is 1.8 degrees, so that 200 steps are required per revolution. These motors can be operated with two windings carrying current at the same time (*two-phase excitation*). In that case the rotor teeth align themselves between the two active stator teeth.

Figure 12.18 shows the actuation sequences and rotor positions for single-winding (single-phase) and two-winding (two-phase) excitation. Note that the stepping angles for the two kinds of excitation are the same but that the rotor positions are offset by half the stepping angle. The two actuation sequences (sometimes called *full-stepping* actuation modes) can be combined into what is called a *half-step* mode in which the number of steps per revolution is doubled. Thus a motor requiring 200 steps per revolution with single or two-phase excitation will require 400 steps per revolution when operated in the half-step mode.

There are other types of stepping motors besides the bifilar permanent magnet type. Some motors use only one winding per pole rather than the two

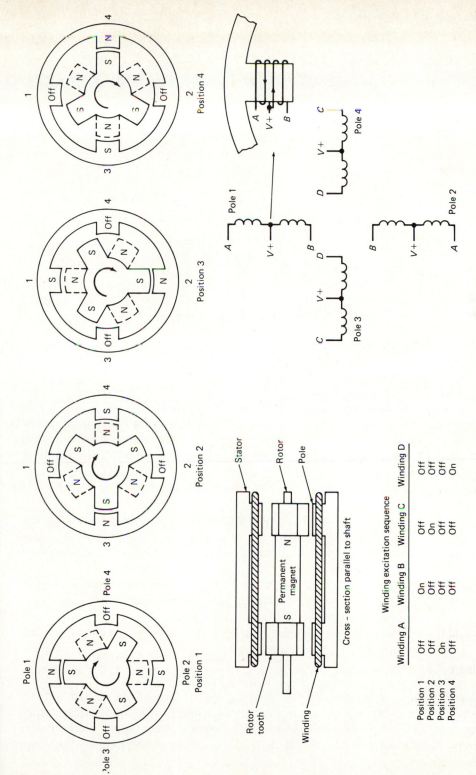

Figure 12.17 Basic stepping-motor operation.

339

Rotor position	Winding A	Winding B	Winding C	Winding D
Single-phase excitation				
0	Off	On	Off	Off
θ	Off	Off	On	Off
2θ	On	Off	Off	Off
3θ	Off	Off	Off	On
Two-phase excitation				
$\theta/2$	Off	On	On	Off
$3/2\,\theta$	On	Off	On	Off
$5/2\,\theta$	On	Off	Off	On
$7/2\,\theta$	Off	On	Off	On
Half-step mode				
0	Off	On	Off	Off
$\theta/2$	Off	On	On	Off
θ	Off	Off	On	Off
$3/2\,\theta$	On	Off	On	Off
$2\,\theta$	On	Off	Off	Off
$5/2\,\theta$	On	Off	Off	On
$3\,\theta$	Off	Off	Off	On
$7/2\,\theta$	Off	On	Off	On

Figure 12.18 Stepping-motor actuation sequences.

windings per pole used in bifilar motors. The drive circuit for these motors must be able to reverse the currents in the stator windings in order to generate both magnetic polarities. *Variable reluctance* stepping motors use a simple iron rotor with no permanent magnet. The rotor is still moved by the magnetic attraction of the rotor to the energized poles of the stator. However, the variable reluctance motor has no residual holding torque when the windings are not energized.

Stepper motors have the following advantages when used as actuators:

- They are easily controlled by computers, since winding currents are simply switched on or off.
- Speed of rotation can be controlled by controlling the time between successive steps.
- Direction of rotation is controlled by the winding excitation sequence.
- Rotor position can be controlled and monitored by counting the number of steps.
- If the motor is left energized once a target position is reached, the rotor will resist movement.

Stepping-motor drive circuits A stepping motor can be controlled directly from a digital output port of a computer with only a few solid-state switches required to control the winding currents. Consider the stepper motor driver circuit shown in Fig. 12.19. Wires A, B, C, and D from the stepper motor are

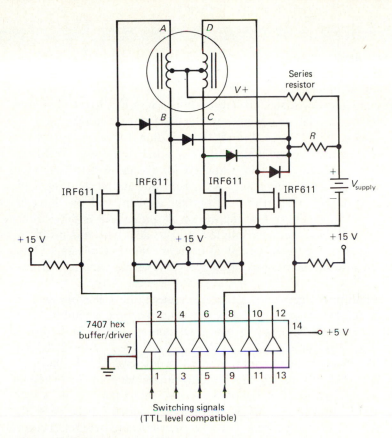

Figure 12.19 Stepper motor drive circuit.

connected to power MOSFETs which act as switches. The $V+$ wire is connected to the motor power supply via a series resistor. When a power MOSFET is switched on, a circuit is completed and current flows through one of the windings.

The power MOSFETs used in this circuit require a gate (control terminal) voltage of between 10 and 15 V to switch on completely. Therefore, if the MOSFETs are to be controlled by a digital output port which operates at TTL voltage levels (i.e., 0 to 5 V), some voltage-level conversion circuits are required. The 7407 hex buffer-driver IC provides the required voltage level conversion. The six buffer-driver circuits in the 7407 have TTL-compatible inputs and open collector outputs. If the open collector outputs are connected to 15 V via pull-up resistors, the output will switch from about 0.2 V to 15 V when the input switches from TTL low (<0.8 V) to TTL high (>2.4 V).

The windings in a stepper motor have a substantial inductance. This inductance presents two problems. First, when a winding is switched on, the equation for the current flowing in the winding is

$$i(t) = \frac{V}{R} \left(1 - \exp \frac{-tR}{L} \right)$$

where V = supply voltage

R = total circuit resistance

L = total circuit inductance

It takes approximately $3L/R$ seconds (three time constants) for the current to reach 95 percent of its final steady-state value. In a typical stepper motor, the winding resistance might be $5\,\Omega$ and the winding inductance $20\,\text{mH}$. Therefore, the time constant, in the absence of any other significant resistance in the circuit, will be $4\,\text{ms}$ and the time required for the current to reach its final value will be about $12\,\text{ms}$. This places an upper limit on the stepping rate of the motor. If the motor is stepped too rapidly, current will not have a chance to build up in the windings, and the motor torque, which is dependent on winding current, will decrease. The simplest way to deal with this problem is to add additional resistance to the circuit to decrease the time constant. The supply voltage must also be increased so that the steady-state winding current remains the same. This is the reason for the series resistor in the $V+$ line in Fig. 12.19. The addition of the series resistor decreases the efficiency of the stepper-motor circuit. The losses in a series resistor become prohibitive in higher-power stepper-motor circuits, so other techniques are used to force current into the windings. Interested readers will find these techniques described in the references on stepper motors cited at the end of this chapter.

The second problem created by the winding inductance occurs when a winding is switched off. As with all inductive loads, the collapsing field will attempt to maintain the flow of current. If no path is provided for the current, damaging voltage spikes will be generated across the switching devices. Therefore, free-wheeling diodes must be included in the driver circuit. The free-wheeling diode circuit used in Fig. 12.19 includes a series resistor to increase the decay rate of the winding currents. If this resistor isn't used, the maximum stepping rate will be limited by the time it takes the winding current to decay.

Stepping-motor limitations Even when the driver circuit is designed to achieve short rise and fall times for the winding currents, there are still limitations on the stepping rates that can be achieved by a stepping motor. A typical stepper-motor torque-speed curve is shown in Fig. 12.20. If the load torque exceeds the *pullout torque*, the motor will not rotate in synchronism with the stepping signals from the controller, and the ability to control rotor velocity and position without feedback will be lost. Since the pullout torque drops as the stepping rate increases, there is a maximum stepping rate that can be achieved for a particular combination of motor, driver circuit, and load.

An additional concern is the presence of resonance frequencies. The combination of the rotor and load inertia and the attractive force between the rotor and the stator teeth is an oscillatory system rather like a mass suspended on a spring. When the rotor is incremented by one step, it oscillates before

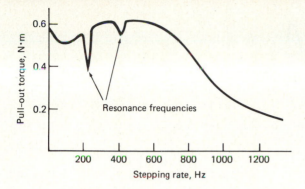

Figure 12.20 Stepper-motor torque-speed curve.

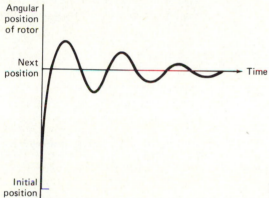

Figure 12.21 Rotor position vs. time when motor is incremented by one step.

reaching its final position, as shown in Fig. 12.21. If the motor is stepped at a frequency equal to the frequency of the rotor oscillation, the rotor has a positive velocity every time it is stepped. As a result, succeeding steps become increasingly oscillatory (see Fig. 12.22). The increased oscillation decreases the motor's pullout torque and can cause the rotor to lose synchronism with the stepping pulses. The same situation can occur at stepping rates which are subharmonics of the oscillation frequency. Usually, a stepper motor can be accelerated or decelerated through the resonance frequencies without losing synchronism, but it cannot be run continuously at these frequencies. The effects of resonance can be decreased by adding mechanical or electrical damping to the stepping-motor drive system. The references listed at the end of this chapter provide more information on solutions to the resonance problem.

Although a stepper motor may have sufficient pullout torque to operate at high stepping rates, it cannot start or stop at these rates. Normally, for a given combination of rotor and load inertia, a *start/stop rate* will be defined which is the highest stepping rate at which the motor can be started or stopped without losing synchronism. Once the motor is started, it is accelerated from the

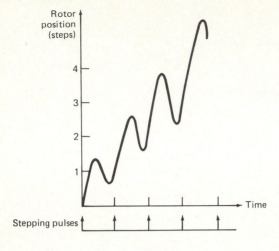

Figure 12.22 Effect of stepping rate equal to rotor oscillation frequency.

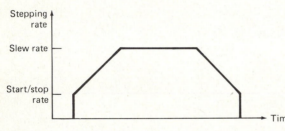

Figure 12.23 Typical stepping-rate profile.

start/stop rate to a maximum *slew rate* defined by the pullout torque characteristics of the motor. When the motor is to be stopped, it is decelerated from the slew rate to the stop/start rate before it is stopped. Figure 12.23 illustrates a typical stepping-rate profile.

Stepping-motor controllers In a typical positioning application where the stepper motor rotor is to be incremented by a number of steps, the motor controller must perform a number of functions. It must generate the proper control signals for the solid-state switches so that the motor windings are actuated in the proper sequence. It must also time the control signals so that the proper stepping-rate profile is followed. Finally, it must plan the stepping-rate profile so that the rotor is brought to a stop after the required number of steps.

The designer has a range of options when deciding how many of the control functions will be performed by microcomputer software (see Fig. 12.24). The microcomputer can be programmed to control the switches in the stepper-motor driver circuit directly. In that case, it performs all the required control operations. Alternatively, an external *translator* circuit can be used to generate the control signals for the switches in the driver circuit. The translator generates the proper winding excitation sequence. Translators are available

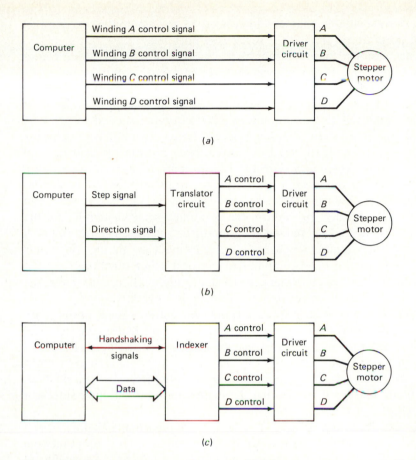

Figure 12.24 Stepper-motor control options. (*a*) Direct computer control. (*b*) Excitation controlled by stepping-motor translator. (*c*) Stepping-rate profile and excitation controlled by stepping-motor indexer.

which can be configured to generate both full-step and half-step sequences. The microcomputer sends a step pulse to the translator to cause the motor to make one step. The microcomputer controls the direction of rotation with another control line. The microcomputer is still responsible for maintaining and planning the stepping-rate profile.

If this is still too much of a burden on the microcomputer, a *stepping-motor indexer* can be used. The indexer accepts commands from the microprocessor specifying the start/stop rate, the acceleration and deceleration rates, the slew rate, and the required final position. It then generates the required stepping-rate profile and informs the microcomputer when the final position has been reached. This may be done by using an interrupt signal or by setting a bit in a status register in the indexer which can be read by the microcomputer. Stepping-motor indexers (sometimes called stepping-motor controllers) are available as plug-in circuit boards for most microcomputer bus systems.

Indexers are also available that are interfaced to the computer via parallel input/output ports or via an RS-232 serial I/O port.

The choice among these options is a classic example of a hardware-software tradeoff. If the designer elects to have the microcomputer control the driver switches directly, hardware costs are minimized but careful programming is required if the motor is to be operated at high stepping rates. For example, if the motor is to be operated at a 2000-step-per-second rate, a new set of control signals must be generated every 500 μs. In addition, while the stepping motor is being accelerated or decelerated, a different time interval must be generated between each step. In order to manage all this, the use of a programmable counter-timer to generate the required time delays is mandatory. A table of stepping-time intervals for acceleration and deceleration can be precalculated in order to reduce the computational demands on the microprocessor. Nevertheless, the stepping-motor driver routines will probably have to be coded in assembly language in order to meet the timing constraints and to leave time for the microcomputer to perform other functions.

An example of a system where a designer might choose to have the microcomputer control the driver switches directly is a printer. Stepper motors are used to move the print head and advance the paper. Since printers are produced in large volumes and selling cost is an important factor, the time and effort to write the software to control the stepper motors can be justified if it saves on hardware costs. In addition, a printer is a predictable system in which the tasks to be performed by the control microcomputer are well-defined. Therefore, the designer can ensure that the microcomputer will have sufficient time to do everything required of it.

The use of a translator reduces the software requirements slightly with a small increase in hardware costs. An indexer reduces the computational requirements on the microcomputer substantially at the expense of a significant increase in hardware costs. Purchased in small quantities, a stepper-motor indexer board will cost a few hundred dollars. However, it removes much of the complexity from the software design. The microcomputer can load the desired stepper-motor position into the indexer and then perform other tasks while the indexer moves the stepper motor to the target position. The use of indexers is particularly attractive in complex systems where several stepper motors must be operated simultaneously and the microcomputer must perform many other tasks besides motor control. Typical examples of such systems include robots and computer-controlled machine tools.

As we have mentioned, stepper motors have many attractive features when used as actuators in microcomputer-based systems. However, they are not suited for all applications. As the required output power increases beyond a few hundred watts or the required stepping rate increases beyond a few thousand steps per second, it becomes more and more difficult to achieve the required performance with a stepper motor without resorting to complex drive circuits and rotor position feedback. Since the principal advantages of stepper motors are their simple drive circuits and ability to operate without feedback, other motors may represent a better choice.

12.3.2 Dc Servomotors

Dc servomotors are dc motors designed for use in actuators where the position or velocity of the actuator is to be controlled. They are characterised by a high ratio of output torque capability to rotor inertia. This large torque/inertia ratio allows the motor to be accelerated and decelerated quickly. This is important when an actuator must rapidly move from one position to another. The magnetic field in a dc servomotor is constant and is generated by permanent magnets in smaller motors (those with output ratings less than a few kilowatts). Since the magnetic field is constant, control of motor torque is achieved entirely by controlling the motor's armature current. Unlike a stepper motor, it is not possible to control the rotor velocity or position of a dc servomotor with any precision without the use of feedback. As a result, dc servomotors often come equipped with feedback sensors such as tachometers (velocity feedback) and shaft encoders (position feedback).

Motor model The operation of a dc servomotor can be understood by examining the electrical and mechanical models of the motor (see Fig. 12.25). The electrical model includes the resistance and inductance of the motor's armature circuit. In addition, the *counter electromotive force* (counter EMF) generated by the movement of the armature in the motor's magnetic field is included. The counter EMF is a voltage that opposes the voltage applied to the motor terminals and is proportional to rotor speed. The equation relating the armature circuit's input voltage and current is

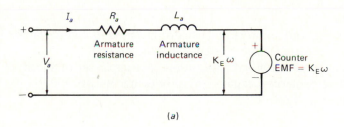

(a)

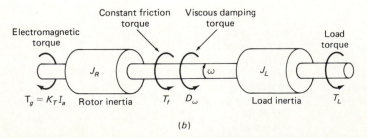

(b)

Figure 12.25 (a) Electrical and (b) mechanical models of the dc servomotor.

$$V_a = L_a \frac{dI_a}{dt} + R_a I_a + K_E \omega$$

where V_a = armature circuit terminal voltage
$\quad I_a$ = armature current
$\quad L_a$ = armature circuit inductance
$\quad R_a$ = armature circuit resistance
$\quad \omega$ = rotor velocity
$\quad K_E$ = motor counter–EMF constant

The mechanical model includes the moments of inertia of the rotor and the load as well as several torques. The friction torque and load torque are constant, independent of motor speed. The viscous damping torque is made up of a number of forces in the motor which are proportional to motor speed. The electromagnetic torque is the torque produced by the flow of the armature current. This torque is directly proportional to the armature current. The equation describing the required electromagnetic torque is

$$K_T I_a = T_{em} = (J_R + J_L) \frac{d\omega}{dt} + D\omega + T_f + T_L$$

where T_{em} = electromagnetic torque
$\quad K_T$ = motor torque constant
$\quad J_R$ = moment of inertia of motor
$\quad J_L$ = moment of inertia of load
$\quad D$ = viscous damping constant
$\quad T_f$ = motor friction torque
$\quad T_L$ = load torque

The characteristics of dc servomotors are usually specified in considerable detail by the manufacturer so that the designer can determine the motor's suitability for a particular motion-control application. A sample set of specifications for a typical dc servomotor capable of producing 300 to 400 W of mechanical power is given in Table 12.3.

Table 12.3 Sample dc servomotor specifications

Peak current	40 A
Continuous current	7.5 A
Torque constant	0.1 N·m/A
Counter-EMF constant	10 V per 1000 r/min
Armature resistance	1 Ω
Armature inductance	200 μH
Armature power dissipation	60 W maximum
Rotor moment of inertia	0.00015 kg·m^2
Friction torque	0.5 N·m
Maximum speed	5000 r/min

Example: robot-arm drive   Using the specifications given in Table 12.3, we can consider the application of this motor in a robotic system. The drive we will consider is for one axis of the robot arm shown in Fig. 12.26. The axis rotates the vertical post that the remainder of the arm is mounted on. Robot-drive systems are generally difficult to analyze, since moments of inertia do not remain constant and velocity profiles vary with the task the robot is performing. We will restrict our attention to a simple case.

The robot arm is to be used in a "pick-and-place" application, where items are repeatedly picked up in one location and placed in another location. The basic specifications for the vertical axis are the following:

1. The axis is to turn 180 degrees for each pick-and-place operation.
2. The time to turn 180 degrees must be 1 s or less.
3. The velocity profile to be followed is trapezoidal, as shown in Fig. 12.27.
4. The duty cycle is 67 percent. That is, the axis is rotating two-thirds of the time.
5. The moment of inertia of the arm and its load is $4 \, \text{kg} \cdot \text{m}^2$.
6. The load torque is $8 \, \text{N} \cdot \text{m}$. This is primarily friction in bearings and gears.
7. The gear reduction between the motor shaft and the axis is 100:1.

The designer must calculate the following quantities to see if a motor is suitable for this type of incremental motion task:

1. Maximum motor velocity required
2. Maximum motor torque required
3. Armature power dissipation

Given the velocity profile, the gear ratio, and the distance to be moved,

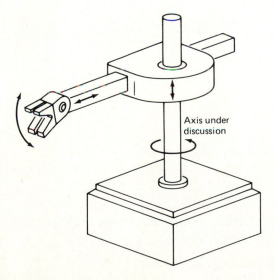

Axis under discussion

Figure 12.26 Robot arm.

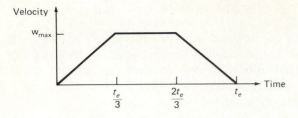

Figure 12.27 Velocity profile for robot axis.

the calculation of the maximum motor velocity is relatively straightforward. The distance moved by the motor is

$$\text{Gear ratio} \times \text{distance moved by axis} = 100 \times \tfrac{1}{2}\text{rev} = 50\,\text{rev}$$

The distance moved must equal the area under the velocity profile. Therefore, the maximum velocity for the simple trapezoidal profile in Fig. 12.27 can be calculated with some basic algebra. The formula is

$$\text{Maximum velocity} = \frac{3 \times \text{distance moved}}{2 \times \text{time required}}$$

$$= \frac{3 \times 50}{2 \times 1}$$

$$= 75\,\text{r/s} = 4500\,\text{r/min}$$

The maximum velocity is less than the specified maximum velocity (5000 r/min) for the motor we are considering.

The maximum motor torque will occur when the motor is accelerating. The acceleration rate is constant and easy to calculate:

$$\text{Acceleration} = \frac{4500\,\text{r/min}}{\tfrac{1}{3}\,\text{s}} = 13{,}500\,\text{r/(min)(s)} = 1413.7\,\text{rad/(s)(s)}$$

When a load is coupled to a motor through a gear ratio, N, the effective moment of inertia of the motor is

$$J = J_R + \frac{1}{N^2}\,J_L$$

Similarly, the total load torque, ignoring viscous damping, is

$$T = T_f + \frac{1}{N}\,T_L$$

With the values given for the motor and the robot-arm system, the calculated values of moment of inertia and load torque are

$$J = 0.00055\,\text{kg} \cdot \text{m}^2$$
$$T = 0.13\,\text{N} \cdot \text{m}$$

Using these values, along with the calculated value for acceleration, in the dc servomotor torque equation, we get

$$\text{Maximum torque} = J\frac{d\omega}{dt} + T$$

$$= 0.907 \, \text{N} \cdot \text{m}.$$

Therefore, the maximum armature current will be

$$\frac{\text{Maximum torque}}{K_T} = \frac{0.9707}{0.1} = 9.1 \, \text{A}$$

This maximum current is considerably less than the 40-A peak current rating for the motor.

The armature power dissipation can be written as

$$\left(\frac{\text{Energy dissipated in armature per movement}}{\text{Time per movement}}\right) \text{duty cycle}$$

The energy dissipated in the armature per movement is

$$R_a \int_0^{t_e} I_a^2(t)\,dt$$

By solving the motor torque equation for I_a, we can express the armature current as

$$I_a = \frac{J(d\omega/dt) + T}{K_T}$$

Inserting this into the formula for the energy dissipated in the armature and performing the integration for the trapezoidal velocity profile, we get

$$\text{Energy dissipated} = \frac{R_a}{K_T^2}\left(\frac{2J^2(d\omega/dt)^2}{3} + T\right)$$

For the system under discussion, the energy dissipated in the armature per movement is 42 J. Therefore, the power dissipation is about 28 W for a 67 percent duty cycle. This power dissipation is safely below the maximum dissipation limit of 60 W for the motor.

We have verified that the motor will operate within its specified limits for this particular robotics application. There is sufficient reserve torque and power-dissipation capability to allow some safety margin if the robot is overloaded. The next stage in the design process is to determine the voltage and current requirements for the servo amplifier that will power the motor. The peak current requirement, 9.1 A, has already been calculated. The peak motor voltage will occur just as the motor reaches maximum velocity during acceleration. At this point the armature current is at its maximum value and the counter EMF is also at its maximum value. The motor terminal voltage is

$$V_a = R_a I_a + K_E \omega = 1 \times 9.1 + \left(\frac{10}{1000} \times 4500 \right) = 54 \text{ V}$$

Allowing some reserve capacity, the servo amplifier output rating should be

Voltage $\quad \pm 60$ V
Current $\quad \pm 15$ A

Interfaces to servomotors Once the motor and servo amplifier have been selected, the interface to the controlling microcomputer must be designed or selected. Three interface options are shown in Fig. 12.28. In the first interface, the computer sends position commands to a conventional analog position control servo loop via a digital-to-analog converter (DAC). These analog controllers are normally in a cascade configuration with an inner velocity control feedback loop and an outer position control feedback loop. The inner velocity loop effectively adds damping to the feedback system and thereby stabilizes it. The design of analog position control servo loops is covered in most texts on servosystems or motor drives. One disadvantage of this interface is that the computer has no control over the velocity profile. Acceleration and deceleration rates and maximum velocities are determined by factors such as amplifier gain, current-limit set-points, and voltage-limit set-points in the analog control loops.

The second interface combines analog and digital control techniques. The velocity control loop is retained and is implemented with analog electronics. However, the position control loop is now implemented with digital circuits. The motor-position sensor is an *incremental shaft encoder* which generates a fixed number of pulses per motor revolution. Therefore, when the motor rotates, a train of pulses is generated. The total number of pulses generated during a movement is proportional to the distance traveled. The pulse rate is proportional to the motor's velocity. The pulses from the encoder are fed to a digital counter which decrements its count every time it receives a pulse from the encoder. The counter also receives pulses from the microcomputer. However, it increments its count every time it receives one of these pulses. The binary count in the counter is sent to a digital-to-analog converter which converts it to a voltage. This voltage is used as a control signal for the analog velocity loop.

To the computer, the interface resembles a stepper-motor translator. Consider a system where the shaft encoder produces 100 pulses per revolution. If the computer wishes to increment the motor by 85.6 r, it sends out a train of pulses, with the total number of pulses equaling

$$100 \times 85.6 = 8560$$

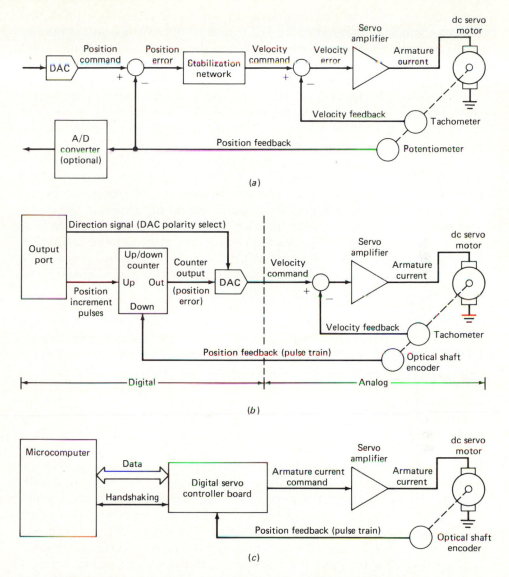

Figure 12.28 Servomotor controllers. (*a*) Analog servo loop. (*b*) Mixed digital/analog servo loop. (*c*) Digital servo loop.

The pulse rate is controlled to achieve the desired velocity profile. In this case, if the pulse rate is kept constant at 1000 pulses per second, the count in the counter will initially increase from zero, since the motor is at a standstill and therefore there are no feedback pulses from the shaft encoder. As the count increases, the DAC output voltage increases, increasing the velocity command. As a result, the motor accelerates until the pulse rate from the encoder equals the pulse rate from the computer. When the two pulse rates are equal, the count in the counter remains at a constant value. As a result, the velocity command remains constant and the motor moves at a constant velocity of 10 r/s. When 8560 pulses have been sent to the counter, the computer stops sending pulses. Pulses are still arriving from the shaft encoder, so the count in the counter now decreases, thereby decreasing the velocity command. As a result the motor begins to decelerate to a stop. When the counter has counted down to zero and the velocity command is zero, the counter will have received 8560 pulses from both the computer and the shaft encoder. Therefore, the motor will have moved the required distance.

The third interface option in Fig. 12.28 is a completely digital servo loop. This resembles a stepper-motor indexer. The only feedback required is position feedback from a shaft encoder. The stabilization normally provided by the velocity feedback loop is provided by digital control algorithms in the digital controller. The computer writes position and velocity profile commands to registers in the servo controller board. The digital servo controller board then controls the movement of the motor. When the motor has reached the required position, the controller informs the computer by generating an interrupt signal or setting a flag bit in an internal status register which can be read by the computer.

Circuit boards implementing the mixed analog and digital servo loop or the completely digital servo loop are available for most common microcomputer bus systems. Controllers are also available in units combined with servo amplifiers. These units can be interfaced to the microcomputer via parallel or serial I/O ports.

SUMMARY

In this chapter we have considered the design of microcomputer output systems which use actuators to control an external process. The actuator usually consists of two elements, a transducer and an amplifier. In actuators which have only two states, either on or off, the amplifier can be reduced to a switch which controls the amplication of power to the transducer. At low power levels, the switch can be a power transistor or power MOSFET which is controlled directly by an output line from a microcomputer output port. At higher power levels, electrical isolation is required between the actuator power supply and the microcomputer. In that case electromechanical or solid-state relays can be used to control power to the transducer.

Transducers which can produce a continuous range of output values require linear amplifiers. These are available in a wide range of output power levels. Depending on their characteristics and intended application, they may be advertised as power operational amplifiers, servo amplifiers, or programmable power supplies. High-power amplifiers may make use of pulse-width modulation in order to reduce their internal-power dissipation. A microcomputer may use pulse-width modulation directly to control a solid-state switch to produce a continuous range of output values from a transducer.

Stepping motors and dc servomotors were considered as examples of actuators that are commonly used in real-time systems. The stepper motor can be controlled relatively easily by a computer. If the stepper motor is operated within its pullout torque limits, its position and velocity can be controlled without the need for position or velocity feedback sensors. The stepper motor is particularly well-suited for low-power and low-speed applications.

When output-power requirements begin to exceed 1 kW or very fast response is required, a dc servomotor should be considered. In position control applications dc servomotors require position feedback sensors and sometimes also require velocity feedback. However, they are capable of high rates of acceleration and high speeds, so they are suited for applications requiring fast response. They are also available with higher output powers than stepper motors. The control of a dc servomotor is more complex than the control of a stepper motor, but complete dc servomotor controllers are commercially available. Therefore, the system designer need not be an expert on dc motor controls in order to use dc servomotors as actuators.

EXERCISES

12.1 A light bulb is to be controlled by a transistor switch. The bulb operates from a 12-V dc source and is rated to dissipate 10 W. The resistance of the cold filament is measured to be 1.3 Ω. What are the average- and peak-current ratings required for the transistor?

12.2 A solenoid with a coil inductance of 0.5 H and a coil resistance of 30 Ω is operated from a 24-V supply. If a free-wheeling diode is used for transient suppression, how long will it take for the coil current to decay to 5 percent of its original value after the switch is opened? A resistor is to be placed in series with the free-wheeling diode to accelerate the decay of current. What value of resistance is required to have the current decay to 5 percent of its original value in 20 ms? What is the peak voltage across the switch?

BIBLIOGRAPHY

Solid-State Switches

Transistors

Millman, J., and C. Halkias: *Integrated Electronics: Analog and Digital Circuits and Systems*, McGraw-Hill, New York, 1972.

Power MOSFETs

Oxner, E.: *Power FETs and Their Applications*, Prentice-Hall, Englewood Cliffs, NJ, 1982.

Thyristors

Grafham, D. R., and F. B. Golden (eds.): *General Electric SCR Manual*, 6th ed., General Electric, Auburn NY, 1979.

Stepping motors

Arcanley, P.: *Stepping Motors*: *A Guide to Modern Theory and Practice*, Peter Peregrinus Ltd, New York, 1982.
Kenjo, T.: *Stepping Motors and Their Microprocessor Controls*, Clarendon Press, Oxford, 1984.

DC servomotors

Electro-Craft Corp.: *DC Motors, Speed Controls, Servo Systems*, 4th ed., Hopkins, MN, 1978.
Leonhard, W.: *Control of Electrical Drives*, Springer-Verlag, New York, 1985.

BOARD-LEVEL DESIGN

BOARD-BASED
MICROCOMPUTER SYSTEMS

The components of a board-based microcomputer system are shown in Fig. 13.1. The microcomputer system is based on a set of printed circuit boards that are plugged into a *backplane* or *mother board*, as shown in Fig. 13.2. Each board performs a function required in the microcomputer system. One board contains the microprocessor, another may contain the memory, and a third contains an I/O interface. Figure 13.3 shows how a typical board-based microcomputer system might be configured. The backplane interconnects the boards via a series of conductors called the *system bus*. The connector between the printed circuit board and the backplane does not provide much support, so a *card cage* is provided to hold the printed circuit boards in place. The card cage has properly spaced *card guides* that hold the edges of the printed circuit boards and guide the boards to the connectors on the mother board. The card cage is installed in a cabinet along with a power supply and possibly a fan for cooling.

Such a system can be purchased as a complete unit from a number of manufacturers. However, it is also possible to purchase each component separately and assemble a system to conform to a particular set of requirements. It is this modularity that makes board-based microcomputer systems attractive. The designer can customize the system for a particular application by choosing the appropriate boards, cabinet, power supply, and cooling technique. In an oceanographic application the designer might choose a set of compact boards using low-power-consumption CMOS integrated circuits. The boards would be installed in a pressure vessel and powered from batteries. In a

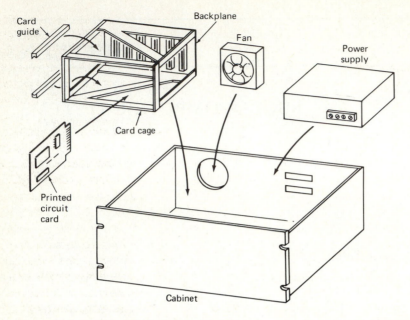

Figure 13.1 Board-based microcomputer system.

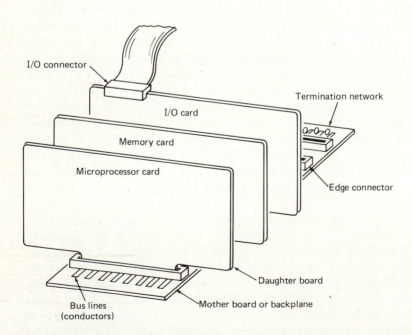

Figure 13.2 Components of a board-level-based microcomputer system.

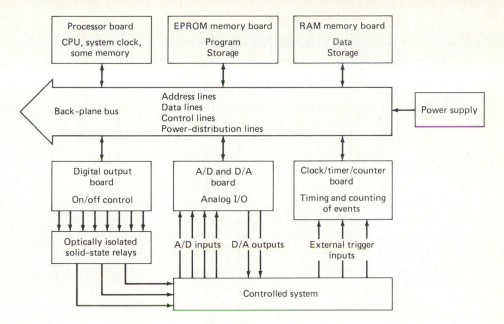

Figure 13.3 Block diagram of a board-based microcomputer system.

biomedical instrumentation application, the designer might choose a set of boards that are based on a powerful 16-bit microcomputer, since considerable numerical computation is required. The boards would be installed in an attractive cabinet (for marketing reasons) along with a conventional power supply and a quiet fan.

Basing a real-time system design on a commercially available microcomputer board system has other advantages besides design flexibility. A popular system will have many manufacturers producing boards for the system. This results in competitive prices, a wide variety of available boards, and multiple sources for popular boards. Designs can be upgraded by replacing the original microprocessor board with a board using a faster or more powerful microprocessor. Designs can also be expanded by adding more memory boards or more I/O boards. In many cases a faulty system can be repaired by systematically swapping boards that are known to be good for the boards in the defective system. This allows personnel who are not highly skilled in electronics to service the equipment.

13.1 THE BACKPLANE BUS

The backplane bus consists of a set of electric conductors that carry digital signals and electric power among the boards plugged into the backplane. A very simple example of such a bus is shown in Fig. 13.4. The assignment of

Connector

Address Ø	1			1				1		1		
Address 1		2			2				2		2	
Address 2	3			3				3		3			Address lines
Address 3		4			4				4		4	
Address 4	5			5				5		5		
Data Ø		6			6				6		6	
Data 1	7			7				7		7		
Data 2		8			8				8		8		Data lines
Data 3	9			9				9		9		
Data 4		10			10				10		10	
System clock	11			11				11		11		
Read/write		12			12				12		12		Control lines
Data strobe	13			13				13		13		
+12 V		14			14				14		14	
−12 V	15			15				15		15		Power-distribution lines
+5 V		16			16				16		16	
Gnd	17			17				17		17		

Figure 13.4 Simplified backplane bus structure.

lines to functions or signals is standardized for a particular bus system so that any board designed for that system can function in any backplane designed for that system. The bus lines are attached to the connectors on the backplane. These connectors have commonly been *edge connectors* (see Fig. 13.5) which use springlike contracts to connect to gold-plated "fingers" on the printed circuit board. This technique is relatively inexpensive, since only one connector is needed. However, these connectors are not very resistant to vibration and

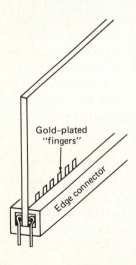

Gold-plated "fingers"

Edge connector

Figure 13.5 Edge connector.

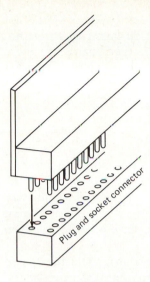

Figure 13.6 Pin and socket connector.

corrosion and do not withstand repeated insertions of the printed circuit board very well. As a result, a plug and socket type of connector (see Fig. 13.6) that is popular in Europe is also gaining in popularity in North America. This connector system does not have the problem of the edge connector but is somewhat more expensive.

The lines on the bus can be divided into five categories:

1. Address lines to select memory locations and I/O registers
2. Data lines
3. Control lines to control data transfers, carry interrupt signals, and arbitrate requests for control of the bus
4. Power-distribution lines
5. Spare lines for future expansion

13.1.1 Address and Data Lines

Microcomputer bus systems have anywhere from 16 to 32 address lines. More recently defined bus systems normally have more address lines than earlier systems because they have been designed with an eye toward the 16- and 32-bit microcomputers with their large address spaces. The number of data lines in microcomputer bus systems ranges from 8 to 32. Again, the more recently defined systems tend to have more data lines than do early systems. Data lines are normally *bidirectional*, meaning that data can flow either from the computer board to another board or from a board back to the computer board on the same lines. The direction of data flow is controlled by a signal on a control line. The S-100 (IEEE S696) bus system has 16 data lines which can be used as two *unidirectional* 8-bit buses (data output and data input) or as a single 16-bit bidirectional data bus.

A backplane bus system with large numbers of data and address lines is not without drawbacks. The boards and backplanes must be larger to accommodate the lines, and connectors are more expensive. Larger boards are more expensive, harder to fit into small enclosures, harder to cool, and less tolerant of vibration.

13.1.2 Lines to Control Data Transfer

The control lines vary considerably from one system to another. All systems have control lines to control the transfer of data from one board to another. The board that initiates a data transfer will be called the *bus master* in the following discussion, while the board that responds to the control signals sent by the master board will be called the *slave board*. Normally the master board is the board that contains the microprocessor while the slave board is usually a memory board or an I/O board.

For a data transfer to occur, the direction of the transfer and the source and destination of the data must be specified. The direction of transfer can be specified by *data direction* lines on the bus which indicate whether a transfer is from the master to the slave (a *write* operation) or from the slave to the master (a *read* operation). The source or destination of data which is to be moved to or from the master board is specified by placing the address of the memory location or I/O register on the address bus. Address decoding circuits on the slave boards ensure that only the addressed memory location or I/O register gains access to the data bus. Many buses also contain a control line which specifies whether the data transfer is between the master and memory or between the master and an I/O device. This allows the bus to accommodate microprocessors which have a separate I/O addressing system of the type discussed in Chap. 2.

Timing of data transfer   Two general methods are used to control the timing of data transfers on the bus. *Synchronous* data transfers occur under control of a system clock and other timing signals generated by the master board (normally the CPU board). *Asynchronous* data transfers rely on *handshaking* signals between the master and slave boards to control the transfer.

In a synchronous data read operation (see Fig. 13.7), the master board places the address of the data on the bus (1) and asserts the DATA READ signal (2), informing the slave board that a read operation is being performed. The master board then reads the data from the bus a fixed time later (4). The slave board must place the data on the bus before this time (3) and hold it valid until the master has read it (5).

In a synchronous write operation (see Fig. 13.8), the master places the destination address (1) and the data (2) on the bus for a fixed length of time while asserting the DATA WRITE signal (3), informing the slave board that a write operation is being performed. The slave board must read the data before

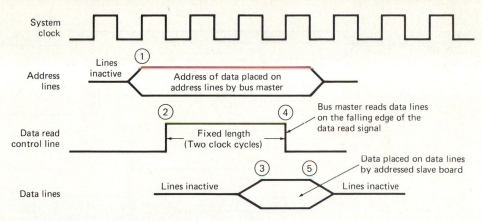

Figure 13.7 Synchronous data read operation.

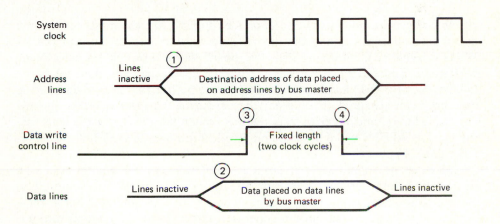

Figure 13.8 Synchronous data write operation.

the master removes the data from the bus. In some systems, the slave board uses the falling edge of the DATA WRITE signal (4) as a *strobe* to latch the data from the bus.

In an asynchronous read operation (see Fig. 13.9), the master places the address of the desired data on the bus (1) and asserts the DATA READ signal (2), informing the slave board that a read operation is being performed. It then waits until the slave board signals that it has placed the data on the bus. The slave board places the data on the bus (3) and then asserts the DATA VALID handshaking signal (4). The master board then reads the data and removes the DATA READ signal (5) and the address of the data from the bus. When the slave board detects that the DATA READ signal has been removed, it removes the DATA VALID signal (6) and the data from the bus.

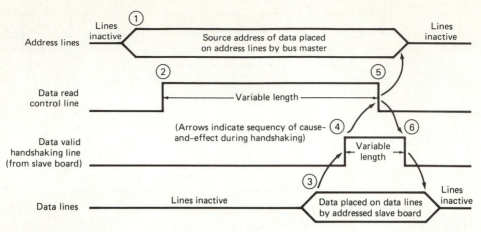

Figure 13.9 Asynchronous data read operation.

In an asynchronous write operation (see Fig. 13.10), the master places the destination address (1) and the data (2) on the bus and asserts the DATA WRITE signal (3), informing the slave board that a write operation is being performed. The slave board reads the data from the bus and then sends the DATA ACCEPTED handshaking signal to the master (4), indicating that the data has been accepted. The master then removes the DATA WRITE signal (5), the address, and the data from the bus. When the slave board detects that the DATA WRITE signal has been removed, it removes the DATA ACCEPT-ED signal (6).

Bus systems employing synchronous data transfers are simpler, since no handshaking lines are required and no additional hardware is required to monitor and control the handshaking lines. However, all the boards on a synchronous bus must be able to operate at the speed required by the bus

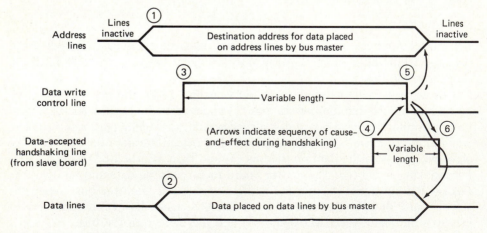

Figure 13.10 Asynchronous data write operation.

master and they must all operate at the same speed for data-transfer operations. On an asynchronous bus, boards can operate at different speeds, since data transfer does not rely on precise synchronization to control signals sent by the master board. A system can include both a fast memory board and a slow I/O board. The system can take full advantage of the memory board's speed while performing memory operations yet still be able to communicate successfully with the I/O board.

13.1.3 Interrupt Control Lines

The interrupt control lines carry *interrupt requests* from a board that generates interrupts to the CPU board and carry *interrupt acknowledge* signals from the CPU back to the interrupting board. In some systems there are multiple interrupt request lines which allow each interrupting board to have its own interrupt request line. This makes the identification of the interrupting board very simple. For example, the CPU board may use a MC68000 microprocessor which has three interrupt input lines IPL_0, IPL_1, and IPL_2 (see Chap. 7). The logical levels on these lines are interpreted as a binary number indicating the priority and source of the interrupt (000 indicates no interrupt). An *encoder IC* can be used to generate these 3-bit binary numbers from up to seven interrupt request lines coming from the system bus (Fig. 13.11). The interrupt acknowledge signal placed on the system bus by the MC68000 CPU after receiving the interrupt will cause the interrupting device to send an *interrupt vector code* to the CPU, as discussed in Chap. 7.

In other bus systems there are only one or two interrupt request lines. In that case the priority of the interrupting device must be determined by some other technique. A common technique is to *daisy-chain* the interrupt request through the boards which generate interrupt requests, as discussed in Chap. 7.

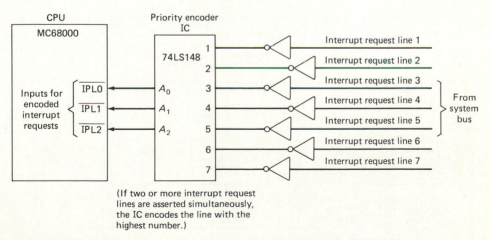

Figure 13.11 Use of multiple interrupt request lines with a Motorola 68000 CPU.

In some cases boards that generate interrupt requests signals cannot be set up to respond to the interrupt acknowledge signal with an interrupt vector code that is appropriate for the microprocessor used in the system. In that case the CPU could read a status register in each board that is capable of generating an interrupt in order to determine if an *interrupt request bit* is set. This is known as *polling*.

13.1.4 Bus Request Arbitration Lines

Bus request arbitration lines deal with situations where there is more than one *bus master* plugged into the backplane. A bus master is a board that initiates data transfers on the bus system. It therefore controls the address lines and many of the control lines in the bus. The bus master is normally the board which contains the CPU, but there are cases where there is more than one possible bus master. High-speed peripheral devices such as hard disks or high-speed analog-to-digital converters may produce data at such a rate that the CPU board cannot handle the job of storing the data. In this case the board controlling the high-speed peripheral device may have special addressing and bus control circuits that allow it to directly access a block of memory and store the data. This is an example of a *direct memory access*, or *DMA*, interface of the type discussed in Chap. 9.

In other situations the computational requirements may be such that more than one microprocessor is required. There may, therefore, be several microprocessor boards plugged into the backplane, all sharing system resources such as memory boards and I/O boards. These are called *multiprocessor* systems. A block diagram of a multiprocessor system is shown in Fig. 13.12. Each CPU board has its own private memory for storing programs and data. Thus the CPU can continue functioning even if it doesn't access to the system bus. CPU boards designed for multiprocessor applications are commonly equipped with a

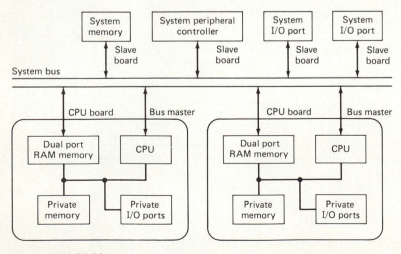

Figure 13.12 Multiprocessor system.

small amount of *dual port* RAM memory. This is memory which has two sets of address, data, and control lines. One set of lines is connected to the system bus while the other set is connected directly to the CPU on the board. Thus the memory can be accessed either from the system bus by another CPU board or by the on-board CPU. Circuitry is included to prevent simultaneous access. The dual port RAM memories are used as *mailboxes* to pass messages among the CPU boards.

Systems with multiple microprocessors or boards performing direct memory access have multiple bus masters, only one of which can have control of the system bus at any time. Control of the system bus must therefore be arbitrated among the various bus masters. The bus request arbitration system consists of lines which allow an inactive bus master to request control of the bus and lines which allow the active bus master to grant control of the bus. Some kind of protocol or logic is normally designed to determine priorities of requests. The active bus master uses the bus request arbitration system to determine whether to relinquish control of the bus. If the priority of the requesting board is high enough, the active bus master disconnects itself from the bus and issues a *bus grant* signal to allow the requesting board to take control of the bus.

13.1.5 Adapting Microprocessor Control Lines to Bus Control Lines

Many bus systems have evolved from buses designed by a single manufacturer for one type of microprocessor. In these cases the timing and number of control signals in the system bus were determined by the control signals produced by the microprocessor for which the bus was originally designed. Some of the most popular bus systems have evolved from systems originally designed around the control signals for the Intel 8080 microprocessor.

This orientation toward a particular microprocessor makes the design of boards containing other microprocessors more difficult than it might otherwise be. The problem is approached in two ways. In some bus systems, most notably the STD (IEEE P961) bus, some control lines are assigned to different control signals for different processors. This results in an inexpensive system but means that boards making use of the processor-specific control lines may not work with all other boards available for the system. The other approach is to rigorously specify the timing and meaning of all control signals and then use logic circuits on each microprocessor board to convert the control signals produced by the microprocessor to those required by the bus specification. This guarantees that all boards meeting the bus specification will work with any other board meeting the specification. However, the logic required to convert the control signals of some microprocessors to those required by the system bus may be complex and expensive.

13.1.6 Power-supply Lines

In a bus-based microcomputer system, the power-supply outputs are normally connected to power-distribution lines on the backplane bus. Each bus conductor and connector contact can carry only a limited amount of current, so it is common for several bus lines to be assigned to a single power-supply output. The voltages distributed on the bus are normally regulated at the power-supply outputs. However, the S-100 (IEEE S696) bus distributes loosely regulated voltages on the bus and relies on regulator circuits installed on each board to regulate the on-board voltages. This reduces power-supply costs and eliminates problems due to voltage drops along bus conductors and across connectors but also results in higher power consumption and more problems with cooling the boards. The trend in bus design is to reduce the number of different voltages distributed on the bus, since most digital circuits can now operate from a single voltage supply. Boards that require other voltages can generate them using compact on-board dc-dc converters.

13.1.7 Bus Electrical Characteristics

Besides standardizing the assignment of signals and functions to bus conductors, a bus standard also specifies the electrical characteristics of the bus.

Bus-driver circuits   Usually the signal levels are defined to be the same as for TTL digital logic. However, the circuits that place signals on the bus (*bus drivers*) have characteristics that are different from those of standard TTL logic gates in two respects:

1. The current-handling capabilities of bus drivers are much higher than those of standard logic gates in order to drive a large number of loads attached to the bus.
2. The outputs of the drivers are designed to avoid *bus contention*.

   Bus contention occurs when standard logic gates are connected to a common line on a bus, because the output of each gate is basically a set of switches which attempt to connect the line either to the supply voltage or to ground depending on their logical state. If one gate attached to the line is attempting to connect the line to the supply voltage while another gate is attempting to connect the line to ground, a high current will flow through the two gates, as shown in Fig. 13.13a, and damage may result. Even if no circuit damage occurs, the voltage of the line is indeterminate and the logical state of the line is therefore also indeterminate. This is bus contention.

Tristate drivers   One solution to this problem is to design the bus drivers so that inactive drivers can effectively be disconnected from the bus. *Tristate* bus drivers, which are discussed in App. B, are examples of this approach. As shown in Fig. 13.13b, an inactive tristate bus driver keeps both its switches open so that there is no connection to the bus line. A tristate bus driver must have an extra input which is used to activate it when it must put a signal on the bus. Bus contention can still occur if two drivers are inadvertently activated at the same time.

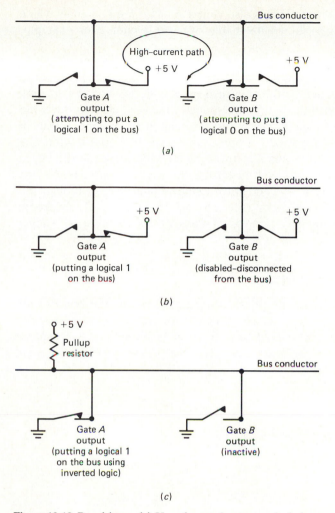

Figure 13.13 Bus drivers. (*a*) Use of normal gates results in bus contention. (*b*) Bus drivers based on tristate output circuits. (*c*) Bus drivers based on open-collector output circuits.

Open-collector drivers The other common solution is to use *open-collector* drivers as shown in Fig. 13.13*c*. These drivers use open-collector output circuits of the type discussed in App. B. A *pull-up resistor* connects the bus to the positive supply voltage so that the bus will be at the supply voltage if all the open-collector bus drivers are in the open-circuit condition. In an open-collector bus system, all the inactive bus drivers are in the logical state that results in an open-circuit condition. The active bus driver then "pulls" the bus line up and down depending on its logical state. If it connects the bus line to ground, the supply voltage will be dropped across the pull-up resistors and so the voltage on the line will be close to zero. Because the inactive bus condition results in the bus voltage being high, it is common for open-collector bus systems to have signals which are *active low*. This means that the signal level is low when the signal is asserted. This is also called *inverted logic*. Signals which are active low are usually written with a line over their name, for example, $\overline{\text{DATA READ}}$.

Maximum data rate The maximum data rate and clock frequencies allowed on the bus may also be defined by the electrical specifications. At fast switching speeds the distributed capacitance and inductance of the conductors affects the switching performance. In some cases fast switching transitions result in oscillations of the signal level called *ringing*, as shown in Fig. 13.14. If the amplitude of the ringing is large enough, the logic may interpret the ringing as valid logical transitions. This will result in errors. *Bus termination networks* can be added to the backplane or plugged into a spare connector to reduce the severity of ringing. These networks (see Fig. 13.15) consist of electronic components which damp the electrical oscillations or clamp them to a level which will not cause errors on the bus.

13.1.8 Bus Physical Characteristics

A bus standard will also define the physical characteristics of the boards and connectors to make sure that all boards designed for the bus system will fit. Board dimensions and connector types for different bus systems have usually been different. This is inconvenient for the user, since a new mechanical design

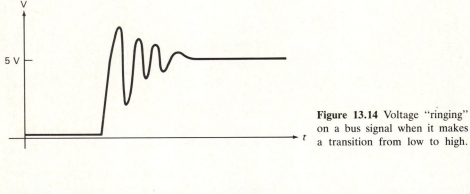

Figure 13.14 Voltage "ringing" on a bus signal when it makes a transition from low to high.

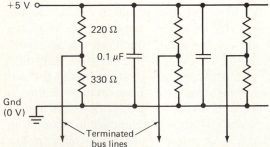

Figure 13.15 Bus termination network.

has to be performed every time a new bus system is used. However, there is now a trend toward standardization using the DIN 41612 pin and socket connector and a family of standardized metric card sizes—the Eurocard family. New bus systems are being designed using these standards, and some old bus systems are being redesigned to incorporate the connector- and card-size standards.

13.2 BOARDS FOR BUS SYSTEMS

Boards for bus systems can be classed into three broad categories:

1. CPU boards
2. Memory boards
3. Peripheral controller and I/O boards

13.2.1 CPU Boards

CPU boards contain the microprocessor, the system clock, and the circuits to generate the bus control signals. Advances in semiconductor technology allow more functions to be placed in an integrated circuit every year. As a result, many CPU boards have become microcomputer boards with memory and basic I/O facilities on the same board as the microprocessor (see Fig. 13.16). The

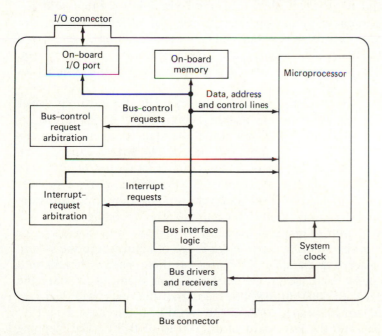

Figure 13.16 Typical CPU board.

system bus is then used to expand the memory system and to add specialized peripheral device controllers and I/O ports. *Single-board* microcomputers are available that do not have any interface to a bus system. These are inexpensive since they do not require the bus interface circuitry and can be used without a card cage and mother board, but they are difficult to expand.

13.2.2 Memory Boards

Memory boards are used to store programs and data. EPROM and EEPROM (electrically programmable and electrically erasable and programmable read-only memory) memory boards are used to hold the program and constants in operational systems where the program is not subject to frequent change. Static and dynamic read/write memory (RAM) boards are available to store data and to store programs during system development. Low-power-consumption CMOS RAM boards are available with rechargeable batteries on the board. These RAM boards will retain their data even when the system is turned off or the power fails.

A popular form of memory board is one that uses static RAM, EPROM, and EEPROM memory devices which are internally organized so that each addressable location contains 1 byte of data. These are commonly called *byte-wide* memory devices. Different chips of this type normally have similar pin assignments and control signals. The designer can choose from a number of combinations of RAM and PROM chips on the same board since the chips are interchangeable and are plugged into sockets on the board rather than soldered in place. Any differences in the pin assignments for different devices are accommodated by on-board jumpers that can be moved to set up a socket for a particular type of memory chip. One standard memory board can thus be used for a variety of system configurations with only the type and number of memory chips on the board changing.

Besides the memory circuits, the memory board will also have bus driver and receiver circuits to interface the memory circuits to the system bus. A board that contains dynamic RAM will usually contain the refresh circuits required by the RAM. However, some dynamic RAM boards expect to be refreshed by circuits on another board. It is wise to check on this point before buying a dynamic RAM board. A memory board will also have *address decoding* circuits to enable specific memory chips when they are addressed. These circuits were discussed briefly in Chap. 8. The decoding circuit will usually be equipped with jumpers or switches so that the user can assign or *map* the memory on the board to a particular range of addresses in the system memory address space. For example, if the user has three memory boards with 16K bytes of memory each, he or she does not want them all to share the address range from 0X0000 to 0X3FFF. Instead the user will want to assign one board to the range 0X0000 to 0X3FFF, the next board to 0X4000 to 0X7FFF, and the last board to 0X8000 to 0XBFFF. The mapping facilities on the memory board allow the user to perform these assignments. Figure 13.17 is a block diagram of a typical memory board.

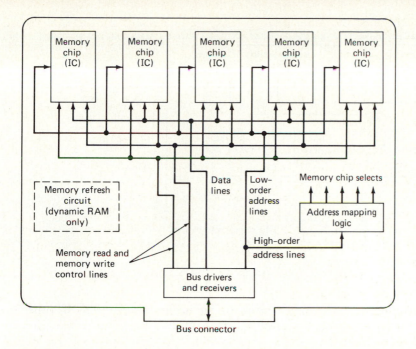

Figure 13.17 Typical memory board.

13.2.3 Peripheral Device Controller and I/O Boards

Peripheral device controller boards control systems that are closely associated with the computer system. Examples include floppy-disk drive controllers, CRT terminal controllers, and real-time clocks. I/O boards act as interfaces between the computer system and the outside world. Examples include analog-to-digital converter boards, serial data receiver/transmitter boards, and general-purpose digital input/output boards. As shown in Fig. 13.18, these boards normally consist of

1. Interface circuits to connect the board to the bus
2. An interface circuit between the board and the controlled device (e.g., floppy-disk drive) or outside world
3. Circuits to perform the interface and control functions
4. Data, control, and status registers
5. Address decoder circuits that decode addresses on the bus and select the appropriate function or register

Each board will usually have several addresses in the memory or I/O address space. These addresses refer to the data, control, or status registers on the board or simply initiate a function when that address is asserted. For example, an A/D converter board could be commanded to start a conversion by writing to a particular address. It is the act of selecting the address that

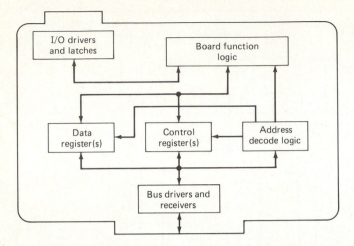

Figure 13.18 Typical I/O board.

initiates the conversion; the data written to the address is ignored. In other A/D boards a specific bit in a control register will have to be set to initiate a conversion.

As in the case of memory boards, the address decoder circuits will usually have jumpers or switches to allow the user to change the addresses to which the board responds. This allows the designer to use more than one copy of a board in a system. It also allows I/O devices to be assigned to a different range of addresses than memory in systems where memory and I/O share the same address space.

13.3 DEVELOPMENT OF BACKPLANE BUS STANDARDS

Most backplane bus standards start as corporate standards defined by individual companies. One manufacturer or a small group of manufacturers supply boards for the system. Normally the only microcomputers available for use on this kind of bus are ones manufactured by the company which defined the bus. Many bus systems remain as corporate standards. Others are abandoned because they do not succeed in attracting a reasonable share of the market. However, some systems become very popular and other companies begin to manufacture boards for these systems. Boards using a variety of microcomputers are introduced, as are specialized peripheral controller and I/O boards. These bus systems attain the status of informal de facto standards.

If an informal bus standard remains popular, it may become a basis for a formal standard developed by a standards-making organization. In North America, the standards-making organization involved in setting standards for microcomputer backplane buses is the Institute of Electrical and Electronic

Engineers (IEEE). The committee setting a formal standard must deal with the following issues:

1. Ambiguities and flaws in the original bus specification
2. Incompatible versions of the standard used by different organizations
3. Proposed extensions to the original specification to add desirable features or to allow the use of wider data and address paths

The committee will be limited in the type of changes they can make, because boards designed to the formal standard must still work with most boards designed to the informal standard. If they do not, then the formal standard is unlikely to be accepted by manufacturers and users.

13.4 SOME IMPORTANT BACKPLANE BUS STANDARDS

13.4.1 STD (IEEE P961) Bus

The STD bus is based on a 56-conductor backplane which supports synchronous transfers of 8-bit-wide data on a bidirectional data bus. The address bus is 16 bits wide, allowing direct access to 64K bytes of memory. The address bus is also used to transfer addresses in the separate I/O address space used by some microprocessors. There are 22 control lines which are used to control data transfers between the CPU and memory locations or I/O registers as well as providing some simple facilities for interrupts and bus request arbitration. In addition, some of the control lines are dedicated to CPU-dependent control signals which are required by some I/O and peripheral controller boards. The remaining 10 lines in the STD bus are devoted to the power bus. They carry the following voltages: +5 V, −5 V, +12 V, and −12 V.

Because the STD bus has control lines which are dedicated to CPU-dependent control signals, not all STD cards will work in every system. Fortunately, a terminology has developed which indicates which cards will work together. A card which is specified to be *STD-bus-compatible* will usually work with any CPU available on the STD bus. A card that will work only with a specific CPU will be specified to be compatible with the particular subset of the STD bus peculiar to that CPU. For example, boards that work only in STD bus systems using the Zilog Z80 microprocessor are said to be *STD-Z80-compatible*. Similarly, boards that work only with STD bus systems using the Motorola 6809 microprocessor are said to be *STD-6809-compatible*. Despite the general acceptance of this terminology, it is still possible to find STD bus boards advertised as STD-bus compatible which in fact work only with certain microprocessors. The best defense is to read the board specifications carefully. Any limitations on which microprocessors the board will work with will usually be given somewhere in the fine print.

STD bus boards use a 4.5-in × 6.5-in card size and use an edge connector to mate with the backplane. A proposal for an alternative format, using the slightly smaller single-width Eurocard with the standard DIN 41612 pin and socket connector, has been made to the IEEE committee which is developing the formal STD bus standard. The relatively small size of the STD boards has some advantages. Each board typically performs only one function, so one is not forced to purchase more functions than required. The small board is easier to cool than larger boards and will fit into smaller, cheaper enclosures. The small size of the board, plus the simple bus interface logic, combine to make STD bus boards relatively inexpensive when compared with boards for other bus systems.

CPU boards based on most of the popular 8-bit microprocessors are available for the STD bus. Microprocessor boards using the newer chips with 16- or 32-bit internal data paths and an 8-bit external data path (e.g., Intel 8088, Motorola 68008, TI 9995) are becoming available. The STD bus has a control line dedicated to memory expansion, so the expanded address space of these newer microcomputers can be used to some extent. In addition, an informal standard has developed for STD bus boards using the Intel 8088 microprocessor that allows the CPU board to access up to 1 Mbyte of memory by first placing the 16 low-order bits of the address on the address bus and then placing the 4 high-order bits on the address bus. Special memory boards that latch the address bus are required.

The STD bus is popular in industrial control and other applications where a relatively simple microcomputer system is required. A wide variety of boards designed to interface with industrial processes are available. In addition, boards using CMOS components for wide-temperature-range operation and good electrical noise immunity are being introduced which should make the system even more attractive for industrial applications. The STD bus, on the other hand, is not well-suited to very high performance applications. For example, the STD bus would not be the best choice in an application requiring the use of one or more high-speed 16-bit microprocessors on the bus, because of its narrow 8-bit data bus, its limited bus arbitration capability, and limitations on bus throughput due to the simple bus interface circuits.

13.4.2 IBM Personal Computer Bus

The system expansion bus (called the PC bus in this section) originally developed for the IBM personal computer (IBM PC) is a good example of how a bus system can develop far beyond its designers' original intentions. The IBM personal computer was designed as a personal home and office computer which would run data-processing programs such as word processors, spreadsheets, and business graphics programs. The computer is based on a mother board which contains an Intel-8088-based microcomputer system. The mother board has a system bus with five connectors. These were originally intended to accommodate boards for I/O interfaces to printers, video terminals, and floppy-disk drives and boards for memory expansion.

However, the system rapidly developed beyond this original data-processing orientation. Several companies began producing data-acquisition boards that contained analog-to-digital converters. Other companies introduced interface boards that allowed the IBM PC to use the IEEE-488 interface bus, which is used to interconnect laboratory instruments (see Chap. 10 for more on the IEEE-488 bus). Scientists and engineers began to move the system into the laboratory and use it to acquire and process data from experiments. Then output boards that could control relays and motors were introduced. The personal computer began to be used in closed-loop applications where it actually controlled experiments or processes.

In order to accommodate these increasingly complex real-time applications, manufacturers introduced add-on boxes that increased the PC bus capacity from its original five positions. Other manufacturers have produced IBM-PC-compatible systems where the microcomputer is taken off the mother board and placed on a plug-in card which is plugged into the PC bus backplane like an I/O or memory card. In addition, special industrial grade systems based on the PC bus are available. These use rugged rack-mountable enclosures, filtered-air-flow cooling systems, and shock-mounted disk drives. We can see that the system has migrated from its original data-processing applications in an office environment all the way to real-time control applications in a factory environment.

The PC bus is based on a 62-conductor backplane. The data bus is 8 bits wide and bidirectional. The address bus is 20 bits wide to accommodate the 20 address lines from the Intel 8088. The control lines are oriented to the control signals of the Intel 8088 microcomputer. Data transfer is synchronous and follows the timing required by the 8088. A line to indicate that a data-transfer operation is to an I/O port is included in order to support the separate I/O address space used by the 8088.

The bus supports six interrupt request lines. These are fed to an interrupt controller IC (Intel 8259A) on the mother board. The interrupt controller provides priority arbitration among the interrupt signals and generates the interrupt vector codes. If two interrupt signals are received simultaneously, the 8259 will first send a single interrupt request signal to the 8088 microprocessor. When the 8088 responds with an interrupt acknowledge signal, the 8259 returns the interrupt vector code of the interrupt source with the higher priority. The interrupt vector codes for the interrupt sources must be written into registers in the 8259 by the 8088 during system initialization. In the IBM PC this is done by a program stored in ROM which runs when power is applied to the system.

The PC bus also supports multiple bus masters to a limited extent. Three DMA request lines and three DMA acknowledge lines are included on the bus for boards which wish to take control of the bus to perform a DMA data transfer. Some of the lines are usually used by the interfaces to floppy-disk and hard-disk drives. There may therefore be only one line left for other boards. Arbitration of requests for the bus on the three DMA request lines is handled by a DMA controller IC (Intel 8237-5) on the mother board. A priority-based arbitration scheme is used in which the highest priority line making a service request will be served first.

There are eight lines devoted to the power supply on the PC bus. They carry $+5$ V, -5 V, $+12$ V, and -12 V.

Boards used on the PC bus originally had to fit inside the IBM PC's cabinet. This constraint determined their size and shape. A board may be a maximum of 33.5 cm long by 10.6 cm high. Some boards may be smaller. They all have a 62-position edge connector to plug into the system bus. The original IBM PC and most of its imitators do not supply a card cage to support the boards plugged into the mother board. Usually they are supported only at one side by a card guide or a bracket. As a result, the system is more susceptible to shock and vibration than are systems designed for industrial use from the outset.

As we have mentioned, the IBM PC and compatible computers are popular choices for data-acquisition applications. A large number of boards and complete data-acquisition systems are available for this task. Many of these come with the signal conditioning circuits required for transducers such as thermocouples and strain gauges. In addition, the software drivers to control the data-acquisition system are often included along with software interface functions for common high-level languages such as BASIC, FORTRAN, and C. The user can access the data-acquisition system by calling these functions from a high-level language program. In many cases, the user can perform all the required data-analysis functions with commercially available statistical analysis or spreadsheet programs and can use commercial available graphics programs to plot the data. The amount of programming required can therefore often be very small.

The PC bus can also be used in real-time control applications. Since the target system and the development system are the same, system development and debugging is simplified considerably. In addition, development software for the IBM PC and similar computers is plentiful and reasonably inexpensive. However, the potential user must make sure that the PC bus computer chosen will be rugged enough for the environment in which it will be installed.

13.4.3 S-100 (IEEE S696) Bus

The S-100 bus represents a step upward in capability from the STD bus and the PC bus. The bus is based on a 100-conductor backplane. The data bus is 16 bits wide and can be used in two modes. It can be used as two unidirectional 8-bit data buses (data in and data out) in order to maintain compatibility with the original S-100 specification or it can be used as a bidirectional 16-bit bus in new designs using 16-bit microprocessors. Data transfers on the bus can be controlled by handshaking signals, so the bus is asynchronous. The address bus, which was originally 16 bits wide, has been expanded to 24 bits. Again, both an I/O and memory address space are supported.

Unlike the STD bus, the control lines on the S-100 bus are the same for all

CPUs. Also the timing of control signals and bus transactions is more rigorously specified than that for the STD bus. This complicates the interface circuits on some CPU boards, since CPU control signals must be converted to the control signals defined for the S-100 bus, but it does allow any S-100 board to work with almost any other S-100 board.

The control bus has well-defined and powerful facilities for interrupt requests and bus arbitration. Nine lines are supplied for interrupt requests. This allows the source of an interrupt to be identified without having to poll every board that could be an interrupt source. Up to 16 *temporary bus masters* can exist on the bus along with a *permanent bus master*. Each temporary master board is assigned a priority. The bus arbitration logic will decide among temporary masters simultaneously requesting bus control on the basis of their priorities.

The S-100 power bus distributes loosely regulated power to the boards. Each board has voltage regulators on board to control the voltage supplied to its circuits. The voltages distributed on the bus are: $+8$ V, $+16$ V, and -16 V.

S-100 boards have dimensions of 5 in by 10 in and use edge connectors. The size of these boards allows multiple functions to be supplied on a single board. For example, S-100 single-board computers incorporating a microprocessor, 64K byte of RAM, a floppy-disk controller, and I/O ports are available.

Microcomputer systems based on S-100 boards are popular as business and personal computers. They are also popular as engineering workstations and microcomputer development systems. The S-100 bus was originally designed around the Intel 8080 microprocessor, and this family of microprocessors (8080, 8085, Z80) is still commonly used. However, other 8-bit microprocessors are supported. CPU boards on most of the 16- and 32-bit microprocessors are also available.

S-100 boards have been less popular in industrial control applications, perhaps because of their reputation as "hobby boards." In fact, the S-100 bus is now more rigorously defined than some "industrial" bus systems and boards are available that have been tested for 200 h prior to shipment, so the prejudice against this system is not justified. A more-valid reason for its relative unpopularity in industrial applications is that the selection of interface boards is more limited than that for some other buses.

13.4.4 Multibus (IEEE S796)

The Multibus backplane bus has about the same capabilities as the S-100 bus. It is based on a backplane consisting of two separate sets of conductors. One set, consisting of 86 lines, carries most of the bus signals. The other set, consisting of 60 lines, is partially assigned to the remainder of the bus signals. The rest of this set is undefined and can be used for special applications. The data bus is 16 bits wide and bidirectional. Data transfers on the bus are performed synchronously. The address bus is a maximum of 24 bits wide, but versions of the bus with 16 and 20 address lines are also defined. Separate address spaces for memory and I/O are supported by the bus standard.

The bus has eight interrupt requests lines so that interrupt requests can be identified without polling all the possible sources. Multiple bus masters are also supported by the bus, although arbitration of simultaneous bus requests is not as convenient as that on the S-100 bus.

Boards for the Multibus are relatively large, measuring 6.75 in by 12 in. Two edge connectors are used, one for the primary 86-conductor bus and another for the 60-conductor auxiliary bus. Because of its large size, each Multibus board normally contains several functions. A criticism sometimes leveled against the Multibus is that it is inflexible, because boards performing individual functions are difficult to purchase at a reasonable price.

The Multibus was originally developed by Intel Corporation but is now an IEEE standard. Intel, however, retains a copyright on the name. Intel supplied good support for the Multibus in terms of boards, engineering support, development systems, and distribution, all of which contributed to making the Multibus quite popular. Other manufacturers also began making Multibus boards, in some cases acting as second sources for Intel boards. This increased the popularity of the Multibus in industrial applications where reliability of supply is important. CPU boards for the Multibus were originally designed around the Intel microprocessors, but CPUs based on other processors (particularly 16-bit microprocessors) are now readily available.

Intel has recently introduced an upgraded version of the Multibus, called Multibus II. This uses the standard Eurocard board sizes and a standard DIN pin and socket connector. The address and data bus widths have been expanded to 32 bits and other performance improvements have been made. Boards for the original Multibus continue to be available.

13.4.5 VME (IEEE P1014) Bus

The VME bus is a relatively recent introduction by a group of semiconductor manufacturers (Motorola, Mostek, Signetics/Philips, and Thomson-CSF). Efforts are underway to develop an IEEE standard for the VME bus. The boards are based on the standardized Eurocard size and use the standard DIN pin and socket connector. The backplane bus has provision to support 32-bit data and address buses. The Eurocard boards are quite small, but the use of the pin and socket connector allows the use of wide data and address buses on the backplane.

The basic VME bus is defined with a 16-bit data bus and a 24-bit address bus on a single width (160 by 100 mm) Eurocard. The basic bus consists of 96 conductors. Bus expansion to 32 bits is provided by use of a double width (160 by 234 mm) Eurocard which has an extra connector. The conductors not required for the bus expansion are dedicated to user I/O functions. Both single-width and double-width cards can coexist on the same backplane as long as proper mechanical support is provided.

Bus transfers on the VME bus are asynchronous. Like the S-100 bus and the Multibus, the VME bus supports multiple interrupt request lines and multiple bus masters. The VME bus has considerable potential in high-performance multiprocessor systems. The user I/O lines on the second bus connector can be used as a dedicated I/O bus, freeing the VME bus for high-speed data transfers between memory and the various bus masters. In addition, an additional bus for interprocessor communication is being developed to allow processors on the bus to pass messages to each other without using the system bus.

At present, the number of boards available for the VME bus is limited. However, the VME bus is likely to be popular for high-performance microcomputer systems in the future, since it uses small boards yet supports very wide data and address paths.

The STD bus, IBM PC bus, S-100 bus, Multibus, and VME bus are by no means the only buses that have been defined. There are buses such as the Motorola Versabus, the DEC Q-bus, and the National Semiconductor CIMBUS which are basically dedicated to one manufacturer's microprocessors. Some other buses, such as the IEEE P896 bus (Futurebus), are in the process of development but do not yet have any manufacturers producing boards for them. Table 13.1 summarizes the characteristics of the five buses that have been described in this section. Figure 13.19 shows the relative sizes and form factors for the boards used in these bus systems.

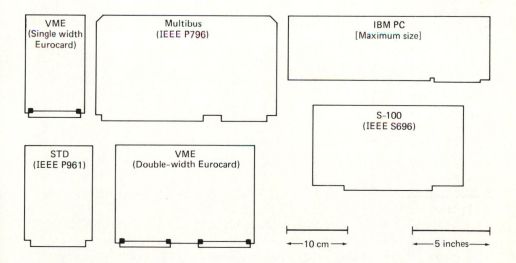

Figure 13.19 Relative sizes of boards for different bus systems.

Table 13.1 Bus comparison

	STD	IBM PC bus	S-100	Multibus	VME
Formal specification (as of early 1984)	Under development (IEEE P961)	None (IBM defines the standard)	IEEE S696	IEEE S796	Under development (IEEE 1014)
Approximate number of vendors	100	100	100	150	50
Microprocessors supported 8-bit data bus	Intel 8080, 8085, 8088; Zilog Z80; National NSC800; MOS Technology 6502; Motorola 6800, 6809, 68008; Texas Instruments 9995	Intel 8088	Intel 8080, 8085, 8088; Zilog Z80; National NSC800; Motorola 6800, 6809	Intel 8080, 8085, 8088, 80188; Zilog Z80; National NSC800; Motorola 6800, 6809 6809	Motorola 68000, 68010; Intel 80186, 80286; also Motorola 68020 and National 32032 (32 bit)
16-bit data bus	Intel 8086, 80186	Intel 8086, 80186	Intel 8086, 80186, 80286; Zilog Z8000; Motorola 68000, 68010; National 16032	Intel 8086, 80186, 80286; Zilog Z8000; Motorola 68010; National 16032	
Address bus width, bits	16 (plus one memory expansion line)	20	16 or 24	24	16, 24, or 32
Data bus width, bits	8	8	8 or 16	8 or 16	8, 16, or 32
Control of data transfer	Synchronous	Synchronous	Asynchronous	Asynchronous	Asynchronous
Theoretical data-transfer rate, Mbytes/s	1	1	6–12 (depending on data bus width)	5–10 (depending on data bus width)	6–24 (depending on data bus width)
Number of interrupt request lines	2	6	10	8	7
Support for multiple bus masters	Poor	Fair	Good	Good	Excellent
Bus voltages	+5, −5, +12, −12, and ground	+5, −5, +12, −12, and ground	+8, +16, −16, and ground (loosely regulated)	+5, −5, +12, −12, and ground	+5, +12, −12, and ground plus +5-V battery backup
Board dimensions, cm	11.4 × 16.5	33.5 × 10.6	13 × 25.4	17.1 × 30.5	16 × 10 or 16 × 23.4
Connector size and type	56-contact edge connector	62-position edge connector	100-contact edge connector	86- and 60-contact edge connectors	96-pin plug and socket connectors (two connectors for double Eurocard format)

13.5 SELECTING A BUS SYSTEM

Besides the five bus systems we have discussed, there are many other bus systems on the market. The designer is faced with difficult choices in choosing among them. The two primary criteria that should be used in choosing a bus system are to minimize risk and to match the bus performance to the performance required by the application.

Risk minimization implies choosing a bus whose basic specifications are clearly defined and are likely to remain unchanged in the future. This gives some guarantee that boards designed for the bus will work together and that boards, and computer systems based on them, won't be made obsolete by changes in the bus standard. Bus systems which are defined by a formal standard issued by an independent body such as the Institution of Electrical and Electronic Engineers (IEEE) provide the best guarantee of this kind of stability.

Even if a bus system is rigorously specified by an independent body, it may not be of much use if only a few vendors are manufacturing boards to be used on the bus. A large number of vendors generally means competitive prices and reduces the risk that manufacturing of boards for the bus system will suddenly stop. The number of different types of boards available for a particular bus system should also be a consideration unless the designer is certain that only a few different boards will be required for all potential applications. At the very least, the designer should verify that boards performing the functions that are certain to be required are available for the bus system under consideration. It is possible, of course, to custom-design boards for a particular bus system, but this should be done only if the potential savings in production costs offset the added design costs or if the function that must be performed by the board is so unusual that no such board is commercially available.

Matching the performance of the bus system to the performance requirements of the application is important because the cost of the boards for a bus system is directly related to the bus performance. The STD bus has a narrow data path, a limited memory addressing range, and very limited support for multiple bus masters, and transfers data between boards relatively slowly. The VME bus, on the other hand, has a wider data path, a large memory addressing range, and excellent support for multiple bus masters, and can transfer data between boards at least six times as fast as the STD bus. However, boards for the VME bus are normally about three times as expensive as are boards for the STD bus. This is because they usually contain high-performance components that can take advantage of the bus performance and because the logic to interface to the faster and more complex VME bus must also be faster and more complex.

In order to minimize costs, therefore, the designer should carefully analyze the performance requirements of the computer system to be designed and then choose a bus system to match these requirements. In many cases the system may start off with relatively modest performance but is designed to be expanded to a higher-performance system. In these cases, of course, the bus system must be chosen with an eye to this future expansion.

In some situations, particular mechanical or environmental constraints will dictate the choice of a particular bus. For instance, size constraints may require that the designer choose a bus system whose cards are smaller than a certain size. In other cases the choice of bus will be dictated by the availability of boards that operate over an extended temperature range or that have very low power consumption.

SUMMARY

Board-based microcomputer systems, which consist of printed circuit boards plugged into a backplane or mother board, represent a convenient and flexible approach to microcomputer hardware design. In most cases, the microcomputer can be assembled from commercially available boards, saving on development time and cost.

Each board usually represents a microcomputer subsystem such as the CPU, memory, or I/O interface. The board is plugged into a backplane bus which carries the signals between boards. The interconnecting lines on the backplane include

1. Address lines
2. Data lines
3. Control lines to control data transfers, carry interrupt signals, and arbitrate requests for control of the bus
4. Power-distribution lines

In order to allow boards from manufacturers to plug into the same bus and operate together successfully, a bus standard must be defined. This may be developed by a manufacturer or group of manufacturers or by a standards-making organization such as the IEEE. Examples of bus standards described in this chapter include the STD bus, the IBM PC bus, the S-100 bus, the Multibus, and the VME bus.

Two primary criteria should be used in choosing a bus system. First, the risk that a bus will become obsolete or that the bus standard will suddenly change should be minimized by choosing a well-established bus with a formally defined standard and a large number of manufacturers supporting the bus. Second, the performance of the bus should be matched to the performance requirement of the application, since the cost of boards is directly related to the bus performance.

EXERCISES

13.1 You are choosing I/O boards for a synchronous bus. The transfer of data is controlled with the timing and control signals shown in Figs. 13.7 and 13.8. The system clock has a frequency of 4 MHz.

1. A board containing input ports has the following timing specifications: Valid data is placed on the data lines a maximum of 700 ns after an address is placed on the system bus and a maximum of 400 ns after the DATA READ signal is asserted.
2. A board containing output ports has the following timing specifications: Data will be read from the bus when the DATA WRITE signal goes low. The address must have been on the system bus for a minimum of 750 ns prior to the DATA WRITE signal's going low. The data must have been on the system bus a minimum of 250 ns prior to the DATA WRITE signal's going low.

Will these two boards work properly in the system? If one or both of them will not work, explain why.

13.2 In the asynchronous bus system shown in Figs. 13.9 and 13.10, do we really need two handshaking lines? Explain.

13.3 What happens on an asynchronous bus if the bus master attempts to transfer data to a nonexistent address (say the board that responds to that address has been removed)? How could we deal with this situation?

13.4 A microcomputer system based on a system bus that has only one interrupt request line has eight boards that can generate an interrupt. The identity of the interrupting board must be determined by polling the status registers of the boards capable of generating an interrupt after an interrupt request is received. Each board that can generate an interrupt is assigned a priority number. In the event of two or more boards' simultaneously generating an interrupt, the interrupt service routine for the board with the highest priority number must be executed first. Figure 13.20 shows the (memory-mapped) addresses of the status registers for the eight boards, the bit in each status register which is set if an interrupt is requested, and the priority number for each board. Write a C function which does the following: Poll the status registers and find the highest priority board that has requested an interrupt. Reset the interrupt request bit for this board. The other bits in the status register must remain unchanged. Return the priority number of this board.

13.5. Given the information presented in this chapter on the STD bus, S-100 bus, Multibus, and the VME bus, which bus would you choose for the following applications:

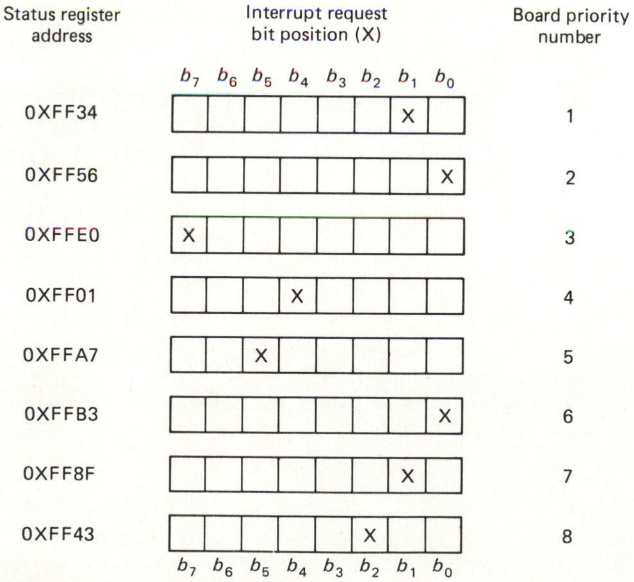

Status register address	Interrupt request bit position (X)	Board priority number

	b_7 b_6 b_5 b_4 b_3 b_2 b_1 b_0	
0XFF34	X at b_1	1
0XFF56	X at b_0	2
0XFFE0	X at b_7	3
0XFF01	X at b_5	4
0XFFA7	X at b_6	5
0XFFB3	X at b_0	6
0XFF8F	X at b_1	7
0XFF43	X at b_2	8

b_7 b_6 b_5 b_4 b_3 b_2 b_1 b_0

Figure 13.20 Status registers.

1. You are designing a portable instrument to be used out of doors. You wish to use the 8-bit CMOS NSC800 microprocessor because of its low power consumption. The system will use less than 32K bytes of memory. There is little chance that this system will ever be expanded to use more memory or a 16-bit microprocessor.

2. You are supervising the design of a new microcomputer-based medical imaging system. Presently you are planning to use the Motorola 68000 microprocessor. It is possible that you may have to go to a multiprocessor system to get the required computational power. In the future it is likely that you will want to upgrade to the Motorola 68020 microprocessor which has a 32-bit data bus.

In each case present your reasoning for your choice of bus system.

13.6 A sales representative from Silicon Gulch Microsystems has heard that you are trying to decide on a bus system to be used in the majority of your company's products. After an exhausting morning of flip charts and overhead transparencies and a three-martini lunch, she convinces you that the new Superbus meets all your technical requirements and that Silicon Gulch Microsystems manufactures some very nice boards for the Superbus. Despite your gratitude for the free lunch, you know that you still have to ask some tough questions before making a decision. What are they?

13.7 The CPU boards in a multiprocessor system could use portions of the system memory for the mailboxes they use to communicate among each other. If this is so, why are dual port RAM memories on each CPU board so popular?

BIBLIOGRAPHY

General

Borrill, P. L.: "Microprocessor Bus Structures and Standards," *IEEE Micro*, vol. 1, no. 1, February 1981, pp. 84–95.

STD bus

Elmore, T.: "Standard Bus for 8-bit Microprocessor Systems," *Microprocessors and Microsystems*, vol. 6, no. 9, November 1982, pp. 455–465.
Ironoak Company: *The STD Bus Buyer's Guide*, 3239 Caminito Ameca, La Jolla, CA 92037, 1983 (updated regularly).
Pro-Log Corp.: *STD Bus Technical Manual and Product Catalog*, 2411 Garden Road, Monterey, CA 93940, August 1982.
Titus, C., J. Titus, and D. Larsen: *STD Bus Interfacing*, Howard W. Sams & Co., Indianapolis, IN, 1982.

IBM personal computer bus

International Business Machines Corporation: *IBM Personal Computer Hardware Technical Reference Manual*, Boca Raton, FL (revised periodically).
Eggebrecht, L.: *Interfacing to the IBM Personal Computer*, Howard W. Sams & Co., Indianapolis, IN, 1983.

S-100 Bus (IEEE S696)

IEEE Standard 696 Microprocessor Bus (*S-100*), available (Computer Society order number 927) from IEEE Computer Society Order Dept., P.O. Box 80452, Worldway Postal Centre, Los Angeles, CA 90080.

Ironoak Company: *The S-100 Bus Buyer's Guide,* 3239 Caminito Ameca, La Jolla, CA 92037, 1983 (updated regularly).

Libes, S., and M. Garetz; *Interfacing to S-100/IEEE-696 Microcomputers*, McGraw-Hill, New York, 1982.

Multibus (IEEE 796)

IEEE Standard 796 Microcomputer System Bus (*Multibus*), available (Computer Society order number 928) from IEEE Computer Society Order Dept., P.O. Box 80452, Worldway Postal Centre, Los Angeles, CA 90080.

Intel Corp.: "Intel Multibus Interfacing," applications note AP-22A, Santa Clara, CA.

Ironoak Company: *The Multibus Buyer's Guide*, 3239 Caminito Ameca, La Jolla, CA 92037, 1983 (updated regularly).

VME Bus

Mostek Corp, Motorola Inc., and Signetics/Philips: *VME Bus Specification Manual*, October 1981.

Signetics Corp.: *VMEbus Compatible Products Directory*, P.O. Box 409, Sunnyvale, CA 94086, July 1983.

EXAMPLE OF A BOARD-LEVEL
SYSTEM DESIGN

In this chapter we bring together the material presented in the previous chapters to design a microcomputer-based system. Working from a description of the desired system and a simple requirements document, a system design is developed using the design techniques described in Chap. 4. Then the selection of the hardware components of the system is discussed. The hardware design is done at the board level using boards for the STD bus described in the previous chapter. Finally, we cover the design of some of the software modules in the system. Most software will be written in the C programming language, but some modules will be written in the assembly language of the microprocessor used in this system.

14.1 A TORQUE WRENCH FOR INDUSTRIAL ROBOTS

The real-time system to be designed is a controller for an automatic torque wrench. As shown in Fig. 14.1, this torque wrench is mounted on the end of a robot manipulator arm. The wrench can grip, release, and rotate the nuts it handles. The wrench is equipped with a sensor to measure the torque applied while it is fastening a nut to a bolt or stud. The purpose of the controller is to control the operation of the torque wrench and coordinate it with the operation of the robot arm so that a set of nuts can be fastened at the proper torque in an automated assembly operation.

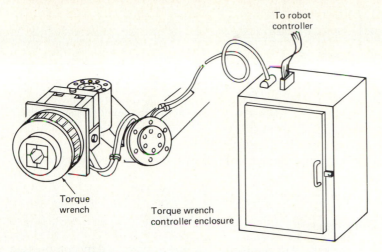

To robot
controller

Torque
wrench

Torque wrench
controller enclosure

Figure 14.1 Torque-wrench controller for an industrial controller.

14.1.1 System Operating Cycle

The sequence for attaching a fastener begins with the robot arm moving the wrench to a pallet containing the nuts. The wrench's open socket is positioned onto a nut and the robot-arm controller sends a message to the torque-wrench controller to grip the nut. The torque-wrench controller causes the socket to close in order to grip the nut and sends an acknowledgement message back to the robot-arm controller. The robot arm then moves the wrench to a stud or bolt on the item being assembled and positions the wrench so that the nut is correctly aligned with the stud or bolt. A message is then sent from the robot-arm controller to the torque-wrench controller to tighten the nut until a specified torque is reached. The torque-wrench controller starts the wrench turning and monitors the torque feedback signal from the wrench. Once the proper torque has been reached, the torque-wrench controller stops the rotation of the wrench and sends an acknowledgement message to the robot-arm controller.

The torque-wrench controller can detect some faults that may occur during the fastening operation. For example, if the nut jams because of misalignment of the threads or a defect in the threads, the torque will rise more rapidly than it would in a normal fastening operation. If the threads are stripped or the nut has been so badly misaligned that it can't be threaded onto the stud or bolt, the torque will not rise to the proper value. If the controller detects a fault in the tightening operation, it sends a fault message to the robot-arm controller, stops the rotation of the wrench, and releases the wrench's grip on the nut.

When the robot-arm controller receives a message that the fastening operation has been completed (either successfully or unsuccessfully), it moves the arm back to the pallet containing the nuts to begin another fastening operation. If the fastening operation just completed is unsuccessful, the

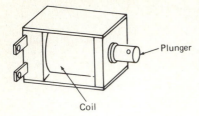

Figure 14.2 Solenoid used to control pneumatic valve.

robot-arm controller activates an indicator to alert a human operator or, alternatively, sends a message about the fault to the computer supervising the entire assembly operation.

14.1.2 System Components

Like many industrial power tools, the torque wrench is powered by compressed air. There are two pneumatic circuits in the wrench. One operates a plunger which causes the wrench's socket to close and open slightly so that nuts can be gripped and released. The second circuit operates a pneumatic motor that rotates the socket so that the fastening operation can be performed. The torque-wrench controller controls the flow of air in these circuits via *solenoid-activated valves*. A solenoid (Fig. 14.2) is similar to the electromechanical relays discussed in Chap. 12 except that the coil is used to attract a plunger rather than an electrical contact. When current is passed through the coil, the plunger is drawn into the coil. When the current is turned off, the plunger is drawn out of the coil by a spring. In a solenoid-activated valve the plunger is used to open and close the valve.

The torque wrench uses a strain gauge bridge (see Chap. 11) to sense the torque applied to the nut. The bridge circuit contains resistors which change resistance when slightly deformed. These resistors are attached to the wrench and cause the bridge to produce an electrical signal proportional to the torque applied by the wrench. The signal has an amplitude of only a few millivolts which must be amplified to be useful. The manufacturer of the strain gauge bridge supplies a signal-conditioning unit for the bridge which supplies the excitation voltage for the bridge and amplifies the bridge signal to an output level suitable for use with analog-to-digital converters.

The torque-wrench controller communicates with the robot-arm controller via an RS-232 *serial interface* of the type discussed in Chap. 10. The two controllers exchange messages consisting of strings of ASCII character codes.

14.2 SYSTEM SPECIFICATIONS

In order to design the torque-wrench controller we need detailed specifications on the nature of the interfaces between the torque-wrench controller and the torque wrench and between the torque-wrench controller and the robot-arm controller. We also need detailed specifications on the actions to be performed by the torque-wrench controller. These specifications are developed through consultation with the designers of the torque wrench, the designers or programmers of the robot arm, and the manufacturing engineers who plan to use the automatic torque wrench in assembly operations. The specifications developed are recorded in the system requirements document which is then used as a reference during the remainder of the design process.

A condensed version of the requirements document is presented here. Only the portions applicable to the aspects of the design discussed in this chapter are included. The document should be read in conjunction with the system block diagram (Fig. 14.3).

Torque-Wrench Controller Requirements Document (Condensed)

A. Interfaces

    A.1. Solenoid valve interface

        There are two solenoid valves to be controlled. Each valve is either completely open or completely closed. It takes approximately 50 ms for a valve to open or close completely. The solenoid coil current must be greater than 500 mA in order to completely open a valve and must be less than 1 A to avoid overheating the coil. A valve is completely closed when the solenoid current is under 50 mA. The coil resistance is approximately 20 Ω. The manufacturer of the solenoid

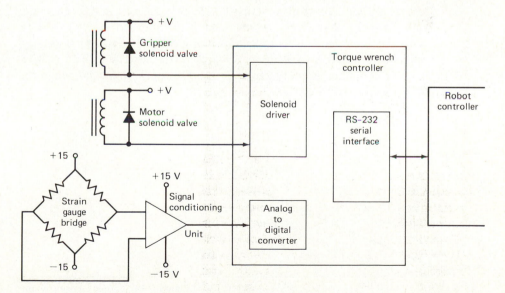

Figure 14.3 System block diagram for torque-wrench controller.

valve recommends that a free-wheeling diode (see Chap. 12) be placed in parallel with the solenoid coil in order to prevent high transient voltages when the coil current is switched off.

A.2. The torque sensor consists of the strain gauge bridge circuit installed in the torque wrench and the signal-conditioning unit. The signal conditioning unit is to be installed in the torque-wrench controller cabinet and must be supplied with the following voltages:

$$+15\,\text{V} \pm 5 \text{ percent at } 20\,\text{mA maximum}$$
$$-15\,\text{V} \pm 5 \text{ percent at } 20\,\text{mA maximum}$$

The torque sensor is linear and has a gain (defined as volts per unit of torque) of $50\,\text{mV/N}\cdot\text{m}$ (newton-meter). Thus a torque of $100\,\text{N}\cdot\text{m}$ will produce a sensor output of 5 V. Although the sensor output is linear, it is subject to an *offset voltage*, as shown in Fig. 14.4. The offset voltage drifts with temperature and time but can be adjusted to remain positive and within 0.5 V of zero.

A.3. Robot-arm controller serial interface characteristics
The specifications for the interface are listed in Table 14.1. This seven-line interface is a commonly used subset of the RS-232-C standard when two pieces of equipment are to be directly interfaced without a modem.

The messages passed between the robot-arm controller and the torque-wrench controller consist of strings of ASCII-encoded alphanumeric characters. A message is terminated by the ASCII "newline" character. This is a nonprinting control code which usually causes a terminal or printer to advance to the next line. The ASCII newline character is represented in the C programming language by the character constant '\n'.

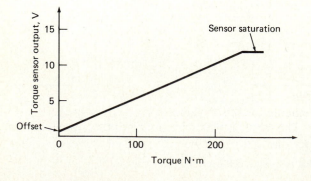

Figure 14.4 Torque sensor imput vs. output characteristics.

Table 14.1 Robot-arm controller serial interface

Interface type: Subset of RS-232-C configured as DTE
Connector type: Female 25-pin subminiature D type
RS-232 lines supported:

AA	Frame ground	Pin 1
BA	Transmitted data	Pin 2
BB	Received data	Pin 3
CA	Request to send	Pin 4
CB	Clear to send	Pin 5
AB	Signal ground	Pin 7
CD	Data terminal ready	Pin 20

Voltage levels: +12, −12
Data format: Asynchronous
8 data bits
1 stop bit
No parity bit
Data rate: 300 bits per second (baud)

Messages from the robot-arm controller to the torque wrench controller are

```
''GRIP\n''          (Close the socket to grip a nut)
''TORQUE xxx\n''    (xxx represent three numeric
                    (0–9) characters standing for
                    the desired torque in newton-meters)
```

As an example, the message "TORQUE 095\n" will cause the torque wrench to tighten the nut to a torque of 95 N · m.

Messages from the torque-wrench controller to the robot-arm controller are

```
''ACK\n''           (Indicates that the operation
                    requested by the robot-arm
                    controller has been successfully
                    completed)
''FAULT 1\n''       (Indicates that the fastening
                    operation was terminated due to
                    the nut jamming)
''FAULT 2\n''       (Indicates that the fastening
                    operation was terminated due to
                    stripped threads or severe
                    misalignment of the nut)
```

B. Performance Requirements

The performance requirements for this design are basically concerned with the fastening operation. The wrench must apply fastening torques up to 150 N · m with an accuracy of ±1.0 N · m. The torque-vs.-time profile for a normal fastening operation is shown in Fig. 14.5. If the required torque is

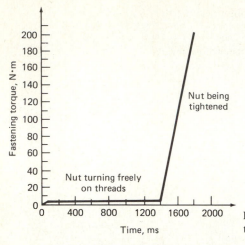

Figure 14.5 Torque-vs.-time profile for a fastening operation.

not reached after 2000 ms, stripped threads or severe misalignment of the nut can be assumed. If the torque rises above a 20 N · m threshold in less than 1000 ms after a fastening operation has begun, a jammed nut can be assumed.

Since the valve which controls the torque-wrench motor takes 50 ms to close, the torque-wrench controller must shut off the current to the solenoid valve before the target torque is reached. Tests of the torque wrench show that the current must be shut off when the torque is 10 N · m less than the target value.

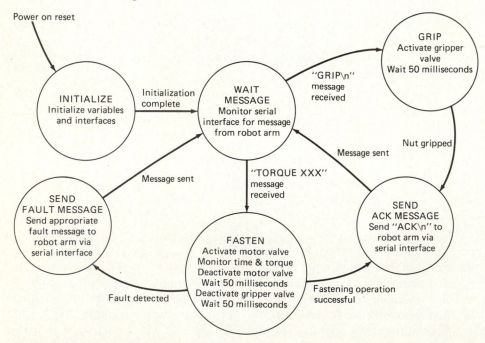

Figure 14.6 ASM diagram of torque-wrench controller.

14.3 BASIC SYSTEM DESIGN

Given the condensed requirements document of the previous section and the system description of Sec. 14.1, we can start designing the system following the guidelines set out in Chap. 4.

14.3.1 Flow of Control: Algorithmic State-Machine Diagram

From the sequence of operations described in Sec. 14.1 an ASM diagram can be developed to describe the flow of control in the torque-wrench controller. The diagram is shown in Fig. 14.6. The sequence of actions is simple and is driven by messages from the robot-arm controller.

14.3.2 Decomposition into Functions:—Tree Diagram

The next stage is to break the design down into a set of functions. This functional decomposition is shown in the form of a tree diagram in Fig. 14.7. Higher-level more-general functions are at the top of the diagram and the lower-level functions operating under the control of the higher-level functions are at the bottom of the diagram. Again we see that the design is really quite simple, requiring no more than 10 to 20 functions.

The sequencing and control functions implement the finite-state machine that sequences the torque-wrench controller through its operations. These sequencing and control functions invoke the functions that do actual work, such as gripping a nut or sending a message to the robot-arm controller. In turn, these functions make use of lower-level functions. For example, the operation of gripping a nut can be divided into the activation of the solenoid valve and the 50-ms delay required to allow the solenoid valve to open. Note that these lower-level functions may involve both hardware and software. The solenoid driver function, for instance, incorporates both the interface circuit to the solenoid valve and the software required to control it.

14.3.3 Estimating Program Length

At this point we can begin to make some initial estimates of program length and the amount of program memory that will be required. On the basis of the rule of thumb provided in Sec. 4.4, it is apparent that the program should be no more than a few hundred lines long if it is written in a high-level language such as C. Thus it should require only about a month for one programmer to develop the software. Since the program is relatively short, the program memory requirements are modest; 8 to 10K bytes of storage should be more than adequate. Similarly, there is no need to store a large amount of data, so a small amount of read/write memory (perhaps 1K byte) will be sufficient for this system.

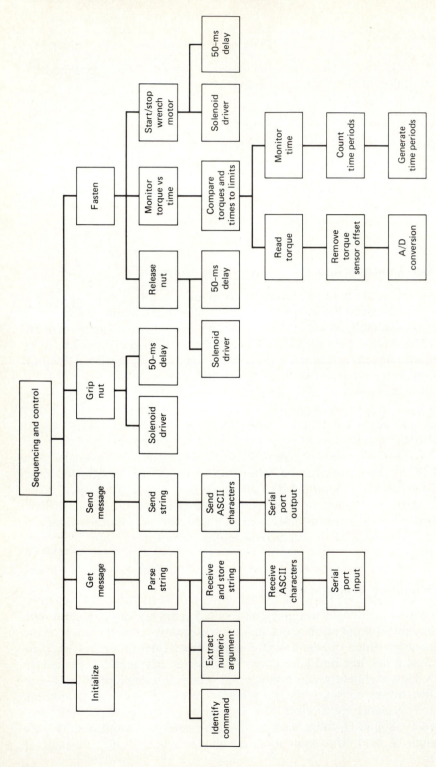

Figure 14.7 Tree diagram of torque-wrench-controller functions.

14.3.4 Module Specification

If this were a larger system, requiring the combined efforts of several programmers and hardware designers, the design would now proceed to the module-specification stage, where the design would be broken up into a set of carefully specified modules that could be assigned to different members of the design team. Then the detailed design of the individual modules would begin. However, this system is simple enough that a less-formal approach can be taken. The specifications for the hardware components of the system will be derived from the system-requirements document and then the hardware components will be selected using these specifications. With the hardware design complete, the system software will be written using the system-requirements document, the ASM diagram, and the tree diagram as guides.

14.4 HARDWARE SPECIFICATIONS

The hardware components that must be selected include a microcomputer, an analog-to-digital converter, solenoid drivers, an RS-232 serial communications interface, a power supply, a mother board and card cage, and a cabinet. Since it is important to monitor the time it takes for a fastening operation in order to detect faults, a hardware timer of some sort may also be required.

14.4.1 Microcomputer and System Bus

This particular application is not particularly demanding on the microcomputer. The amount of memory that must be addressed will be less than 16K bytes at most, so the microcomputer won't require the ability to address large amounts of memory. No complex computations are required and the system's speed of response is limited by the inherent sluggishness of electromechanical components such as solenoid valves and robot arms. Therefore, strong "number-crunching" capabilities and fast instruction-execution speeds are not primary considerations in the selection of the microcomputer. As a result, a simple 8-bit microcomputer should prove more than adequate in this design.

Given that only one 8-bit microcomputer is required, there is no need to use an expensive high-speed system bus that can handle multiple microcomputers and large amounts of memory. Instead, a less-expensive system bus offering more modest performance should be selected. The STD bus system, which has a large number of reasonably priced cards available and is oriented toward industrial applications of this type, is a good choice.

14.4.2 Analog-to-Digital Converter

The analog-to-digital (A/D) converter must be specified for input voltage range, resolution, accuracy, and conversion time.

Voltage range   The voltage from the strain gauge amplifier can range from 0 V at a torque of $0 \, \text{N} \cdot \text{m}$ with no offset voltage to 8 V at a torque of $150 \, \text{N} \cdot \text{m}$ with the maximum offset voltage of 0.5 V. The input voltage range of the A/D converter must be greater than or equal to this range. In order to make the best use of the converter's resolution, the converter's input voltage range should match the range of the input signal fairly closely. An input voltage range of 0 to 10 V is common for many A/D converter systems and is quite close to the 0- to 8-V range of the strain gauge signal. Therefore, we will specify the input voltage range of the A/D converter as 0 to 10 V.

Resolution   As discussed in Secs. 9.3 and 9.4, an A/D converter's resolution must be specified because the converter produces a fixed-length binary code which can take on only a finite number of discrete values. Therefore, the binary number produced by the converter is usually only an approximation of the magnitude of the continuous quantity at the converter's input. In this case, since torque must be controlled to $\pm 1.0 \, \text{N} \cdot \text{m}$, the A/D converter must resolve to at least $2 \, \text{N} \cdot \text{m}$. Since the converter's input range is 0 to 10 V, which is equivalent to 0 to $200 \, \text{N} \cdot \text{m}$, the resolution as a fraction of the input range must be at least 1/100. The resolution of an A/D converter producing binary codes of length n bits expressed as a fraction of the input range is

$$\text{Resolution} = \frac{1}{2^n}$$

A converter producing 6-bit codes will have a resolution of 1/64 and a converter producing 7-bit codes will have a resolution of 1/128. Thus the A/D converter used in this system must produce at least 7-bit codes. Since there are other sources of inaccuracy in the system, it will actually be necessary to choose an A/D converter system which produces longer codes.

Accuracy   In order to account for the other inaccuracies in the A/D converter system, such as nonlinearity and gain errors (see Secs. 9.3 and 9.4), it is necessary to determine an overall accuracy specification which the converter must meet. In the present system, the A/D converter must measure torque to an accuracy of at least $\pm 1.0 \, \text{N} \cdot \text{m}$ over an input range of 0 to $200 \, \text{N} \cdot \text{m}$. The accuracy of the converter must therefore be ± 0.5 percent or better over the temperature range that is expected. The A/D converter will be mounted in a cabinet which is in turn installed in a heated factory, so a reasonable temperature range is 10 to 50°C.

It must be pointed out here that the preceding analysis has been carried out under the implicit assumption that the torque sensor is perfectly linear and accurate (with the exception of offset). A complete error analysis must include the effects of error sources in the sensor. The analysis of the effects of sensor errors is described in Chap. 11. To keep this present example reasonably simple, we will continue to assume that the torque sensor is ideal.

Given an overall accuracy specification for the A/D converter, the resolution specification is actually redundant, since the resolution is incorporated in the accuracy specification. However, A/D converters are usually categorized and advertised by the length of the binary code they produce, while their overall accuracy is usually listed only in the fine print on the specification sheet. A knowledge of the required resolution in terms of the length of the binary codes produced by the converter is therefore useful in quickly identifying converters which may be suitable. Once some candidate converters have been found, their specification sheets can be examined to determine if their overall accuracy is sufficient for the application.

Conversion time The conversion time of an A/D converter is the time between the moment the converter is commanded to start a new analog-to-digital conversion and the moment the new binary code is available at the output of the converter. The conversion time is important in the present application, since it is necessary to sample the torque often enough so that there is no possibility of exceeding the desired final torque by more than the allowable error between samples. From Fig. 14.5 we can see that torque rises at a rate of $0.365 \, \text{N} \cdot \text{m/ms}$ during a normal fastening operation. If torque is sampled once every 2 ms, there is no possibility of overshooting the target torque by more than the allowable error of $1.0 \, \text{N} \cdot \text{m}$. The A/D converter will be specified to have a conversion time of less than 2 ms. This means that the converter will probably have to be of the successive approximation type rather than the slower dual slope type (see Sec. 9.4).

14.4.3 Solenoid Drivers

The solenoid drivers are output devices which have an interface both to the microcomputer and to the solenoid valves. A simplified diagram of a solenoid driver is shown in Fig. 14.8. The interface to the computer is a register that can be addressed by the microcomputer. The microcomputer writes the desired state of the solenoid valve into the register. For example, a binary 1 stored in the register causes the valve to open and a binary 0 stored in the register causes the valve to close. Since the register is a data-storage element, the valve remains in one state, either open or closed, until the microcomputer writes to the register again and changes its contents.

The driver's interface to the solenoid valve is basically a switch that opens and closes a circuit that allows current to flow through the solenoid coils. The switch is opened or closed depending on the contents of the register in the driver. The solenoid coils are specified to have a resistance of $20 \, \Omega$ and to carry a current of 0.5 to 1 A when the valve is open. Therefore, by applying Ohm's law ($E = IR$), we can see that the voltage applied to the solenoid coils must be from 10 to 20 V. The switch in the solenoid driver must be capable of blocking at least this much voltage when open and must be capable of conducting at least 1 A when closed. In order to allow for the overshoots and spikes that can

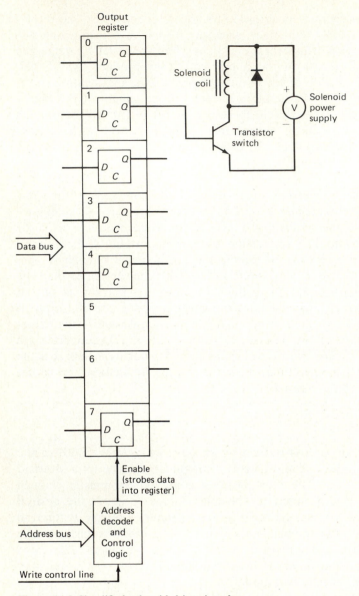

Figure 14.8 Simplified solenoid driver interface.

occur in actual circuits, the specification should be made very conservative—for example, a blocking voltage of at least 30 V and a current-carrying capacity of at least 2 A.

The design of output interface circuits such as solenoid drivers was discussed in Chap. 12. However, as we will see in the next section, it is often possible to purchase an appropriate interface off the shelf without the need to do any detailed electronic circuit design.

14.4.4 Serial Communications Interface

The serial interface must meet the signal level standards for RS-232C (see Chap. 10) and must support at least the seven-line subset of the RS-232C control and data lines that is supported by the robot-controller serial interface (see Table 14.1). It must also be able to operate at a data rate of 300 bits/s.

14.4.5 Power Supply

Complete specification of the power supply must wait until all the remaining electrical components are chosen so that the voltage and current requirements can be determined. However, it is already apparent that the power supply must supply ± 15 V for the strain gauge amplifier and 10 to 20 V for the solenoid coils. The ac line voltage to the power supply may be either 115 or 220 V, depending on the factory in which the system is installed.

14.4.6 Timer Function

Specification of the mother board, card cage, and cabinet must also wait until the remainder of the hardware is chosen. However, it is possible to make some design decisions about the timing functions required in the system. The system must keep track of the time required to fasten a nut, since the time required for the fastening torque to rise can be used to check for faults in the fastening operation. The system must also start the A/D converter at 2-ms intervals in order to monitor the torque during the fastening operation. Therefore, the system timer must have a resolution of at least 2 ms and must keep track of time for a period at least as long as the longest fastening operation, which is 2000 ms.

These specifications are difficult to meet with a completely software-based timer since the system must simultaneously perform other tasks while keeping track of the time. One approach to the design of the timer function would be a hardware timer which sends an interrupt pulse to the microcomputer every 2 ms. This is an example of a real-time clock of the sort discussed in Sec. 9.5. The microcomputer responds to the interrupt by starting an A/D conversion and incrementing an internal variable. This variable can be made to represent the total elapsed time in milliseconds for a fastening operation if it is reset to zero at the beginning of each fastening operation.

The system must also generate 50-ms delays while the solenoid valves open and close. These delays can be generated fairly easily in software, since no other tasks have to be performed during the delay interval. However, a hardware timer could also be used to advantage in this case, since software timing loops must be calibrated on the target system, which can be difficult to do.

14.5 HARDWARE SELECTION

Table 14.2 summarizes the specifications developed for the hardware components of the system. Using these specifications, buyer's guides, and vendor catalogs, it is possible to select the hardware components. In many cases it will be possible to find several suitable alternatives for each component, so that the choice can be based on nontechnical factors such as cost, delivery time, or vendor reputation. In the case of the design presented in this chapter, we have chosen boards that we used in previous projects in instances where there was a choice among several suitable boards. Therefore, it should be clearly understood that the boards we have chosen are not necessarily the "best" by any objective measure.

14.5.1 Microcomputer Board

Prior to starting the search for a microcomputer board it is necessary to decide on the microprocessor that will be used. The specifications call for an 8-bit microprocessor, but do not specify any particular model. There are no particular requirements for extremely low power consumption or operation over an extreme temperature range, so we do not need to consider these factors when choosing the microprocessor. The factors that should be considered are the availability of development software such as assemblers and compilers and previous experience with the microprocessor. In this example we will choose the Intel 8085 microprocessor for the design because we have the required

Table 14.2 Specification summary

Microcomputer board
   STD bus board using an 8-bit processor. Must have at least 8K bytes of EPROM memory and 1K byte of RAM memory. Must have interrupt input available for real-time clock. Preferably it should also have a programmable counter on board that can be used as a real-time clock.
A/D converter board
   Must be STD-bus-compatible.
   Number of input channels: 1.
   Input voltage range: 0 to 10 V.
   Resolution: >7 bits.
   Accuracy: ±0.5 percent of full-scale range over a temperature range of 10 to 50°C.
   Conversion time: <2 ms.
Solenoid driver board
   Must be STD-bus-compatible.
   Number of outputs: 2.
   Blocking voltage: >30 V.
   Current rating: >2 A.
Serial interface board
   Must be STD-bus-compatible.
   Must be RS-232C-compatible.
   Number of channels: 1.
   Communications mode and data rate: Asynchronous at 300 bits/s.

development software and have used this microprocessor extensively in the past. Another design group might choose the Zilog Z80 microprocessor or the Motorola 6809 microprocessor based on past experience and development facilities and the resulting design would be as good as the one based on the Intel microprocessor.

A number of STD bus microcomputer cards using the 8085 microprocessor are available. Most of them have sockets on the board for EPROM and RAM memory chips and many also include some additional circuitry such as I/O

SPECIFICATIONS

Word Size:	8-bit data bus (Up to 24K bytes)
System Clock (T State)	333ns at 3.0 MHz or 250ns at 4.0 MHz and external clock control
Memory Capacity:	Six 24-pin sockets provide up to 12K of RAM or 24K of EPROM
Memory Addressing:	On-board decode allows on-board memory to be configured on any 2K/4K boundary within the 64K memory address field.
Memory Speed:	350ns access (max.)
I/O Addressing:	On-board programmable timer

PORT
PORT ADDR

(HEX)	FUNCTION
7B	Cntl. Reg.
7C	CTC Channel 0
7D	CTC Channel 1
7E	CTC Channel 2
7F	Mode Register

Interrupts:	Multi-level vector interrupt with interrupt request originating from customer specified I/O only. Timer/counter channels interrupt CPU through RST 7.5, RST 6.5, RST 5.5 inputs.
Interface:	All Address, Data and Command Signals are TTL compatible.
Power Requirements:	+5 VDC ±5% at 1.5 amps max.
Card Dimensions:	Height: 6.5 Inches (16.51 cm) Width: 4.48 Inches (11.38 cm) Thickness: 0.062 Inch (0.158 cm) Assembly Thickness: 0.442 Inch (1.123 cm)

ENVIRONMENTAL CHARACTERISTICS

Operating Temperature:	0° to 55°C
Storage Temperature;	−40° to 80°C
Relative Humidity:	0% to 90% (without condensation)

Figure 14.9 Intersil ISB-3111 microcomputer board: condensed specification.

ports or programmable counter-timers. In our application we have a requirement for a hardware timer, so we have narrowed our search to boards that have a programmable counter-timer included on the board. Of these, the Intersil ISB-3111 has been selected. It has a large-scale integrated circuit on board which contains three counter-timers which can be read and controlled by the microprocessor. The counter-timers can be fed the clock pulses from the quartz crystal clock circuit used by the microcomputer, so very precise and stable monitoring of time is possible. Jumpers are provided on the board so that the output signals from the counter-timers can be fed to the interrupt inputs of the 8085 microprocessor. In addition, this board has six 24-pin sockets for EPROM and RAM memory chips. Using readily available chips, we can put 16K bytes of EPROM and 2K bytes of RAM on this board which will be more than sufficient for this application. The condensed specifications for the ISB-3111 board are shown in Fig. 14.9.

14.5.2 Analog-to-Digital Converter

There are many analog-to-digital converter cards available for the STD bus. Even after weeding out those which do not meet the specifications for input voltage range, resolution, accuracy, or conversion time, there are still several candidates. The card chosen is the Analog Devices RTI-1226. This board, the basic specifications of which are given in Fig. 14.10, has a resolution of 10 bits and a maximum error of ± 0.1 percent of the full-scale input range at a temperature of 25°C. The maximum error due to drift in gain and offset with change in temperature is

$$\Delta T \left(\frac{\text{gain error}}{T} + \frac{\text{offset error}}{T} \right) = (50°C - 25°C)(\pm 50\,\text{ppm}/°C + \pm 25\,\text{ppm}/°C)$$

$$= 1875\,\text{ppm}$$

This maximum error will occur when the A/D board temperature has risen to the maximum extreme of the temperature range (50°C). When the board temperature drops to the minimum extreme (10°C), the error will be less. If we add the maximum error at a fixed temperature of 25°C (± 0.1 percent) to the error due to change in temperature (± 0.1875 percent), we get a total error of ± 0.29 percent, which is within our specification of ± 0.5 percent.

Other important features of this board include the ability to set the input range to 0 to 10 V by an on-board jumper and a total conversion time of 40 µs. The board is equipped with an input multiplexer that allows up to 16 analog input signals to be switched into the A/D converter. This considerably exceeds our requirements, but the other boards considered were also equipped with multiple input channels, so we must accept this surplus capability.

Input

Number of input channels	16 single-ended or 8 differential (jumper-selectable)
Input overvoltage protection	±35 V with power applied, ±20 V with power off
Input impedance	$>10^8 \Omega$
Input current	±50 nA
Input voltage ranges	0 V to +10 V, ±5 V, ±10 V (jumper-selectable)
Analog input connector	3M #3493, 34 pin
A/D resolution	10 bits (1024 counts)
A/D output codes	Binary, offset binary, two's complement (jumper selectable)
Total conversion time	40 µs
System throughput	25,000 conversions per second
Common mode voltage (CMV)	±10 V minimum
Common mode rejection (CMR)	60 dB
Linearity	±1/2 least-significant bit (LSB)
Differential nonlinearity	±1/2 LSB
Total system error (adjustable to zero)	±0.1 percent of full-scale range (FSR)
Temperature coefficient	
Gain	±50 ppm/°C of FSR
Offset	±25 ppm/°C of FSR

Interface parameters

Compatibility	Meets all electrical and mechanical STD bus specifications
Implementation	Memory mapped I/O, compatible with all CPU types
Address Selection	Three contiguous bytes in a 16-byte block jumper selectable in any one of 256 locations in 64K of memory space
Power requirement	+5 V ±5 percent at 750 mA (on-board dc/dc converter generates an isolated ±15 V to power the data-acquisition components)

Temperature

Operation	0 to +70°C
Storage	−55°C to +85°C

Figure 14.10 Analog Devices RTI-1225 A/D board: condensed specification.

14.5.3 Solenoid Drivers

STD bus boards designated specifically as solenoid drivers are not listed in any of the buyer's guides. However, there are a large number of what are called "industrial" output boards. These consist of a series of switches which can be controlled by the microcomputer. The switches, which are usually transistors or relays, can carry several amperes of current and block voltages ranging from 30 to 200 V. These boards are used to control electrical loads such as pilot lamps, small motors, and relay or solenoid coils. The switches are normally electrically isolated from the digital circuits which interface to the STD bus. Therefore, even if a switch is damaged by an electrical transient or overload in the load circuit, the damage will be limited to the switch and will not propagate back into the microcomputer.

The solenoid driver we have chosen actually consists of two boards (see Fig. 14.11). A Pro-Log 7507 general-purpose interface card plugs into the STD bus and provides the interface to the microcomputer. The Pro-Log 7507 card in turn is connected by a ribbon cable to a Gordos PB-8 I/O module board, which is mounted outside the STD bus card cage. Up to eight standard I/O modules, available from several companies, can be plugged into the Gordos PB-8 board. These modules provide electrically isolated digital input and output to industrial loads. As discussed in Chap. 12, output modules are available to control both ac and dc loads. The dc output module (see Fig. 14.12) contains an optically isolated transistor that can carry up to 3 A of current and block up to 60 V. The Gordos PB-8 board has screw terminals which allow the modules to be connected to their loads.

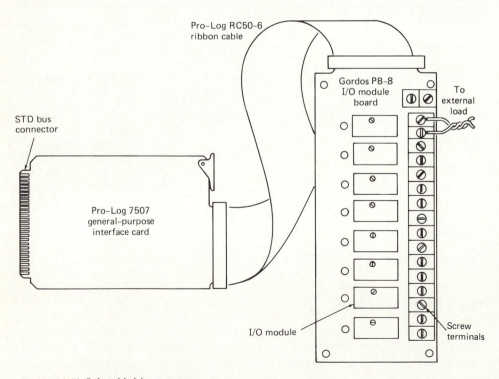

Figure 14.11 Solenoid driver components.

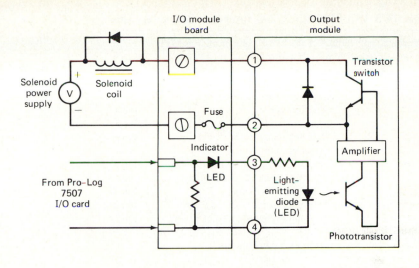

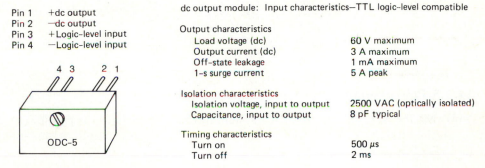

Pin 1	+dc output
Pin 2	−dc output
Pin 3	+Logic-level input
Pin 4	−Logic-level input

dc output module: Input characteristics—TTL logic-level compatible

Output characteristics
  Load voltage (dc)   60 V maximum
  Output current (dc)   3 A maximum
  Off-state leakage   1 mA maximum
  1-s surge current   5 A peak

Isolation characteristics
  Isolation voltage, input to output   2500 VAC (optically isolated)
  Capacitance, input to output   8 pF typical

Timing characteristics
  Turn on   500 μs
  Turn off   2 ms

Figure 14.12 Dc output module characteristics.

14.5.4 Serial Communications Interface

Serial communications cards are among the most common I/O cards for the STD bus system. We have chosen the Intersil ISB-3700 dual-channel synchronous/asynchronous communications card for this design. The ISB-3700 card has two independent serial interfaces, each of which can be used either as a RS-232-C interface or a 20-mA current loop interface. For our purposes we are concerned only with the characteristics of the RS-232-C interface. These characteristics are presented in Table 14.3.

The ISB-3700 supports a larger subset of the RS-232-C signals than does the serial interface for the robot-arm controller. It also allows the designer to select from a wide range of data formats and data rates. The primary task in designing the interface will be to set up the ISB-3700 properly so that its characteristics match those of the serial interface for the robot-arm controller. The second serial interface could be used to connect a video terminal or a printer to the torque-wrench controller for debugging or data-logging purposes.

Table 14.3 ISB-3700 RS-232-C interface characteristics

Interface type: Subset of RS-232-C configurated as DTE
Connector type: Female 25-pin subminiature D type
Lines supported:

BA	Transmitted data	Pin 2
BB	Received data	Pin 3
CA	Request to send	Pin 4
CB	Clear to send	Pin 5
CC	Data set ready	Pin 6
AB	Signal ground	Pin 7
DB	Transmitter signal element timing	Pin 15
DD	Receiver signal element timing	Pin 17
CD	Data terminal ready	Pin 20
CE	Ring indicator	Pin 22
DA	Transmitter signal element timing	Pin 24

Voltage levels: +12, −12
Data format: Synchronous or asynchronous
        5, 6, 7, or 8 data bits
        1, 1.5, or 2 stop bits
        Parity enabled or disabled, odd or even
Data rate: 50, 75, 110, 134.5, 150, 300, 600, 1200, 1800
        2000, 2400, 3600, 4800, 7200, 9600, 19,200 baud

14.5.5 Power Supply

Now that the STD bus cards for the system have been chosen, it is possible to determine the specifications for the power supply. The power requirements for the STD bus cards are

ISB-3111 microcomputer card	+5 V dc ±5 percent at 1.3 A
RTI-1226 A/D converter card	+5 V dc ±5 percent at 0.75 A
7507 general-purpose interface card	+5 V dc ±5 percent at 1.0 A
ISB-3700 serial communications card	+5 V dc ±5 percent at 0.72 A
	+12 V dc ±10 percent at 0.13 A
	−12 V dc ±10 percent at 0.05 A

In addition it is necessary to supply +15 V ±5 percent at 20 mA and −15 V ±5 percent at 20 mA to the strain gauge amplifier and 10 to 20 V at a current of 1 to 2 A to the solenoid valves.

Switching and linear power supplies Power supplies can be divided into two categories based on the technique used to convert and regulate the power. *Linear supplies* use a conventional transformer to convert the input ac voltage to a lower ac voltage which is then rectified and regulated to the required dc output voltages. *Switching supplies* immediately rectify the input ac voltage and then convert the resulting dc voltage to a high-frequency ac voltage using transistor switches. This high-frequency power (20 kHz or higher) is then

GENERAL SPECIFICATIONS

Dielectric Withstand	Input/Output = 2500V ac (1 minute)
	Input/Safety GND = 1500V ac (1 minute)
	Output/Safety GND = 500V ac (1 minute)
Ambient Temp Range	0 to +70°C operating
	−20°C to +85°C storage
Temperature Coefficients	+5V output = 0.02%/°C
	Other outputs = 0.05%/°C
Efficiency	70% typ
Full Output Rating	0 to +50°C (derated 50% @70°C)
Cooling	Convection
Hold-Up Time	20ms min at full load and nominal input voltage
Transient Response	Output voltage returns in less than 1 ms max following a 50% load change
Mean Time Before Failure	Over 35,000 hrs calculated per MIL-HDBK-217 @ 25°C ambient and typical load

INPUT SPECIFICATIONS

Input Voltages	90V to 130V (115V nominal), 180V to 260V (230V nominal), jumper selectable
Input Frequency	47−440Hz

OUTPUT SPECIFICATIONS

Line Regulation	0.5% max (±10% change)
Load Regulation	+5V output = 2% max
	Other outputs = 5% max (at 50% to 100% of rated load)
Cross Regulation	0.2% max on +5V output when any other output changes from 50% to 100% of rated load
	5% max on other outputs when +5V output changes from 50% to 100% of rated load
	1% max on other outputs when any other output excluding +5V output changes from 50% to 100% of rated load
Output Voltage	+5V: adjustable from +4.5V to OVP trip point
Adjustment	Other outputs: when +5V is set at 5.00V, they will be within ±5% of nominal output voltage
Noise and Ripple	1% p-p typ (2% p-p max)
Overvoltage Protection	+5V output = 6.2 ± 0.4V
Overcurrent Protection	Maximum current cannot be drawn from all outputs at the same time. The overcurrent protection feature will reduce all output voltages to a safe dissipation level when the average power rating exceeds 125% of maximum power. The overload feature will also protect against short circuit on any output.

Figure 14.13 Analog Devices ADSC power-supply specifications.

passed through a small transformer and rectified to produce the required dc output voltages. Regulation in the switching power supply is carried out by the same circuit that converts the input power to a high frequency.

Switching power supplies are generally smaller, lighter, and more efficient than are the equivalent linear supplies. The greatest benefit of the increased efficiency is reduced heat production and thus easier cooling of the equipment. In addition, switching power supplies operate properly over a wider input voltage range than do linear supplies. This is very useful in areas where voltage dips, or "brownouts," occur on power lines. However, switching power supplies do produce electrical noise at their outputs at frequencies which are multiples of the frequency at which the internal transistors operate. While this electrical noise is on the order of a few tens of millivolts in most cases, it can affect sensitive analog circuits such as transducer amplifiers that are amplifying signals which themselves have amplitudes of only a few millivolts. Digital circuits are much less sensitive to the noise produced by switching power supplies.

STD bus power supply   The power-supply market is very competitive, so there are many power supplies available that meet the requirements for this application. We have chosen an Analog Devices ADSC 40-31 switching power supply to supply the power for the STD bus cards and the solenoid coils. It supplies $+5\,$V at up to $4\,$A, $+12\,$V at up to $3\,$A, and $-12\,$V at up to $1\,$A. The $+12$-V output will be used to power the solenoid coils. Figure 14.13 summarizes the other specifications for this supply.

Strain gauge amplifier power supply   The power for the strain gauge amplifier must be relatively noise-free, since the amplifier handles very low level signals. Several manufacturers sell low-noise dc-dc converters which accept input power at $5\,$V or $12\,$V dc and supply output power at $+15\,$V dc and $-15\,$V dc. We have chosen an Intronics DCE 12/12/150 dc-dc converter which we will supply from the -12-V output of the Analog Devices ADSC 40-31 switching power supply. It will deliver $+15\,$V and $-15\,$V at currents up to $150\,$mA. The output ripple and noise is specified to be less than $1\,$mV rms.

14.5.6 Packaging

The design decisions to be made about packaging the system include choosing a card cage and mother board, selecting an enclosure, and determining whether the system will require cooling. STD bus card cages are available from both manufacturers of STD bus cards and independent suppliers. We chose a Pro-Log BRO4-T card cage and mother board which will accept up to four STD bus cards. The card cage comes equipped with a power cable which is terminated with a nine-pin plug. Pro-Log sells a matching socket and cable assembly ending in nine bare wires that can be connected to the power supply.

The choice of enclosure and the design of the cooling system are interrelated. A small enclosure is easy to transport and install but may require forced air cooling, since there is not sufficient surface area to allow cooling entirely by natural convection. A larger enclosure allows cooling entirely by natural convection and so can be sealed completely, which is very desirable in industrial environments.

Since the torque-wrench controller will be installed on the factory floor, a totally sealed enclosure should be used. The STD bus cards dissipate a total of 21 W of heat. The power supply, which is 70 percent efficient, dissipates about 14 W when both solenoid valves are activated. The strain gauge amplifier and its power supply dissipate less than 1 W. Thus the total heat to be dissipated by the enclosure is about 36 W. The air temperature within the enclosure must be kept below 50°C to ensure proper operation of the microcomputer boards. In a modern factory using industrial robots, the air temperature should not rise above 35°C. The enclosure should therefore be sized so that the temperature differential between the inside and the outside will be less than 15°C when 36 W of heat is being dissipated from the exposed sides of the enclosure.

Detailed analysis of heat transfer from a sealed enclosure is beyond the scope of this book. Interested readers should refer to a text on thermal design for electronic equipment such as the one listed at the end of this chapter. However, as a rough rule of thumb, a cabinet can transfer on the order of 0.01 to 0.03 W/in^2 of exposed surface area when the temperature difference between the inside and outside is 15 to 20°C. The upper end of the range ($0.03 \ W/in^2$) is achieved if the cabinet is painted to maximize heat transfer by radiation and a fan is used inside the cabinet to circulate air more rapidly than is possible by simple natural convection.

If the measures just described are used to maximize heat transfer, the minimum exposed surface area for the enclosure is

$$\frac{36 \ W}{0.03 \ W/in^2} = 1200 \ in^2$$

Based on this required surface area, we chose a standard NEMA 12 industrial enclosure (oil- and dust-tight) with dimensions of 20 by 20 by 10 in. The total surface area of the enclosure is 2000 in^2. This allows some leeway in mounting the enclosure. However, the enclosure should be mounted to maximize the surface area exposed to open air. It should also be mounted so that the largest surfaces are oriented vertically to ensure that the efficiency of convective heat transfer is maximized. In addition, the microcomputer card cage should be mounted near the bottom of the cabinet where the air will be the coolest. The internal fan should be mounted so that it forces air up between the cards in the card cage in order to avoid "hot spots" in the card cage.

14.6 HARDWARE CONFIGURATION

In this system, choosing the proper components constitutes most of the hardware design. However, some wiring does have to be performed, and some jumper connnections must be made, before the hardware is ready to operate. In addition, a programmer's model of the hardware must be developed to assist in the design of the system software.

14.6.1 Power Supply

The schematic diagram for the power supply wiring is shown in Fig. 14.14. The input circuit consists of a line cord, power switch, fuse, and transient suppressor. The transient suppressor is a nonlinear resistor, called a metal oxide varistor (MOV), whose resistance drops dramatically when a high voltage is placed across it. When a high-voltage transient from a lightning strike or some other cause occurs on the power line, the transient suppressor absorbs the energy in the transient, protecting the power supply.

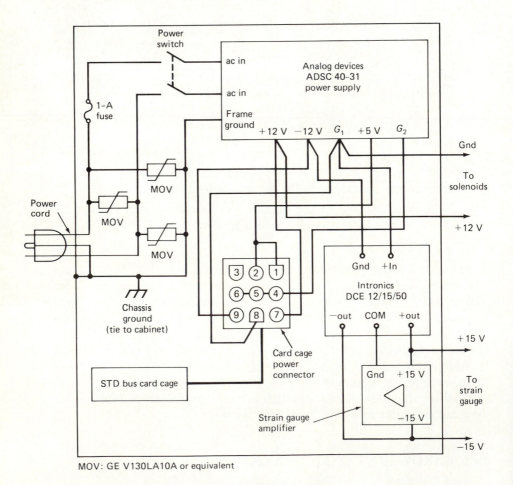

MOV: GE V130LA10A or equivalent

Figure 14.14 Wiring diagram for system power supply.

14.6.2 Microcomputer Board

Memory configuration The microcomputer board requires the greatest number of jumper connections before it is ready to install. The first set of jumper connections configure the six sockets for the on-board memory chips. The sockets must be configured for the range of memory addresses which will cause the memory chip in a particular socket to be selected. In addition, each socket must also be configured for the particular type of memory chip to be used in that socket.

In this design we will use three $4K \times 8$ EPROM memory chips to store the program and one $2K \times 8$ static RAM chip to store variable data, thereby allowing some margin for future expansion. (Remember that in this context, 1K equals 1024, so we are referring to chips with 4096 and 2048 memory locations, respectively, with each location containing 8 bits.) The EPROM chips, known as 2732s, are available from several manufacturers, including Intel and Mostek. The static RAM chip, known as a MK4802, is available from Mostek Corporation.

The 8085 microprocessor begins execution at memory address 0000 after the reset operation which occurs when power is applied. Thus, one EPROM chip should occupy the address space from 0 to 4K. Program development is easier if the program memory is kept as one contiguous block, so the remaining two EPROM chips will be assigned to succeeding 4K blocks of the address space. Finally, the static RAM chip will be assigned to the 2K block of addresses just above the addresses devoted to the EPROMs.

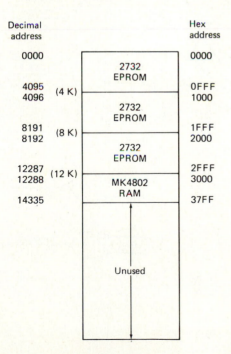

Figure 14.15 System memory map.

Figure 14.15, a *memory map*, shows where the memory chips are located in the address space of the 8085 microprocessor.

Figure 14.16 is a simplified schematic diagram of the circuit used to assign the memory chips to particular address ranges. The five most-significant bits of the memory address bus (ADR_{11} to ADR_{15}) are decoded into lines which become active when different 2K, 4K, and 16K segments of the memory address space are addressed. For example, if the address bus contained the address 19000 decimal, the lines marked 2–4K, 0–4K, and 16–32K would all become active—that is, go from a high state to a low state. By logically ANDing these lines, a memory chip enable pulse can be generated when a particular 2K, 4K, or 16K segment of memory is addressed. If the socket contains a 2K × 8 memory chip, a 2K segment line and a 16K segment line must be ANDed together. If a 4K × 8 memory chip is used, a 4K segment line and a 16K segment line must be ANDed together in order to assign the chip to a unique segment of the 64K address space. The low-order address lines of the address bus (i.e., ADR0 to ADR11) are sent to the sockets. The enabled

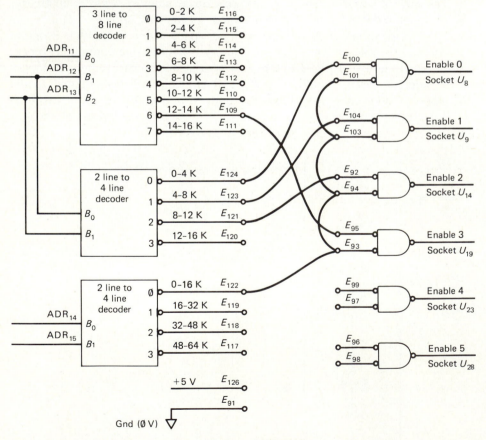

Figure 14.16 Memory chip select circuit for Intersil ISB-3111.

Function	U8		U9		U14		U19		U23		U28	
	From	To	From	To	From	To	From	To	From	To	From	To
2716 EPROM	E_3	E_2	E_9	E_8	E_{15}	E_{14}	E_{21}	E_{20}	E_{27}	E_{26}	E_{33}	E_{32}
	E_{38}	E_4	E_{41}	E_{10}	E_{44}	E_{16}	E_{47}	E_{22}	E_{50}	E_{28}	E_{53}	E_{34}
	E_5	E_6	E_{11}	E_{12}	E_{17}	E_{18}	E_{23}	E_{24}	E_{29}	E_{30}	E_{35}	E_{36}
2732 EPROM	E_{37}	E_2	E_{40}	E_8	E_{43}	E_{14}	E_{46}	E_{20}	E_{49}	E_{26}	E_{52}	E_{32}
	E_{38}	E_4	E_{41}	E_{10}	E_{44}	E_{16}	E_{47}	E_{22}	E_{50}	E_{28}	E_{53}	E_{34}
	E_5	E_6	E_{11}	E_{12}	E_{17}	E_{18}	E_{23}	E_{24}	E_{29}	E_{30}	E_{35}	E_{36}
4118 and 4801 RAM	E_1	E_2	E_7	E_8	E_{13}	E_{14}	E_{19}	E_{20}	E_{25}	E_{26}	E_{31}	E_{32}
	E_5	E_4	E_{11}	E_{10}	E_{17}	E_{16}	E_{23}	E_{22}	E_{29}	E_{28}	E_{35}	E_{34}
	E_{39}	E_6	E_{42}	E_{12}	E_{45}	E_{18}	E_{48}	E_{24}	E_{51}	E_{30}	E_{54}	E_{36}
4802 RAM	E_1	E_2	E_7	E_8	E_{13}	E_{14}	E_{19}	E_{20}	E_{25}	E_{26}	E_{31}	E_{32}
	E_{38}	E_4	E_{41}	E_{10}	E_{44}	E_{16}	E_{47}	E_{22}	E_{50}	E_{28}	E_{53}	E_{34}
	E_{39}	E_6	E_{42}	E_{12}	E_{45}	E_{18}	E_{48}	E_{24}	E_{51}	E_{30}	E_{54}	E_{36}
34000 ROM	E_3	E_2	E_9	E_8	E_{15}	E_{14}	E_{21}	E_{20}	E_{27}	E_{26}	E_{33}	E_{32}
	E_{38}	E_4	E_{41}	E_{10}	E_{44}	E_{16}	E_{47}	E_{22}	E_{50}	E_{28}	E_{53}	E_{34}
	E_5	E_6	E_{11}	E_{12}	E_{17}	E_{18}	E_{23}	E_{24}	E_{29}	E_{30}	E_{35}	E_{36}

Figure 14.17 Memory device jumper configuration.

memory chip will decode the low-order portion of the address on these lines in order to select the particular memory location being addressed.

With the jumper connections shown in Fig. 14.16, the memory chip in socket U8 is enabled when memory locations 0000 decimal to 4095 decimal are addressed, since lines 0–4K and 0–16K are ANDed together to form the enable signal. Socket U_9 is enabled for the next 4K segment in the address space and socket U_{14} for the next 4K segment after that. These three sockets will be used for the 2732 EPROM memory chips. Socket U_{19} will be enabled for addresses in the 2K range between 12288 decimal and 14335 (12–14K). This socket will be used for the MK 4802 static RAM chip.

The required jumper connections to configure a socket for the particular type of memory chip to be installed are listed in Fig. 14.17. Jumper connections are made on the microcomputer board by connecting a short piece of wire between the jumper posts. For instance, to configure socket U_8 for a 2732 EPROM, wires are connected between jumper posts E_{37} and E_2, E_{38} and E_4, and E_5 and E_6.

Counter-timer configuration The next portion of the microcomputer card to be configured is the counter-timer. The counter-timer is a single integrated circuit, the Intel 8253. A block diagram of the 8253 is shown in Fig. 14.18. This chip contains three 16-bit counters which are interfaced to the microcomputer. The microcomputer can perform read and write operations to the counters by using I/O mapped input/output instructions (see Sec. 2.5). On the ISB-3111

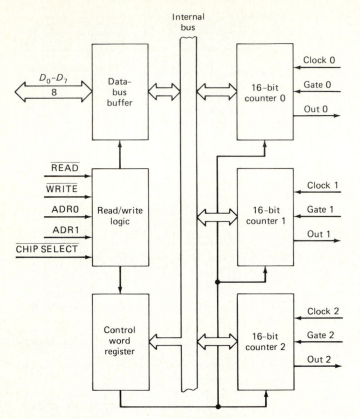

Figure 14.18 Block diagram of Intel 8253 counter-timer IC.

microcomputer board, the internal registers of the 8253 counter-timer chip are assigned the following addresses in the I/O address space:

Control word register	7F (hexadecimal)
Counter 0	7C
Counter 1	7D
Counter 2	7E

The microcomputer can load a counter with an initial value, start the counting operation, and read the value of the count as counting progresses. Once a counter is started, it decrements the count every time the digital signal at the counter's clock input goes from a low state to a high state. The counter can be programmed, by writing the appropriate control word to the control word register, to cause its out line to go from a low state to a high state when the counter value reaches zero. The gate input to the counter enables counting when it is high and disables counting when it is low. The gate inputs are all held permanently high on the ISB-3111 microcomputer board, so counting can never be disabled by these inputs.

We can create a real-time clock from one of the counters in the 8253 by connecting a constant frequency clock signal to the clock input of the counter and connecting the out line of the counter to an interrupt input of the microcomputer. The microcomputer initializes the counter with a value chosen so that interrupts will occur at intervals equal to the basic resolution of the real-time clock. For example, if the clock signal has a frequency of 250,000 Hz and the desired resolution of the clock is 2 ms, then the counter is loaded with

$$0.002 \times 250,000 = 500$$

The microcomputer will be interrupted when the counter has counted down from the original value to zero. At every interrupt the microcomputer updates the variables keeping track of the elapsed time and reloads and restarts the counter.

Interrupt system configuration The microcomputer card has a 250,000-Hz clock signal available at one of its jumper posts. This will be connected to the jumper post for the clock input to counter 0 of the 8253 counter-timer chip. The out line from counter 0 is brought to a jumper post which can be connected to the jumper post for one of the 8085 microprocessor's five interrupt inputs. One of these interrupt inputs (INTR) is already dedicated to interrupts from other cards on the STD bus and one of the other inputs (TRAP) is normally used to warn the microprocessor of faults, such as loss of system power, so there are three inputs left to choose from. The INT7.5 interrupt input has the feature that it latches the interrupt signal, which would complicate our design somewhat. The INT6.5 input will therefore be chosen for the counter-timer interrupt. The software needed to set up the counter-timer and handle the interrupts is described later in Secs. 14.8 and 14.9. The 8085 interrupt system will be examined in greater detail at that point.

14.6.3 Analog-to-Digital Converter Board

Input stage The first step in configuring the A/D converter board is to connect the output of the torque sensor amplifier to the signal inputs on the board. The signals are brought onto the board through a 34-pin connector mounted on the end of the board opposite to the STD bus connector. The output of the torque sensor amplifier is connected to the pins for channel 0 of the multiplexer. The multiplexer and amplifier on the board can be configured for what are called *single-ended*, or *differential*, inputs. Differential inputs are useful when there is a common voltage difference between the two output leads of a sensor and the analog ground, or common point, of the A/D converter. If one of the output leads of the sensor is at the same potential as the analog ground, as it is in this case, then the board can be set up for single-ended inputs. This is done by jumper connections between jumper posts on the board. The use of differential inputs is discussed in detail in Sec. 11.4.

Voltage range and data format   With the input stage properly configured, the next step is to choose the A/D input voltage range and the A/D output code. This A/D board has input ranges of 0 to +10 V, ±5 V, and ±10 V, which can be selected by making the appropriate jumper connections. This application requires that the board be configured for an input range of 0 to +10 V. This board can be configured to produce an A/D output code in straight binary, offset binary, or twos complement format (see Sec. 9.3). Since the input to the A/D converter is always positive, the straight binary-code format is the appropriate choice.

Address selection   The final stage in the configuration of the A/D card is to select the addresses used for I/O operations between the A/D card and the microcomputer. The A/D card has three 8-bit registers which can be accessed by the microcomputer. They are addressed using memory-mapped I/O (see Sec. 2.5) and have sequential addresses. The starting address for the three address block can be selected by making appropriate jumper connections on the board. When the board is delivered it is prewired so that the three registers are at locations FFFB, FFFC, and FFFD in the memory address space. Since the memory installed on the microcomputer board occupies only the address space from 0000 to 37FF, there is no conflict if the boards are used as delivered.

14.6.4 Solenoid Driver Interface

The configuration of the Gordos I/O rack portion of the interface consists of plugging in the output modules, connecting the wires to the solenoids to the screw terminals on the I/O rack, and connecting a ribbon cable (Pro-Log RC50-6) between the I/O rack and the Pro-Log 7507 interface card plugged into the STD bus. In this case the two output modules are plugged into positions 0 and 1 on the I/O rack. The solenoids are then wired up to the screw terminals corresponding to these two module positions. The solenoid connected to position 0 controls the grip-release valve and the solenoid connected to position 1 controls the wrench motor valve.

The Pro-Log interface card supplies +5-V power to the I/O rack via the ribbon cable. The card can be configured so that the power can be turned on or off under software control, or so that the power is always supplied to the rack as long as power is supplied to the card. In this case the card will be set up in the latter mode. I/O operations to the interface card are done using I/O-mapped input/output instructions. The card has four 8-bit registers that can be accessed by the microcomputer. When the card is first delivered, these registers have contiguous addresses starting at 50 (hexadecimal). Since these addresses do not interfere with any other I/O-mapped addresses in the system, they can be left unchanged.

14.6.5 Serial Interface

Electrical interface   The design of the electrical interface between the two serial interfaces consists of the following steps:

1. Select a connector to be installed on the cabinet of the torque-wrench controller.
2. Decide which lines from the ISB-3700 must be connected to this connector.
3. Decide which connector pins the lines are to be connected to.
4. Design the cable that connects the connector on the torque-wrench controller cabinet to the serial interface connector on the robot-arm controller.

The serial interface on the robot-arm controller uses a female 25-pin subminiature D-type connector (sometimes called a DB-25 connector). These connectors are the type most commonly used for RS-232-C interfaces. A male 25-pin subminiature D-type connector will therefore be used on the torque-wrench controller, thereby allowing the use of a standard interconnection cable with a male connector on one end and a female connector on the other end.

The two serial interfaces are both configured as DTE (data terminal equipment). Therefore, a null modem connection (see Chap. 10) will have to be made so that they can communicate with each other. The null modem connection can be made in the wiring between the connector on the ISB 3700 card and the connector mounted on the cabinet of the torque-wrench controller. The required connections are shown in Fig. 14.19. The connections ensure that the four data control lines—DTR, DSR, RTS, and CTS—will work properly. Since the serial interface in the robot-arm controller does not support the data set ready (DSR) line, there is no need to connect the data terminal ready (DTR) line from the ISB 3700 interface. The remaining lines on the ISB-3700 interface are used for either the 25-mA current loop interface, modem control, or synchronous communication. Therefore, they do not have to be connected to the connector on the cabinet either.

The registers in the ISB 3700 occupy a block of contiguous addresses in the I/O address space. The starting address of the block can be selected by a jumper on the board. We will set the starting address to A0 (hex).

14.7 SOFTWARE DESIGN I: HIGH-LEVEL ROUTINES

We will develop the software for the torque-wrench controller in a top-down fashion starting with the high-level functions at the top of the tree diagram in Fig. 14.7. As mentioned at the beginning of this chapter, most of the program will be written in the C programming language with a few short functions written in the assembly language of the Intel 8085 microprocessor. Even though we use C in this example, the concepts presented are applicable to other high-level programming languages as well.

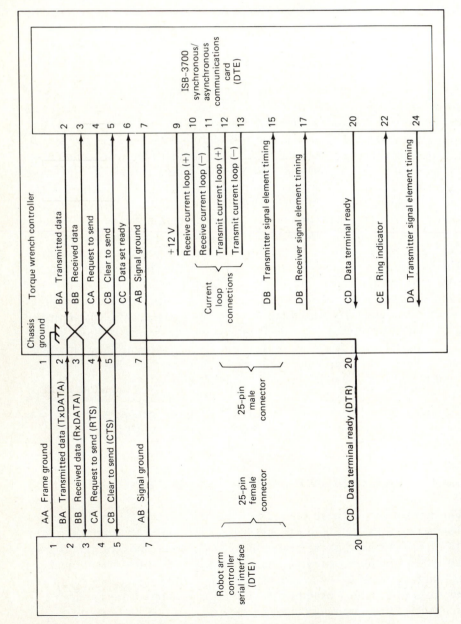

Figure 14.19 Connection diagram for serial interface between robot-arm controller and torque wrench controller.

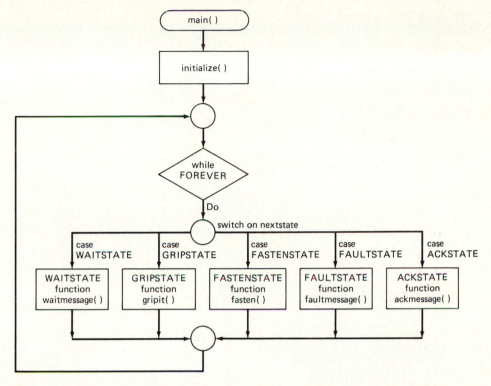

Figure 14.20 State sequencer flowchart.

14.7.1 State Sequencer

The first function to design is the state sequencer which performs the sequencing and control of program flow. The state sequencer is the first C function executed after the system is started up. Therefore, it is given the name main( ), to indicate to the compiler that it is the first function to be executed. A flowchart for this function, which is listed below, is given in Fig. 14.20.

State Sequencer Function

```
#define FOREVER          1

#define WAITSTATE        1   /* Constants used to indicate */
#define GRIPSTATE        2   /* next state to be entered by */
#define FASTENSTATE      3   /* state machine */
#define FAULTSTATE       4
#define ACKSTATE         5

#define FAULT1           1   /* Constants indicating the nature */
#define FAULT2           2   /* of a fault condition */
#define FAULT3           3
```

```
main( )
{
int param, nextstate /* param is used to transfer parameters among */
                     /* states. */
initialize( ) /* Do required initializations such as */
              /* I/O interfaces to a defined state */

nextstate = WAITSTATE;

/* Enter State Sequencer */

while(FOREVER) /* loop continuously */
     {
     switch (nextstate){

            case WAITSTATE:
            nextstate = waitmessage(&param);
            break;

            case GRIPSTATE:
            nextstate = gripit( );
            break;

            case FASTENSTATE:
            nextstate = fasten(param, &param);
            break;

            case FAULTSTATE:
            nextstate = faultmessage(param);
            break;

            case ACKSTATE:
            nexstate = ackmessage( );
            break;

            default:
            param = SYSTEMFAULT;
            nextstate = faultmessage(param);
            }
     }
}
```

The operation of the state sequencer is straightforward. Depending on the value of the variable nextstate, the switch statement selects a function to execute. This function performs the activities required for a particular state and determines if a transition to another state is required. If a transition is required, the function passes control back to the state sequencer and returns a new value for the variable nextstate which will be used on the next loop through the state sequencer to invoke the function for the new state. The variable param is used to pass parameters between states. For instance, the WAIT state uses this variable to pass the required fastening torque to the FASTEN state.

14.7.2 WAIT State

The function for the WAIT state, waitmessage( ), monitors the serial interface for ASCII characters sent from the robot-arm controller and assembles the received characters into a message string. Once a complete string has been received, the function determines which message was sent and on that basis selects the appropriate next state for the system. If the message received was TORQUE xxx\n, the function also extracts the required fastening torque from the message and puts it in the variable param to be passed to the FASTEN state.

The C language code for waitmessage( ) is given below:

```
waitmessage(torque)
  int *torque:        /* Pointer to variable which will be used to
                         pass torque setpoint to FASTEN state */
{
char string[15];      /* Array of characters used to hold messages
                         received from robot controller */

getmessage(string);   /* Assemble a message from the robot controller */

if(*string == 'G')    /* Check to see if first character in message
                         is G */

return(GRIPSTATE);    /* If so, message must be ''GRIP\n'' so next
                         state must be the GRIP state */

if(*string == 'T')    /* Check to see if first character in message
                         is T */
  {                   /* If so, message must be ''TORQUE xxx\n'' */
/* Extract the fastening torque from the message and convert it from a
   string of ASCII numeric characters stored in the array string[ ] to a
   binary integer stored in the variable pointed to by torque */
  *torque = 0;
  while(*string != '\0'  /* while(not end of string) */
    {
    if(*string >= '0' && *string <= '9') /* if(character is a numeral) */
      *torque = *torque * 10 + (*string - '0');
    string++;           /* point to next character in message */
    }
  return(FASTENSTATE); /* next state is FASTEN */
}
return(WAITSTATE);      /* This is the default if the message is not
                           one of the two legal messages */
}
```

The only tricky section in the function above is the conversion of the torque set-point from its representation as a string of ASCII characters to a binary integer. Examination of the ASCII character set (App. A) shows that the codes for the ASCII numeric characters form a contiguous increasing positive sequence. Thus the test

$$*string >= '0' \&\& *string <= '9'$$

determines whether the character pointed to by string is a numeric character. Also, the expression

$$*string - '0'$$

returns the numeric value of the ASCII numeric character pointed to by string. For instance, if the character is '7', then the expression will return the integer 7, since the ASCII code for the character '7' is 55 and the ASCII code for the character '0' is 48. To convert a string of numeric characters to an integer, the pointer to the string is incremented to point to each character in succession, and the following expression is applied to each character to build up the integer:

$$integer = integer * 10 + (*charpointer - '0')$$

The waitmessage( ) function calls another function, getmessage( ), to assemble a message string. The code for the getmessage function is given below.

```
getmessage(stringptr)
  char *stringptr;      /* pointer to array where string is to
                          be stored */
{
while((*stringptr = serialin( )) != '\n')
/* Get character from serial interface and store in array string while
   not end of message ('\n') */
    {
    string++;  /* point to next free location in array */
    }
*string = '\0';    /* insert standard end of string character */
}
```

The function serialin( ) is an I/O driver function which checks whether a character is available at the serial interface and returns it to the calling function. The design of this function is discussed in Sec. 14.8.

14.7.3 GRIP and ACK States

The function for the GRIP state, gripit( ), is very simple, since it makes use of two lower-level functions which will be described in Sec. 14.8. The listing of gripit( ) is given below

```
gripit( )
{
grip( );    /* activate gripper solenoid */
wait50( ); /* 50-ms delay while valve is activated */
return(ACKSTATE);  /* return next state to state sequencer */
}
```

Similarly, the function for the ACK state, ackmessage( ), is also very simple;

```
ackmessage( )
{
sendmessage(''ACK'\n''');  /* send ''ACK'\n''' to robot controller via
                              serial interface */
return(WAITSTATE);  /* return next state to state sequencer */
}


sendmessage(stringptr)
char *stringptr;
{
while(*stringptr != '\0')  /* while not end of string */
    {
    serialout(*string);  /* pass character to serial output */
    string++;            /* point to next character */
    }
}
```

The function sendmessage( ) (above) is passed a pointer to a character string. It disassembles the string and passes each character to the I/O driver function handling serial output (serialout( )).

14.7.4 FASTEN State

The FASTEN state function, fasten( ), is the most complex function in the program. The flowchart for this function is presented in Fig. 14.21. The listing is given below.

```
#define LIMIT 102        /* 20 N·m threshold indicating a jammed nut */
                         /* scaled to value produced by A/D converter*/
fasten(torque, result)
int torque;              /* target torque passed to function */
int *result;            /* pointer to variable where flag indicating
                            success or failure of fastening operation
                            is to be stored */

{
external int clocktick; /* declaration of flag variable used by real-time
                            clock to indicate that 2 ms have elapsed */
long int temp;          /* temporary variable used in scaling torque */
int setpoint;           /* scaled and offset target torque */
int time;               /* elapsed time (milliseconds) */
int value;              /* torque measurement from A/D */
temp = torque - 10;     /* generate torque setpoint */
temp = (temp * 1024)/200;
setpoint = temp + gettorque( );   /* gettorque returns the
                            torque sensor offset since no torque
                            is applied */
time = value = 0;        /* initializations */
startwrench( );
startclock( );
```

```
while((time>1000 && time<2000) && (value<setpoint) || (time<=
1000 && value<=LIMIT))
/* while(torque<setpoint and no faults have been detected) */
{
while(!clocktick)        /* clocktick is set to 1 by the interrupt
                            service routine for real-time clock */

{
/* wait */;
}
clocktick=0;             /* reset clocktick */
time += 2;               /* update elapsed time */
value=gettorque( );      /* get torque measurement from A/D */
}
stopclock( );            /* stop the real-time clock */
stopwrench( );           /* stop the wrench motor */
wait50( );               /* wait 50 ms for the valve to close */
release( );              /* release the nut from the wrench */
wait50( );               /* wait 50 ms for the valve to close */

if(time<=1000)           /* indicates that a nut jammed */
  {
  *result=FAULT1;

  return(FAULTSTATE);    /* return next state to state sequencer */

  }
if(time>=2000)           /* indicates stripped or misaligned fastener */
  {
  *result=FAULT2;
  return(FAULTSTATE);    /* return next state to state sequencer */
  }
return(ACKSTATE);        /* successful fastening operation, return
                            next state to state sequencer */

}
```

The fasten( ) function is passed an integer representing the desired fastening torque in newton-meters. For a number of reasons, this integer cannot be used directly as the set-point which is compared with the measurement from the torque sensor. First, the torque value at which the motor must be shut off is 10 N·m less than the target torque passed to the function. Second, the torque measurement returned from the A/D converter is not in units of newton-meters and contains an offset due to the torque sensor offset voltage. Therefore, the first action in the fasten( ) function is to convert the original target torque passed to it to a usable set-point.

Subtracting 10 from the target torque passed to the function results in the torque at which the motor must be shut off. This new target torque must now be scaled to correspond with the measurements from the torque sensor:

$$\text{Scaled torque} = \text{target torque} \times \frac{\text{full-scale A/D output}}{\text{full-scale A/D input}}$$

$$\text{Scaled torque} = \text{target torque} \times \frac{1023}{200 \, \text{N} \cdot \text{m}}$$

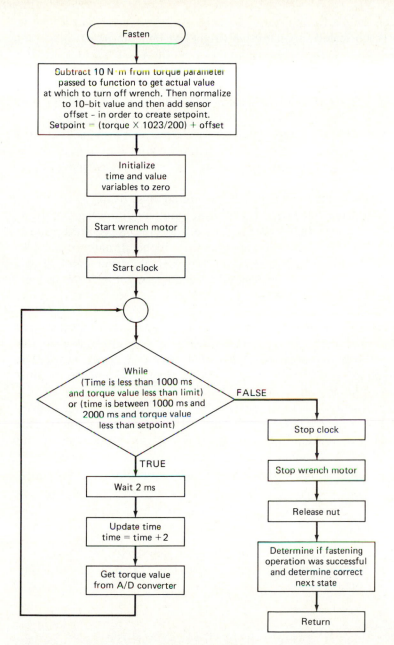

Figure 14.21 Flowchart for fasten( ) function.

Finally, the torque sensor offset is added to the scaled value to produce the final set-point. Since the torque sensor offset drifts gradually over time, it is necessary to measure the offset just prior to the fastening operation to get an accurate value. This is done by measuring the torque sensor output when the wrench motor is off and no torque is being applied. The value measured is the sum of the offsets in the torque sensor and the A/D converter and it can be added directly to the scaled target torque to get the final set-point.

The fastening operation can now commence, so the wrench motor and the real-time clock are started. The real-time clock interrupts the microcomputer every 2 ms. The interrupt service routine for the real-time clock sets a flag variable, clocktick, to 1 to signal the fasten( ) function that 2 ms has elapsed. The fasten( ) function stays in a loop waiting for the flag variable to be set to 1. When it detects that the variable has been set to 1, it resets it to 0. Then it adds 2 ms to the variable keeping track of elapsed time (time) and gets a torque measurement from the A/D converter using the lower-level-function get-torque( ).

The measured torque and the elapsed time are compared with limits to see if the desired torque has been reached or if some fault condition has occurred. For example, if the elapsed time exceeds 2000 ms and the measured torque is still less than the set-point, then a fault due to stripped threads or a badly misaligned fastener is assumed to have occurred. On the other hand, if less than 1000 ms has elapsed and the measured torque exceeds the 20-N·m limit given in the requirements document, then a fault due to jammed nut is assumed.

If no fault condition is detected and the measured torque is still less than the set-point, the function loops back and waits for the next interrupt from the real-time clock. If a fault condition is detected or the measured torque is greater than or equal to the set-point, the real-time clock and the wrench motor are stopped and the nut is released.

The fasten( ) function places an error code in the variable pointed to by the pointer result if a fault was detected during the fastening operation. This error code is used in the FAULT state to send the appropriate message to the robot-arm controller. The fasten( ) function also returns the code for the next state to the state sequencer so that the correct state transition will be made.

14.7.5 FAULT State

The function for the FAULT state, faultmessage( ), simply sends a message to the robot-arm controller describing the type of fault. The listing of fault-message( ) is given below.

```
faultmessage(flag)
int flag;                /* variable containing code which
                            indicates nature of fault */
{
switch(flag)
        {
        case FAULT1:
        sendmessage(''FAULT1\n'');
        break;
```

```
        case FAULT2:
        sendmessage(''FAULT2\n'');
        break;

        default:
        sendmessage(''SYSTEM FAULT\n'');
        }

return(WAITSTATE);    /* return next state to state sequencer */
}
```

The "SYSTEM FAULT\n" message is sent only if there is an error in the software. It is useful for debugging the system but should never occur once the system is running properly.

14.8 SOFTWARE DESIGN II: I/O DRIVER ROUTINES

In this section the lower-level functions which control the input/output interfaces will be described. The programmer's model for each interface will be discussed and then the functions associated with the interface will be presented.

14.8.1 Solenoid Drivers

The programmer's model for the solenoid driver interface is shown in Fig. 14.22. It consists of one of the 8-bit registers in the Pro-Log 7507 interface card. The register is located at I/O address 0x52. Each bit position corresponds to an I/O module. When output modules are used, as they are in this system, writing a 1 to the bit position turns the module on and writing a 0 to the bit position turns the module off. In this system only two modules, 0 and 1, are used, with module 0 controlling the grip-release valve and module 1 the wrench motor.

Unfortunately it is not possible to access individual bits in the register directly. Instead a complete 8-bit code must be sent to the register. As a result, it is necessary to ensure that changing the state of one bit does not alter any of the other bits in the register. This can be done by maintaining a variable which contains the same bit pattern as the output register. When a bit is to be changed, the contents of the variable are first changed using an appropriate masking operation (see Apps. A and E) which changes only the state of the desired bit and then the updated contents are sent to the output register.

	D_7	D_6	D_5	D_4	D_3	D_2	D_1	D_0
I/O address 52H	Module 0	Module 1	Module 2	Module 3	Module 4	Module 5	Module 6	Module 7

Figure 14.22 Programmer's model of Pro-Log 7507 interface card (solenoid driver).

The four solenoid driver functions are listed below

```
#define GRIP 0x80        /* mask used to set bit D7 to 1 */
#define RELEASE 0x7F     /* mask used to set bit D7 to 0 */
#define START 0x40       /* mask used to set bit D6 to 1 */
#define STOP 0xBF        /* mask used to set bit D6 to 0 */
#define SOLENOIDS 0x52   /* I/O address for solenoid driver register */

int solstate;            /* external variable containing solenoid
                            driver bit pattern */

grip( )                  /* activate gripper valve */
{
solstate |=GRIP;         /* OR solstate with GRIP mask to set bit D7
                            to 1 */
output(SOLENOIDS, solstate);   /* send contents of solstate to solenoid
                                  driver register */
}

release( )               /* deactive gripper valve */
{
solstate &=RELEASE;      /* AND solstate with RELEASE mask to set bit
                            D7 to 0 */
output(SOLENOIDS, solstate);
}

startwrench( )           /* start wrench motor */
{
solstate |=START;        /* OR solstate with START mask to set bit D6 to 1
                            */
output(SOLENOIDS, solstate);
}

stopwrench( )            /* stop wrench motor */
{
solstate &=STOP;         /* AND solstate with STOP mask to set bit
                            D6 to 0*/
output(SOLENOIDS, solstate);
}
```

The function output( ), which is used in the functions above to write data to the solenoid driver register, is supplied with the C compiler for the 8085 microprocessor. The function makes use of the 8085's I/O-mapped output instruction to send the data passed to it (solstate) to the specified I/O address (SOLENOIDS). All 8085 I/O operations work on 8-bit data only, so the integer solstate will be truncated before being sent to the solenoid driver register. However, only the lowest 8 bits of solstate are used in any event, so the truncation has no effect on the proper operation of these functions.

14.8.2 Analog-to-Digital Converter

The programmer's model for the Analog Devices RTI-1226 A/D converter board is shown in Fig. 14.23. There are three 8-bit registers which are accessed

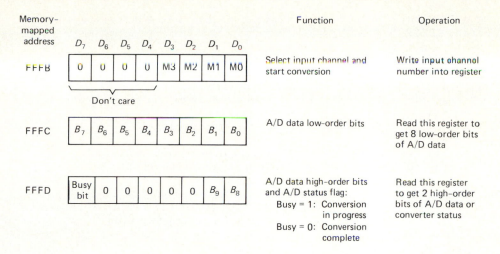

Memory-mapped address	D_7	D_6	D_5	D_4	D_3	D_2	D_1	D_0	Function	Operation
FFFB	0	0	0	0	M3	M2	M1	M0	Select input channel and start conversion	Write input channel number into register
		Don't care								
FFFC	B_7	B_6	B_5	B_4	B_3	B_2	B_1	B_0	A/D data low-order bits	Read this register to get 8 low-order bits of A/D data
FFFD	Busy bit	0	0	0	0	0	B_9	B_8	A/D data high-order bits and A/D status flag: Busy = 1: Conversion in progress Busy = 0: Conversion complete	Read this register to get 2 high-order bits of A/D data or converter status

Figure 14.23 Programmer's model of Analog Devices RTI-1225 A/D card.

using memory-mapped I/O operations. An A/D conversion is started by writing the number of the desired input channel (0 to 15) to the register at address 0xFFFB. This causes the multiplexer to connect the selected analog input to the A/D converter and initiates the A/D conversion. When the conversion is complete, the low-order 8 bits of the data are read from the register at address 0xFFFC. The remaining 2 bits are stored in the two low-order bit positions (D_0 and D_1) of the register at address 0xFFFD.

It is possible to determine when a conversion is complete and data is ready by monitoring bit D_7 of the register at address 0xFFFD. This bit, called the BUSY bit, is set to 1 when a conversion is in progress and is reset to 0 when the conversion is complete.

The function gettorque( ) which gets a torque measurement from the A/D converter is listed below.

```
#define MUXADDR    0xFFFB    /* address of register used to select
                                input channel */
#define ADLO       0xFFFC    /* address of register which contains
                                low-order 8 bits of A/D data */
#define ADHI       0xFFFD    /* address of register which contains
                                high-order 2 bits of A/D data and
                                BUSY bit */
#define BUSYMSK    0x80      /* mask to isolate BUSY bit */
#define DATAMSK    0x03      /* mask to isolate two high-order data
                                bits */
#define CHANO      0x00      /* input channel number of torque sensor */

gettorque( )
{
char *mux, *adlo, *adhi;  /* pointers to the registers on the A/D card */
int lowpart, hipart;
```

```
mux = MUXADR;                    /* set up pointers to A/D registers */
adlo = ADLO;
adhi = ADHI;

*mux = CHANO;                    /* select input channel and start
                                    conversion */
while (*adhi & BUSYMSK == 1)  /* monitor BUSY bit to see if conversion
                                    is complete */
/* loop until BUSY bit = 0 (i.e. conversion complete) */
    ;
lowpart = *adlo;                 /* get low-order 8 bits of data */
hipart = *adhi & DATAMSK;   /*get high-order 2 bits of data and
                                mask them off */
return((hipart =<<8) + lowpart);   /* shift and add low and high parts
                                       to get 10-bit value and return it
                                       to calling function */
}
```

Note that the two portions of the 10-bit data from the A/D converter are initially stored in two separate variables, lowpart and hipart. The variable hipart is shifted left by eight positions to place the two high-order bits of the A/D data in the correct positions and then lowpart, which contains the 8 low-order bits, is added to it to form the complete 10-bit number.

14.8.3 Serial Interface

The ISB 3700 serial interface board uses a large-scale integrated circuit, the Signetics 2651 programmable communications interface, to perform most of the required serial communications functions. Most serial interface boards rely on chips of this type, which are called UARTs (universal asynchronous receiver/transmitter) or USARTs (universal synchronous/asynchronous receiver/transmitter). The programmer's model of a particular board will depend largely on the characteristics of the UART or USART chip used on that board.

To the programmer, one of the serial interfaces on the ISB 3700 consists of nine 8-bit registers which can be accessed by the microcomputer's input/output instructions. The nine registers are

1. *Received-data holding register*. This register contains the last character received by the serial interface. The microcomputer reads received data from this register. Nondata bits such as stop bits and parity bits are stripped from the received character before it is loaded into this register.
2. *Transmitted data holding register*. This register contains the next character to be transmitted by the serial interface. The microcomputer writes data to be transmitted to this register. Nondata bits, such as start, stop, and parity bits, are added to this data by circuits in the interface and then it is transmitted serially.
3. *Mode register* 1. This register is used to control the data format used by the serial interface. The microcomputer writes a mode control byte to this

register when the interface is initialized. Table 14.4 shows the format of the mode byte. In all the following tables, bit 7 is the most-significant bit. When a bit pattern is shown within a bit field, the left-most bit is the most significant.

4. *Mode register* 2. This register is used to control the data transmission rate (the baud rate) and the source of clock signals that control the data-transmission rate. The microcomputer writes a mode control byte to this register when the interface is initialized. Table 14.5 shows the format for the mode byte.

Table 14.4 Mode register 1 format

Bit field	Function	
0	Mode/baud rate factor	00 = synchronous mode, 1 × baud rate
		01 = asynchronous mode, 1 × baud rate
1		10 = asynchronous mode, 16 × baud rate
		11 = asynchronous mode, 64 × baud rate
2	Character length	00 = 5 bits; 10 = 7 bits
		01 = 6 bits; 11 = 8 bits
3		
4	Parity control	0 = disabled; 1 = enabled
5	Parity type	0 = odd parity; 1 = even parity
6	Asynchronous mode:	00 = invalid; 10 = 1.5 stop bits
	stop bit length	01 = 1 stop bit; 11 = 2 stop bits
7		

Note: 1 × baud rate, 16 × baud rate, etc., refers to the required clock frequency if an external clock is used. If the internal clock is used in asynchronous mode, a 16x clock rate is used automatically.

Table 14.5 Mode register 2 format

Bit field	Function		
0	Baud-rate selection	0000 = 50 baud	1000 = 1800 baud
		0001 = 75 baud	1001 = 2000 baud
1		0010 = 110 baud	1010 = 2400 baud
		0011 = 134.5 baud	1011 = 3600 baud
2		0100 = 150 baud	1100 = 4800 baud
		0101 = 300 baud	1101 = 7200 baud
3		0110 = 600 baud	1110 = 9600 baud
		0111 = 1200 baud	1111 = 19,200 baud
4	Receiver clock select	0 = external clock; 1 = internal clock	
5	Transmitter clock select	0 = external clock; 1 = internal clock	
6			
	Not used		
7			

5. *Command register*. This register controls operation of the serial interface. The microcomputer writes a control byte to this register when the interface is initialized. Table 14.6 shows the format of the control byte.
6. *Status register*. This register contains a data byte that indicates receiver and transmitter conditions and modem/data set status. The microcomputer reads data from this register to check the status of the interface. Table 14.7 shows the format of the status byte.

Table 14.6 Command register format

Bit field	Function	
0	Transmit control	0 = disable; 1 = enable
1	DTR control	0 = force DTR high(on); 1 = force DTR low(off)
2	Receive control	0 = disable; 1 = enable
3	Asynchronous mode: force break	1 = send a break signal; 0 = no signal
4	Reset error flags	0 = normal 1 = reset error flags in status register
5	RTS control	0 = force RTS high(on); 1 = force RTS low(off)
6	Operating mode	00 = normal operation; 10 = local loop back 01 = asynchronous mode: echo mode; 11 = remote loop back
7		

Table 14.7 Status register format

Bit field	Function	
0	Transmitter ready	0 = transmit register full (BUSY) 1 = transmit register empty (READY FOR DATA)
1	Receiver ready	0 = receive register empty (NO DATA) 1 = receive register full (DATA AVAILABLE)
2	Transmitter empty/DSR change	0 = normal 1 = transmitter has no new characters to send or DSR line has changed state
3	Parity error detect	0 = normal; 1 = parity error
4	Overrun error	0 = normal; 1 = receiver has loaded new data into receive register before previous data was read by microcomputer
5	Framing error	0 = normal; 1 = wrong number of stop bits received
6	UNUSED	
7	Data set ready status	0 = DSR input is high (ON); 1 = DSR input is low (OFF)

7. *Interrupt vector register*. This register is used to store an interrupt vector that is sent to the microprocessor after the ISB 3700 has generated an interrupt request signal and has received an interrupt acknowledge signal from the microcomputer. The microcomputer initializes this register with an interrupt vector number if the ISB 3700 is to be used as part of a vectored interrupt system.
8. *Interrupt control register*. This register is used to enable or disable the generation of interrupt request signals by the ISB 3700. The register has the following format:

Bit 0	Channel 1	0 = interrupt disable 1 = interrupt enable
Bit 1	Channel 2	0 = interrupt disable 1 = interrupt enable
Bit 2 to bit 7 not used		

When interrupt requests are enabled, they are generated when new data is available in the received-data register or when the transmit-data register is ready to accept another character.
9. *External status register*. This register contains additional status information used when the ISB 3700 is being used with a modem.

With the card jumpers set up so that the base address is A0, as described in the previous section, the addresses for the registers of channel 1 of the serial interface are

Received-data holding register	A0 (read only)
Transmit-data holding register	A0 (write only)
Status register	A1 (read only)
Mode register 1	A2 (write only)
Mode register 2	A2 (write only)
Command register	A3 (write only)
Interrupt vector register	All addresses from A8 to AB
External status register	All addresses from B0 to B3
Interrupt control register	All addresses from B4 to B7

The received-data and transmit-data registers share the same address. The received-data register is accessed when a read operation is performed and the transmit-data register is accessed when a write operation is performed. The two mode registers also share the same address. In this case, a defined sequence must be followed in accessing the two registers. Data must be written to mode register 1 first, then data can be written to mode register 2.

Initializing the serial interface   To initialize a serial interface channel in the ISB 3700, the microcomputer must write control bytes to the two mode registers and the control register. A C function, initserial( ), which initializes channel 1 of the ISB 3700 so that it can communicate with the robot-arm controller, is listed below:

```
#define BASE            0xA0         /* I/O base address for serial I/O board */
#define SERDATAREG      BASE         /* Tx and Rx data register address */
#define SERSTATUSREG    BASE+1       /* Status register address */
#define SERMODEREG      BASE+2       /* Mode register address */
#define SERCOMREG       BASE+3       /* Command register address */
#define SERINTCNTRL     BASE+0x14    /* Interrupt control register address */

#define MW1             0x4D         /* Mode register 1 control word. Sets
                                        interface up for 1 stop bit, no parity,
                                        8 data bits, asynchronous mode */
#define MW2             0x35         /* Mode register 2 control word. Selects
                                        internal clock for data rate timing and
                                        sets data rate to 300 baud. */
#define CMD             0x15         /* Command register control word. Selects
                                        normal operation, forces RTS and DTR
                                        high (i.e., asserted ON), resets error
                                        flags, and turns on transmitter and
                                        receiver */
#define INTOFF          0            /* Interrupt control register control word.
                                        Disables interrupts */
initserial( )
{
output(SERMODEREG, MW1);  /* Write control byte to mode register 1 */
output(SERMODEREG, MW2);  /* Write control byte to mode register 2 */
output(SERINTCNTRL, OFF); /* Write control byte to interrupt control
                             register */
output(SERCOMREG, CMD);   /* Write control byte to command register */
}
```

The function which initializes the serial interface is very simple. The difficult part in designing the function is determining the proper control words to load into the registers. Try to derive the control words yourself, using the tables defining the functions of the individual bits in the ISB 3700 registers.

Sending and receiving data via the serial interface   Flowcharts for simple functions to send and receive data are shown in Figs. 14.24 and 14.25. These functions are listed below:

```
#define RXRDYMSK    2     /* Masks off receive register status
                             bit in status register */
#define NODATA      0

serialin( )               /* gets one character from serial interface
                             and returns it */
{
int status;
int data;
status = NODATA;
```

```
while((status & RXRDYMSK) == NODATA)     /* Wait until the receiver has a */
        {                                /* character */
        status = input(SERSTATUSREG);
        }
data = input(SERDATAREG);
return(data);
}

#define TXRDYMSK    1     /* Masks off transmit register status bit
                             in status register */
#define NOTREADY    0

serialout(data)             /* sends one character to serial interface */
char data;
{
int status;
status = NOTREADY;
while((status & TXRDYMSK) == NOTREADY)    /* Wait until the transmitter
                                             is ready */
        {
        status = input(SERSTATUSREG);
        }
output(SERDATAREG, data);
}
```

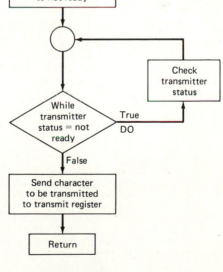

Figure 14.24 serialin( ) flowchart.

Figure 14.25 serialout( ) flowchart.

These functions are extremely crude. They perform no error checking, and more importantly, they force the microcomputer to wait until the serial interface has received a character or is ready to transmit a character. They are adequate for the torque-wrench-controller design because the microcomputer has no other functions to perform when it is using the serial interface.

However, in most applications it is not feasible to have the microprocessor continuously checking the serial interface status register to see whether it can transmit or receive data. Instead, the serial interface is set up to interrupt the microprocessor when a character is received or the transmitter is ready to send another character.

14.8.4 Real-Time Clock

The programmer's model for the Intel 8253 counter-timer IC on the microcomputer board is shown in Fig. 14.26. Only the portions used in the present design are shown. The control register is used to select the operating modes for the counters while the counter load register is used to load the initial value into the counter which is then counted down. The bits in the control register are divided into four fields, each of which controls a particular aspect of a counter's operation. The SC field (bits D_6 and D_7) selects the counter to which the

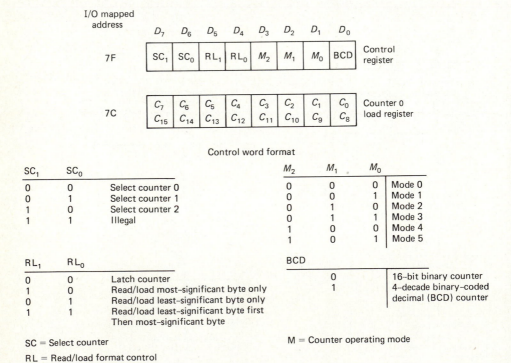

Figure 14.26 Programmer's model of Intel 8253 counter-timer IC.

remainder of the control word will be applied. The RL field (bits D_4 and D_5) determines how the counter load register will be used. Since this register is only 8 bits long whereas the counter can contain 16-bit numbers, it is possible to set up the register to load its contents into the most-significant 8 bits of the counter or into the least-significant 8 bits or to load the first 8 bits received into the least-significant 8 bits of the counter and then load the next 8 bits received into the most-significant 8-bit positions in the counter. The M field (bits D_1, D_2, and D_3) set the counter mode. The counter mode determines such things as whether the counter counts down once or repetitively from the initial count and the functioning of the output line and gate input line of the counter. Finally, the BCD field (bit 0) determines whether the counter will count in binary or in binary-coded decimal (BCD) format.

For this design we wish to use counter 0, so the SC field of the control register will be set to 00. It will be necessary to load the counter with 16-bit values. Therefore, the RL field should be set to 11 so that the 16-bit value can be sent to the counter in two successive write operations to the counter load register. In this application, counting is most readily done in binary, so the BCD field will be set to 0.

The detailed description of the five possible counter modes for the Intel 8253 is given in the data sheet listed in the references at the end of this chapter. The mode that meets the requirements of this design most closely is mode 0: interrupt on terminal count. In this mode the counter output line is initially low after the mode has been set by writing to the control register. After the 16-bit count value is loaded into the counter, the output line will remain low but the counter will start counting the clock pulses sent to it. When the counter value reaches zero (the terminal count), the output line goes high and stays high until a new count value is loaded into the counter. A real-time clock can be implemented by using the counter output line to interrupt the microprocessor and designing the interrupt service routine to reload the count into the counter load register to cause the output line to go low again and to start a new timing interval. To select mode 0, the M field in the control register must be set to 000.

The functions which control the real-time clock are listed below:

```
#define CONTROLWORD    0x30    /* CONTROL register contents to
                                  set up counter 0 in desired
                                  configuration */

#define HICOUNT        0x01
#define LOCOUNT        0xF4    /* To generate a 2-ms delay with a
                                  250,000-Hz clock frequency, the ini-
                                  tial count must be 500 decimal or 1F4
                                  hexadecimal. HICOUNT is the most-
                                  significant 8 bits of this count and
                                  LOCOUNT is the least-significant 8
                                  bits. */

#define CNTRLREG       0x7F    /* I/O mapped address of CONTROL
                                  register */
```

```
#define CNTROLD        0x7C    /* I/O mapped address of counter 0
                                  LOAD register */

int clocktick;                 /* definition of external variable
                                  used to indicate that a real-time
                                  clock interrupt has occurred */

initclock( )           /* real-time clock initialization */
{
output(CNTRLREG,CONTROLWORD);  /* send control word to 8253 CONTROL
                                  register */
}

startclock( )    /* start real-time clock */
{
output(CNTROLD,LOCOUNT);
output(CNTROLD,HICOUNT);  /* send low and high bytes of count
                             to 8253 Counter 0 LOAD register,
                             thereby starting the counter */
enable( );    /* enable 8085 interrupt system so
                 interrupts are recognized */
}

stopclock( )    /* stop real-time clock */
{
disable( );     /* disable 8085 interrupt system so that interrupts
                   are ignored */
}

clockint( )     /* real-time clock interrupt service routine */
{
clocktick=1;    /* set clocktick flag to signal other functions
                   that the interrupt has occurred */
startclock( );  /* start the real-time clock again to generate
                   next 2-ms interval */
}
```

The Intel 8253 is initialized by the initclock( ) function which sets up the control register. This is done during the INITIALIZE state within the initialize( ) function. At this point the counter is still inactive since no initial count has been loaded into it. In addition, the 8085 interrupt system is disabled and therefore won't recognize any interrupt signals at its interrupt input terminals, because the 8085 interrupt system is automatically disabled when the microcomputer is turned on. The interrupt system must be explicitly enabled by an interrupt enable instruction before it will respond to interrupt signals.

When it is necessary to start the real-time clock, the count must be loaded into counter 0 to start the counting operation and the interrupt system must be enabled so that the interrupt generated on the terminal count will be recognized. The function startclock( ) performs these two operations. The function enable( ), which enables the interrupt system, is an assembly-language function which is discussed in the next section.

The real-time clock can be stopped at any time by disabling the interrupt system. The counter will continue to count down to the terminal count and the output line will then go high, but it will be ignored by the microcomputer, since interrupts are disabled. Therefore, the function stopclock( ) simply calls the assembly-language function disable( ) (discussed in Sec. 14.9) which disables the 8085 interrupt system. The clock can be started again at any time by calling the startclock( ) function.

The interrupt service routine (clockint( )), executed when a real-time clock interrupt occurs on the terminal count, must do three things. It must set a flag variable, clocktick, to 1 to signal other functions that the interrupt has occurred. Then it must reload counter 0 with the initial count to restart the counter and it must reenable the interrupt system, since the interrupt system is automatically disabled when an interrupt is recognized. These two operations can be performed by calling the function startclock( ).

The function wait50( ), which generates the 50-ms delays required to allow the solenoid valves to open and close, can be designed so that it uses the real-time clock. All it must do is start the clock, count 25 of the clock interrupts which occur every 2 ms, and then stop the clock. The listing for this function is given below.

```
wait50( )
{
int count;
clocktick = 0;
startclock( );
for(count = 0; count < 25, count++)
        {
        while(!clocktick)
        /*wait*/;
        clocktick = 0;    /* reset flag */
        }
stopclock( );
}
```

14.9 SOFTWARE DESIGN III: ASSEMBLY-LANGUAGE FUNCTIONS AND SYSTEM INITIALIZATION

Although the bulk of the software for this design can be written in the C programming language, a small amount of programming must be done in the assembly language of the 8085 microprocessor. There are two areas where assembly-language programming is required. When the microprocessor begins execution, a small amount of initialization must be done in assembly language before execution can be transferred to the C program. The 8085 interrupt system is not directly accessible by statements in the C language, so assembly-language functions must be written to control the interrupt system. In addition, a small amount of assembly-language code is required to interface interrupt service routines written in C [such as the clockint( ) function in the previous section] with the 8085 interrupt handling mechanism.

14.9.1 Programmer's Model of the Microprocessor

The programmer's model of the Intel 8085 microprocessor's internal registers is shown in Fig. 14.27. It is similar to the model for the Intel 8085 discussed briefly in Chap. 2 (see Fig. 2.7). The accumulator is a data register that acts as a source or destination register for many 8085 instructions, particularly those involving arithmetic or logical operations. The flag register contains bits which store information related to the result of an arithmetic or logical operation. For example, one bit indicates whether the result of an operation is zero or nonzero. The three register pairs—BC, DE, and HL—can be used to store either data or addresses. Each pair can be used as a unit to hold a 16-bit value (memory addresses in particular) or each register in the pair can be accessed separately to store 8-bit data. The program counter register contains the address of the program instruction currently being executed.

Stack operations In common with most other microprocessors, the 8085 makes use of a stack in memory to store the return address when a subroutine is called (see Sec. 7.3). The stack is also used to pass subroutine parameters and to store the system state during interrupt processing. The stack pointer register contains the address of the top of the stack. The contents of a register pair can be stored on the stack by a PUSH instruction. For instance, the assembly-language instruction

PUSH B

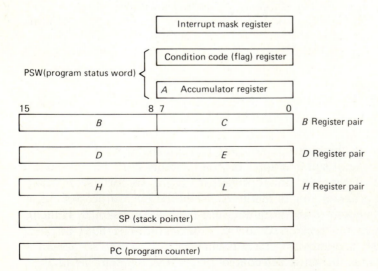

Figure 14.27 Programmer's model of Intel 8085 microprocessor.

stores the contents of the B and C registers on the stack. The contents of the top of the stack can be stored in a register pair by a POP instruction. For the purposes of stack operations, the accumulator and flag registers are grouped into a pair called the PSW register pair. The stack pointer is decremented as data is pushed onto the stack and incremented as data is popped off the stack. As a result, the stack grows from high-memory addresses to low-memory addresses.

The contents of the stack pointer can be accessed by means of a special instruction, SPHL, which swaps the contents of the stack pointer register and the HL register pair. Therefore, the stack pointer can be initialized by loading the desired value into the HL register pair using normal 8085 data move instructions and then executing SPHL to move the value to the stack pointer register.

Interrupt system As mentioned previously, there are five interrupt inputs to the 8085:

> TRAP
> INT7.5
> INT6.5
> INT5.5
> INTR

Of these, the TRAP input is always enabled and can never be disabled; it is said to be *nonmaskable*. The remaining four inputs can be enabled or disabled as a group. Special instructions are included in the instruction set to allow the assembly-language programmer to enable or disable this group of interrupt inputs. The EI instruction enables the group and the DI instruction disables the group. We will refer to this group enable and disable as enabling and disabling the interrupt system even though the TRAP input isn't affected. The INT7.5, INT6.5, and INT5.5 inputs can also be individually disabled, if the entire group is already enabled, through the use of interrupt mask bits in the interrupt mask register.

Figure 14.28 shows the interrupt mask register in detail. The purpose of the individual bits differs depending on whether the register is being read from or written to. In the read mode, the 3 bits devoted to pending interrupts ($I_{7.5}$, $I_{6.5}$, and $I_{5.5}$) indicate whether the corresponding interrupt input has a pending interrupt when interrupts are disabled. The interrupt enabled flag is set to 1 when the interrupt system is enabled and is reset to 0 when the interrupt system is disabled. The interrupt mask status bits ($M_{7.5}$, $M_{6.5}$, and $M_{5.5}$) indicate whether the corresponding interrupt input has been disabled. If the bit is a 1, the corresponding input is disabled, otherwise the input is enabled, provided that the entire interrupt system is enabled.

In the write mode, the mask set enable bit and the three interrupt mask bits ($M_{7.5}$, $M_{6.5}$, and $M_{5.5}$) are used to selectively disable the corresponding

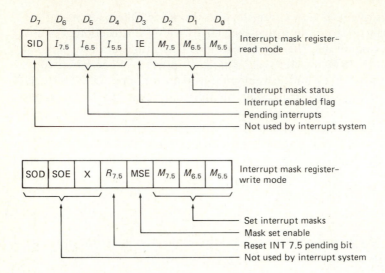

Figure 14.28 Intel 8085 interrupt mask register.

interrupt inputs. In order to change the interrupt mask bits, the mask set enable bit in the word written to the register must be 1. Then a 1 written to an interrupt mask bit will disable that particular interrupt input. For instance, writing a 1 to $M_{7.5}$ will disable the INT7.5 interrupt input and interrupt signals received at that input will be ignored. Writing a 0 to a interrupt mask bit will enable the corresponding interrupt input, provided that the entire interrupt system is enabled.

The assembly-language programmer can read the interrupt mask register by the use of a RIM (read interrupt mask) instruction which copies the contents of the interrupt mask register into the accumulator. The interrupt mask register can be written to by use of the SIM (set interrupt mask) instruction which copies the contents of the accumulator into the interrupt mask register.

14.9.2 System Initialization: Stack Pointer

When power is applied to the Intersil ISB-3111 microcomputer card, a circuit on the card forces the Intel 8085 microprocessor to perform a reset operation. After a reset, the program counter register contains 0000, the interrupt system is disabled, and the three interrupt mask bits in the interrupt mask register are set to 1, disabling their corresponding interrupts. The state of the other registers is unspecified. It is from this initial condition that the program must begin execution. The first operation performed should be the initialization of the stack pointer register, since the program cannot even call a subroutine without using the stack. The stack pointer should initially point to the highest address in RAM memory, since the stack will grow down. After that the program can jump to the C function main( ) which will call its own initialization function to initialize the rest of the system. The assembly-language code required is listed below.

```
EXTERNAL  main       ;indicate to linker that ''main'' will be
                     ;found in another module

STKTOP EQU  37FF     ;highest address in RAM, top of stack

CSEG                 ;assembler directive indicating that
                     ;the following is a program segment

START:LXI H,STKTOP   ;load address of top of stack into HL reg. pair
      SPHL           ;swap contents of SP reg and HL reg pair
      JMP main       ;jump to start of C program
```

The code above must be linked with the remainder of the program so that it is located at location 0000 in memory where execution begins after the power-on reset operation.

14.9.3 System Initialization: Interrupt System

The next order of business is to provide some assembly-language functions that can be called from the C program to control the enabling and disabling of the interrupt system. The functions to enable and disable the entire system are straightforward:

```
PUBLIC enable
PUBLIC disable      ;let the linker know that these labels
                    ;will be used in other modules

enable: EI          ;enable interrupt system
        RET         ;return to calling function

disable: DI         ;disable interrupt system
         RET        ;return to calling function
```

As long as the relocatable object code for these functions is linked with the relocatable object code of the C program, they can be called by the C program just as if they were functions written in C. Note, however, that the C compiler documentation should be studied carefully for any special requirements in linking assembly-language functions to C programs. Some compilers, for instance, expect a prefix on the assembly-language label for the function. For example, an assembly-language function called

```
enable( )
```

in the C program might have to be labeled as

```
cenable:
```

in assembly language.

The assembly-language function to set the interrupt mask register is somewhat more complicated than the previous functions, since a parameter (the desired register contents) must be passed from the calling function. Most C compilers for the 8085 microprocessor pass parameters on the stack in a fashion similar to that described for Motorola 68000 C compilers in Chap. 7. Thus, when the assembly-language function is first entered, the top of stack will contain the return address of the calling function, and below it will be the parameters. The order of the parameters on the stack varies from compiler to compiler and must be determined by referring to the compiler documentation. Since the function under discussion has only one parameter, the order isn't important. The function to set the interrupt mask register is listed below.

```
PUBLIC setmsk
setmsk: POP H      ;pop return address off stack and store in
                   ;HL register pair
        POP D      ;pop mask parameter off stack and store in
                   ;DE register pair
        MOV A,E    ;move mask (8 bits only) from E reg to
                   ;accumulator
        SIM        ;Set Interrupt Mask instruction
        PUSH D     ;push parameter back on stack
        PUSH H     ;push return address back on stack
        RET        ;return to calling function
```

The parameter is pushed back onto the stack in order to leave the stack pointer unchanged upon return to the calling function. Most C compilers expect the stack pointer to be unchanged across a function call. It is wise to check the compiler documentation to see if there are any other special requirements.

Although it isn't required for this design, a function to allow the interrupt mask register to be read is useful in many cases. This function must return a value (the contents of the interrupt mask register) to the calling function. Most C compilers for the 8085 appear to use the HL register pair for returning values from a function. Therefore, the function should look like this:

```
PUBLIC  rdmsk

rdmsk:  RIM        ;Read Interrupt Mask instruction
        MOV L,A    ;move result from accumulator to L reg
        MVI H,00   ;clear H register by loading 00
        RET        ;return to calling function
```

14.9.4 Interrupt Handling

The interrupt service routine for the real-time clock interrupt, clockint( ), was listed in the previous section. However, some additional programming must be done to ensure that this routine is entered properly and that the interrupted function is reentered correctly. In the Intel 8085, an interrupt on the TRAP, INT7.5, INT6.5, or INT5.5 lines causes the processor to disable the interrupt system, push the contents of the program counter (the address of the next instruction in the function being interrupted) onto the stack, and then jump to one of the following addresses:

$$
\begin{array}{ll}
\text{TRAP} & \text{0x24} \\
\text{INT5.5} & \text{0x2C} \\
\text{INT6.5} & \text{0x34} \\
\text{INT7.5} & \text{0x3C}
\end{array}
$$

These interrupt inputs can be said to be *autovectored* inputs, since the interrupt vector is predetermined. The response to an interrupt on the INTR line is similar except that the interrupting device must place an interrupt vector on the data bus which determines the address to which the processor will jump.

The interrupt handling sequence for the INT6.5 input used for the real-time clock is shown in Fig. 14.29. Once the Intel 8085 microprocessor has responded to the interrupt and has jumped to location 0x34, the program at

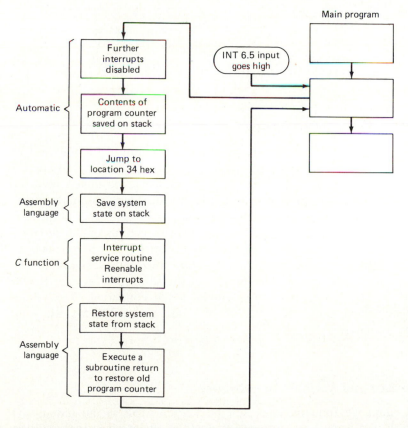

Figure 14.29 Interrupt handling procedure for INT6.5.

that location must save the system state so that the interrupted function can resume properly once the interrupt service routine has executed. Then the interrupt service routine can be called. Upon return from the interrupt service routine, the system state must be restored to that which existed when the microprocessor was interrupted; and interrupts must be reenabled, if that wasn't already done in the interrupt service routine. Finally, a subroutine return instruction is used to recover the program counter contents that were saved at the beginning, and the interrupted program resumes execution. The required assembly-language code is shown below:

```
EXTERNAL clockint

ORG   34H                ;must start at address 34H for INT6.5
      PUSH PSW
      PUSH B
      PUSH D             ;save processor state (i.e. registers)
      PUSH H             ;on stack
      CALL clockint ;call interrupt service routine
      POP H
      POP D
      POP B              ;restore processor state from stack
      POP PSW            ;note that EI (enable interrupt system)
                         ;isn't required since it was enabled in
                         ;clockint( )
      RET                ;return to interrupted function
```

14.9.5 Linking the Software Modules

This completes the assembly-language programming required for this design. The entire assembly-language portion of the program can be placed in one file and assembled into a single relocatable object module to be linked with the relocatable object modules produced by the compilation and assembly of the C-language portions of the program. The modules must be linked so that the assembly-language code will precede the C-language portions of the program and the linker must be instructed to locate the code starting at location 0000. In addition, the linker must be instructed to place the data segment of the program (i.e., addresses for variable storage) at the starting address of RAM. The product of the linking operation will be a file consisting of machine-language code which can be loaded directly into the EPROMs using an EPROM programmer.

14.9.6 Interface and Variable Initialization

One final C-language function must still be written. This is the initialize( ) function which performs the required initializations of interface registers and program variables. It is listed below.

```
#define INTMSK 0x0D     /* interrupt mask register contents */

initialize( )
{
external int solstate; /* declare solstate as external */

solstate = 0;               /* initialize solstate */
release( );
stopwrench( );              /* ensure that solenoids are inactive */
initclock( );               /* initialize real-time clock */
initserial( );              /* initialize serial interface */
setmsk(INTMSK);             /* unmask INT6.5 input */
}
```

SUMMARY

We have presented the design of a small real-time microcomputer system based on commercially available circuit cards. This is a particularly simple system, but the design approach presented can be applied to more complex systems. The hardware design consists mainly of determining the specifications for the components in the system and then choosing boards that meet these specifications. Little or no design at the level of individual integrated circuits is required.

The software is written primarily in a high-level programming language. A small amount of assembly-language programming is required to initialize some microcomputer registers and to give the high-level-language program access to the microcomputer interrupt system. The assembly-language routines are sufficiently short and simple so that no great expertise in assembly-language programming is required to write them.

EXERCISES

14.1 Using the technique described in Sec. 4.4, make an estimate of the length of the program for the torque-wrench controller. How does the estimate compare with the actual length of the program?

14.2 Condensed specifications for three A/D converter boards for the STD bus are given in the table below.

	Board A	Board B	Board C
Resolution	8 bits	12 bits	16 bits
Input voltage range	0 to 5 V, 0 to 10 V, ±5 V, ±10 V (jumper selectable)	0 to 10 V, ±10 V (jumper selectable)	0 to 5 V, 0 to 10 V, ±10 V (jumper selectable)
Conversion time	33 μs	28.5 μs	10 ms
Accuracy at 25°C	±0.4 percent of full-scale range (FSR)	±0.03 percent of FSR	±0.0075 percent of FSR
Temperature coefficient			
Gain	±20 ppm/°C of FSR	±30 ppm/°C of FSR	±10 ppm/°C of FSR
Offset	±35 ppm/°C of FSR	±50 ppm/°C of FSR	±15 ppm/°C of FSR

Which of these boards would be suitable for the torque-wrench controller? Explain your reasoning.

14.3 Describe the jumper connections required if four 2716 EPROMs ($2K \times 8$) were used instead of the three 2732 EPROMs. Draw the memory map for this new configuration.

14.4 Describe in detail the software and hardware changes required if the INT5.5 interrupt input rather than the INT6.5 input is used for the real-time-clock interrupt.

14.5 A possible cause of incorrect torque-wrench operation is loss of air pressure in the pneumatic circuits. A pressure sensor is to be installed to monitor the air pressure. The sensor plus its amplifier is linear and has a gain of 30 psi/V. There is no offset voltage and the maximum output is 10 V. The pneumatic circuit pressure is to be monitored by the torque-wrench controller prior to a gripping or fastening operation. If the pressure is below 100 psi, the torque-wrench controller sends a fault message to the robot-arm controller and does not attempt to carry out the gripping or fastening operation. Modify the torque-wrench-controller design to meet these requirements and present complete documentation of the hardware and software changes required.

14.6 Write a C-language function to initialize channel 1 of the ISB-3700 serial I/O board. The base address of the channel 1 registers is C0 hex. The interface must be initialized to the following state:

1. Asynchronous operation with internal clock
2. 7 data bits
3. 1 stop bit
4. Odd parity enabled
5. Data rate of 1200 bits/s (baud)
6. Channel 1 interrupt enabled
7. Interrupt vector is 02 hex

14.7 Write a more sophisticated version of the function serial_in( ) that checks for parity, overrun, and framing errors. The function should call an error-handling function, rxerror(type), if an error is detected. A code indicating the type of error should be passed to rxerror( ). You do not have to write the rxerror( ) function.

14.8 The nuts which the automatic torque wrench cannot tighten properly will be tightened by human workers. In order to speed up this manual rework, a marking system is to be installed on the torque wrench to mark the fasteners which haven't been tightened properly. The marking system consists of a paint sprayer mounted on the torque wrench. The sprayer is controlled by a solenoid valve with characteristics identical to those already used on the torque wrench. A burst of paint with a duration of 100 to 300 ms is enough to make a clear mark on the fastener that can be spotted by the human workers. Add this new system to the torque-wrench-controller design. Document the addition thoroughly.

14.9 During the debugging of the torque-wrench controller, a video display terminal is connected to the serial I/O port to simulate the robot-arm controller. A second terminal is connected to the second serial I/O port in the system to allow the system to display debugging information. Develop a test plan for the system. Describe how the program will be "instrumented" to assist in debugging. Include listings of the instrumented functions. The torque wrench may not be available for use during debugging. Write a "stub" function that simulates the torque sensor data that would come from the torque wrench. How could you verify that the solenoid drivers are working properly in the absence of the torque wrench?

BIBLIOGRAPHY

Microcomputer board

Intel Corp.: *Intel Component Data Catalogue*, 3065 Bowers Ave., Santa Clara, CA 95081. (Contains data sheets for 8085 microprocessor and 8253 counter-timer IC.)

Intersil Inc. Systems Division: *ISB-3111 8085 Based Processor Card: Product Specification*, 1275 Hammerwood Ave., Sunnyvale, CA 94086.

Leventhal, L. A.: *8080A/8085 Assembly Language Programming*, McGraw-Hill, New York, 1978.

A/D board

Analog Devices Inc.: *RTI-1226 User's Manual*, P.O. Box 280, Norwood, MA 02062.

Solenoid driver boards

Gordos International Corp.: *Gordos Off-the-Shelf Distributor Catalog*, 250 Glenwood Ave., Bloomfield, NJ 07003.

Pro-Log Corp.: *7507 General Purpose Interface Card User's Manual*, 2411 Garden Road, Monterey, CA 93940.

Serial interface board

Intersil Inc. Systems Division: *ISB-3700 Dual Independent Channel, Synchronous/Asynchronous Communications Card: Product Specification*, 1275 Hammerwood Ave., Sunnyvale, CA 94086.

Power supplies

Pressman, A. I.: *Switching and Linear Power Supply, Power Converter Design*, Hayden, Rochelle Park, NJ 1977.

Chryssis, G.: *High Frequency Switching Power Supplies: Theory and Design*, McGraw-Hill, New York, 1984.

Thermal design

Steinberg, D. S.: *Cooling Techniques for Electronic Equipment*, Wiley, New York, 1980.

Software

Manx Software Systems: *Aztec C Version 1.05 User's Manual*, P.O. Box 55, Shrewsbury, NJ 07701.

Supersoft Inc.: *Supersoft C Version 1.2 User's Manual*, P.O. Box 1628, Champaign, IL 61820.

(These two C compilers are suitable for developing the software for this design. Several other suitable compilers are also available.)

Microsoft Corp.: *Utility Software Manual*, 10700 Northup Way, Bellevue, WA 98004.

(Documentation for MACRO-80 Assembler and LINK-80 Linker/Loader which are used in conjunction with the above C compilers.)

IMPLEMENTATION TOOLS

FIFTEEN

DEVELOPMENT SYSTEMS

Microcomputer development systems can be used for both software and hardware development. Figure 15.1 shows the components of a complete microcomputer development system. The function of the various components and possible configurations of these components will be discussed in this chapter.

Because the focus of this book is on the development of board-level systems in which commercially available boards are integrated into a system, the hardware-development requirements take on a minor role compared with the software-development requirements.

The essential capability that a software-development system must have is to be able to write and correct (*debug*) the software. This usually requires the following components:

1. Suitable peripheral devices for software development
2. An operating system that supports the development system software
3. Software for writing, editing, and saving programs
4. Software for compiling and assembling programs
5. Software for linking relocatable object modules for both software and hardware for loading programs into the target system memory
6. Software and hardware for debugging programs

Each of these capabilities will be discussed in turn in the following sections. Other software-development capabilities for jobs such as word processing and project management will be discussed in Sec. 15.7.

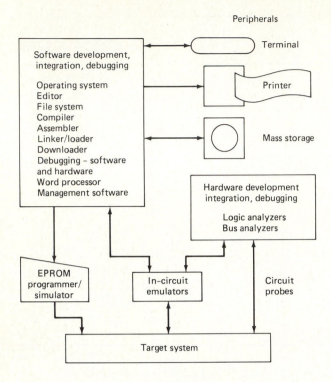

Figure 15.1 Components of a microcomputer development system.

15.1 PERIPHERAL DEVICES

Although there are many different configurations for a software-development system, there are certain necessary *peripheral devices* (see Sec. 3.4) in addition to the software components.

In order to support effective software development, a development system must be equipped with CRT terminals for the programmers to use in entering and editing their programs.

With the exception of some development approaches for very small projects, the amount of software required for even the few software development functions listed above is considerable. This software cannot all be conveniently held in the solid-state program memory (PROM or ROM) of a development system. Thus a *mass storage* system is required—usually a high-speed nonremovable disk (e.g., a Winchester-type disk system). Also, the programs that are developed must be backed up on some form of removable storage medium for security and archiving. Usually a lower-speed removable disk system such as a floppy disk would be added to the system for this purpose. Finally, the system requires a line printer to allow the production of program listings and other printed documents.

15.2 OPERATING SYSTEM

An operating system acts as the basis for the software on the development system. The operating system is a program that forms an intermediary between the user and the development-system hardware (see Sec. 3.5 and Chap. 16). It controls the printers, the terminals, and the mass storage system. The operating system usually manages a file system that allows the user to assign a name to a collection of data such as a program and then store this named collection (called a *file*) on the mass-storage system. The user can then recall the file by giving the name of the file and the appropriate command to the operating system. Operating systems will be discussed in more detail in Chap. 16.

15.3 EDITOR

An editor program is used to create the data that constitutes a file. The editor allows a user to enter data (usually text) into the system from a terminal and then correct or alter the data as required. Editors come in three basic types:

- Line-oriented
- Screen-oriented
- Language-oriented

Early editors were designed for use with electromechanical communication terminals (e.g., Teletype KSR-33). Such editors were *line-oriented*, viewing the entered data as a series of numbered lines. In order to make a change, the user must specify the line number where the change is to be made.

More recently designed editors are oriented to video terminals. They make use of the *cursor*, a lighted block or dash on the screen which shows where the next character will be entered. The user can move the cursor around the screen by using the keys on the terminal keyboard. Changes, insertions, and deletions are made by moving the cursor to the place in the entered text where the alteration is desired. These *screen-oriented* editors are easier to use than line-oriented editors and immediately show the results of an editing operation.

Since in software development the editor is used to enter the programs that are being developed into the system, some development systems have *language-oriented editors* designed to be used with a particular programming language. These editors check the program for errors in *syntax* (rules for program construction) as it is being entered. An example of such an error in a C program would be the omission of a semicolon at the end of a statement. This checking in the editor speeds up the development process, since corrections are made more quickly when the programmer doesn't have to wait for the compiler to find the syntax error. Language-oriented editors are also able to perform automatic program formatting.

15.4 OBJECT-CODE GENERATION

Once the program has been entered into the development system, it must be translated into *object code* (machine code) that can run on the target system. If the development system and the target system use the same type of computer, *native* compilers and assemblers can be used to do the translation. These produce object code that will run on both the development and on the target system.

If the development system and the target system are based on different types of computers, *cross-compilers* and *cross-assemblers* must be used to do the translation. These programs run on the development-system computer but produce code for the target-system computer.

15.5 SOFTWARE INTEGRATION

The process of software integration consists of two steps. First, if the software has been developed and translated into machine language as a set of modules, the program modules must be *linked* together into one program. This program must be *relocated* to the range of memory addresses it will occupy in the target system.

Some simple software-development systems based on *absolute assemblers* do not allow the linking and relocation of machine-language modules. Absolute assemblers produce machine language that will operate only in a specified area of memory. The programmer specifies the starting memory address for the machine-language program when the assembly-language program is written. In addition, the entire program must be assembled at the same time. Modules cannot be assembled separately and then linked together. The use of an absolute assembler makes the independent development of software modules by different programmers rather difficult. Also, machine language generated by an absolute assembler is difficult to interface to the machine language generated by a high-level language translator. Therefore, absolute assemblers should be avoided for all but the simplest software-development projects.

The second step of software integration is to transfer the linked and relocated machine-language program to the target system. This can be accomplished in a number of ways. Two of these are

- Downloading
- EPROM programming

15.5.1 Downloading

Let us assume that the target system can be connected to the development system as if it were a terminal. The bytes of object code can then be serially transferred or *downloaded* to the target system by the development system. In order to receive the program we will assume here that the target system is equipped with a *monitor program* that allows it to act like a terminal to receive the data from the development system. The monitor must store the downloaded program in the RAM memory locations specified in the downloaded data. Such monitor programs are stored in ROMs and are available for many microcomputer boards.

A good example of such a monitor is "TUTOR," for the Motorola MC68000 Educational Computer Board. It is contained in two 64K-bit ROM memories and controls downloading and debug functions for the board.

Machine-language programs that are downloaded to a target system are often encoded into an *ASCII hexadecimal* format. Each byte of the machine-language program is first converted to two hexadecimal digits and these hexadecimal digits are then converted to their ASCII character code equivalents:

$$\begin{array}{cccc} & 1010 & 0100 & 01000001 \quad 00110100 \\ 10100100 & & & \\ & A & 4 & \text{"A"} \qquad \text{"4"} \end{array}$$

$$\text{Binary} \longrightarrow \text{Hexadecimal} \longrightarrow \text{ASCII code}$$

This ASCII-encoded data is transferred as a series of *records*. The ASCII encoding allows the data to be directly viewed on a display terminal. Each record also contains the starting address of the block of memory where the data in the record is to be stored in the target system's memory. This address is determined by the linker in the host system.

Checksum   Error checking may also be included in the record. One error-checking technique is the use of a *checksum*. The checksum of N bytes can be computed as the sum of the N bytes (mod 2). The checksum is the last data sent in the record. Upon receiving the data, the target system can compare the transmitted checksum with the sum it generates on the data received. If there is a difference, then a transmission error has occurred.

The following is an example of the object code of a small program that has been converted to a hexadecimal downloading format called "S-records." This particular format has been specified by Motorola, but there are many other protocols such as "TEKHEX" defined by Tektronix for use in the Tektronix development systems.

Example: Program object code in Motorola S-records   Each line of the following sequence of S-records follows immediately after the previous line in the downloading sequence.

The S_0 record is a header record and contains the ASCII-coded file name.

The S_1 records contain the starting address (in hexadecimal) for loading the program in target memory, and all the instructions in the program are encoded with a hexadecimal digit for each 4 instruction bits (e.g., the 16-bit MC68000 instruction MOVEQ #1,d0 has a hexadecimal object code of 7001).

The S_9 record contains the starting address for execution of the program. Other types of records are described in Motorola documentation.

S0060000786571AB

    S0 = type of record = S_0 record = header record
    06 = number of hex digit pairs to follow in record
  0000 = zeros
  78657 = ASCII file name
    1AB = *checksum* = the ones complement of the sum of all bytes except S_0

S123100024 . . . 4020

    S1 = (only beginning and ending printed here)
    23 = type of record = S_1 record = address + data
  1000 = number of hex digit pairs to follow in record
    24 = RAM address for loading the following data
. . . 40 = 20 (base 16) hex digit pairs of data
    20 = checksum

S11B1020FFFC70022D40FFF82E2EFFFCDEAEFFF82D47FFF
C4E5E4E752B

    – interpreted as in the previous S_1 record

S90500001000EA

    S9 = type of record = S_9 record = end-of-file + execution address
    05 = number of hex digit pairs to follow in record
  0000 = zeros
  1000 = program starting address
    EA = checksum

15.5.2 EPROM Programming

In some cases the target system does not have the software or the communication facilities to accept the program directly from the development system. In this case the program must be transferred in an EPROM to the target system for testing. Regardless of how the testing takes place though, the program must usually be stored in EPROM or some other permanent form of storage once the program is completely tested and debugged. The development system is usually equipped with an *EPROM programmer*. The development system transfers a machine-language program to the EPROM programmer which then stores the program in an EPROM plugged into a socket on the programmer. The programmed EPROM can then be transferred to a socket on a memory board in the target system, resulting in the transfer of the program from the development system to the target system.

EPROM simulator   Transferring programs via EPROMs is very inconvenient during the test and debug phase, since it takes 15 to 20 min to erase and reprogram an EPROM and this must be done every time a change is made in the program. *EPROM simulators* are designed to eliminate this problem. An EPROM simulator consists of RAM memory attached by a cable to the development system which can then write programs and data into the RAM. The RAM memory can also be accessed by the target system through a cable to a plug whose pin connections exactly duplicate the pin arrangements of an EPROM in the target system. This plug is inserted into an EPROM socket in the target system. Thus the development system can rapidly load machine-language programs into the simulator which looks like an EPROM memory to the target system.

15.6 SOFTWARE DEBUGGING

Debugging a program can be performed using either the development or the target system. As much debugging as possible should be done on the development system because it normally has facilities such as terminals, printers, and disk drives which may not exist on the target system. Also, if testing and debugging is performed on the development system, the time-consuming transfer of programs from the development system to the target system after every change can be avoided.

Debugging in the development system If the development system and the target system use the same type of processor, then testing can be performed by running the machine-language program that was produced by the compiler or assembler on the development system itself.

On the other hand, if the development system and target system are based on different microprocessors, the situation becomes more complex. If the software is written in a high-level language, testing can proceed using one or both of the following methods:

1. The program can be compiled using the native compiler of the development system processor, assembled using the native assembler, and run on the development-system (DS) processor to check the logical flow of the high-level program (see Fig. 15.2).
2. A cross-compiler and cross-assembler can be used that run on the development system processor but produce code for the target system processor. Code written in the assembly language of the target system can also be linked into this software. There are two ways that the resulting code can be tested in the development system itself (see Fig. 15.3):

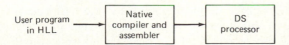

Figure 15.2 Checking a high-level-language program using the processor in a development system.

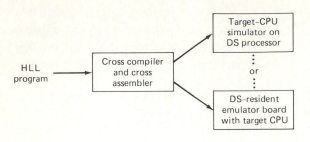

Figure 15.3 Using a target CPU or a simulator to check a high-level-language program.

- Use a *simulator program* which emulates the target processor on the development system processor. That is, a simulator will execute the machine code instructions for the target machine by using combinations of the development system native instructions that achieve the same result.
- Use an *emulator processor* on a board in the development system. The emulator processor is capable of executing the machine code of the target processor although the board will not in general have the same non-CPU facilities.

In either of these cases, complete testing and debugging of the software on the development system is not usually possible. In most real-time systems, the software interacts with the hardware interfaces of the target system (e.g., A/D converters). This interaction is often hard to simulate on the development system. Also, the target system will operate at a different speed than the development system, so time-sensitive programs must be eventually tested on the target system.

Finally, if method 1 is used, differences between the compilers for the target system and the development system will not become apparent until the program is tested on the target system itself. Even if the method 2 is used, the differences between the simulator and the target processor, or between the emulator system and the target system, can lead to performance differences. These differences will also not become apparent until the program is tested on the target system itself.

Debugging in the target system   Unfortunately, the facilities for debugging in the target system are usually rather crude. After downloading the software to the target system, the programmer should ideally be able to completely control the execution of the program, starting and stopping the program at any point, and displaying any variable. The control should be exercised at the level in which the program is written. For example, if the program is written in a high-level language, it should be possible to command the program to start or stop at a certain line number in the high-level language or to display a variable by giving the name of a variable. Simple debugging systems (e.g., Motorola TUTOR monitor) do not approach this level of sophistication, since program starting and stopping points must be specified by the memory address where the program statment is stored. Variables to be displayed must also be specified by the address where the variable is stored. The programmer must determine these addresses from listings of the machine language produced by the translation program. This is obviously a tedious and error-prone process.

Assembly-langauge symbolic debuggers allow reference to assembly-language line numbers and to assembly-language variable names in debugging commands. For programmers using assembly language, these features make the debugger quite convenient, but high-level programmers must still translate back to the original program.

To avoid such translation, *high-level language symbolic debuggers* for some languages that are available for microcomputer development systems allow the programmer to trace C or Pascal variables directly.

Symbolic debugging can be carried out in the target system with the assistance of an *in-circuit emulator* system. One arrangement for this purpose (shown in Fig. 15.4) has an in-circuit emulator with a probe that replaces the target processor. The emulator now controls the target system with a processor of the same type as in the target system. The emulator contains a symbol table

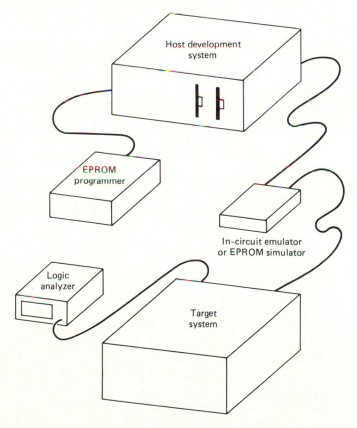

Figure 15.4 A host-based software and hardware development system.

and other information downloaded from the development system to enable symbolic debugging.

The emulator can also control the allocation of blocks of target system memory such that the user program can reside

1. Completely in the emulator, where it can be easily modified
2. Completely in the target system, after it has been debugged
3. Partially in the target memory for debugged portions of the program and partially in the emulator for the rest

In-circuit emulators will be discussed further in the next section.

15.7 OTHER SOFTWARE-DEVELOPMENT-SYSTEM CAPABILITIES

It has been stated that the development effort for a significant software package can be divided into the following major components [1]:

- 25 percent: Software development time
- 50 percent: Debugging and integration
- 25 percent: Administration and project management

So far we have dealt with the first two components, but what sort of facilities are required for the third?

In a major software development much of the project management software can be resident on the development system. Some of these facilities include

- Planning and performance tools (e.g., PERT charts), which can be used to develop contract proposals, allocate personnel resources, locate potential development problems, and assess project progress as milestones are passed.
- Accounting packages, which are used to charge computer usage and programming time to appropriate accounts.
- Word-processing capabilities, which can be used in software development to construct a user's manual. The integration of the word-processing capabilities and the development capabilities allows the programmer to have easy access to the program text as well as the documentation text to simplify creation and modification of each.
- Shared databases. Since many software projects require a number of contributors all working on one aspect of the project, it is necessary to define and share common databases. An example of such a database in C is the use by all team members of a common set of definitions (e.g., the C-language definition "#define CONSTANT 1"). These can all be located in a file and included in the compilation by using "#include".

Other capabilities of a comprehensive development system include optimizing compilers and a "make" command that recompiles only the files containing programs that have been modified since the last time the software system was linked together. The resulting object code for all source files is then relinked to form a new version of the program. The total "make" operation must be specified to the system by a description file or "makefile". The use of the makefile for the generation of executable code has the advantage that it documents and automates the often complex compiling, assembling, and linking of many functions to form a complete program. This capability is available, for example, in UNIX-based development systems.

A well-known maxim states that 80 percent of the execution time of a program can be associated with 20 percent of the code. This leads one to the conclusion that in any program under development, the 20 percent had better be as efficient as possible. In real-time systems the optimization of the program may be necessary to meet performance-time specifications. However, in a complex software system, it may not be clear to the designer just where all the components of critical code are located. Hardware and software tools are available for making this assessment for some development environments.

In the UNIX-based development systems, a C program can be "profiled" to produce a histogram of the amount of time spent in each routine in a program. This is achieved by the compiler's adding special function calls to the user program (upon request).

A combined hardware-software approach has been taken in the Hewlett-Packard HP64000 development system. In this case, a combination of a hardware analyzer that observes bus addresses and development-system software that determines either the percentage of time the program spends in a named user routine or the percentage of time spent in a particular memory address range. The advantage of this approach is that a program can be profiled in its real-time environment.

15.8 EXAMPLES OF AVAILABLE DEVELOPMENT SYSTEMS

Many possible solutions are available for software development. The choice of a particular approach depends on a large number of factors, some of which will be discussed in the next section. The approaches will be presented from the most primitive to the most sophisticated.

Single-board development systems For very small software developments, single-board combined development and execute environments are available using BASIC or FORTH.

For example, STD bus cards with a FORTH interpreter are available with interfaces for a terminal and an I/O port. The STD bus boards for similar types of development are available with BASIC interpreters.

These systems can be inconvenient from a number of standpoints, including ease of program documentation and ability to track down real-time bugs (the execution time cannot be predicted accurately); but for small systems for use in laboratories, the approach may have some advantages.

Target system with removable development boards   Another method of doing software development that does not require the purchase of a complete development system is to augment the target system (see Fig. 15.5) with the

- Cards necessary to provide operating-system support for software development (terminal and printer interfaces, mass storage, extra memory, EPROM programmer)
- Software to carry out the required level of development (operating system, editor, native compiler, assembler, and linker)

Once the project software is complete, the development tools can be removed and used in the next system.

Personal-computer-based target system   In some applications it is possible to use a personal computer as the "target" computer, in which case, the software development takes place in the personal computer. Such systems are suitable for clean laboratory-type environments (see Fig. 15.6). The usual peripherals (floppy disk, display, keyboard) are augmented by I/O interface boards (A/D, D/A, etc.) for data acquisition and control. Examples of such systems are the IBM, Hewlett-Packard, and Digital Equipment Corp. personal computers.

Personal-computer-based development systems   At the lower end of the true development system ladder are personal computers that can be used as software development systems for external target systems. There are a number of systems available that are produced by microprocessor manufacturers.

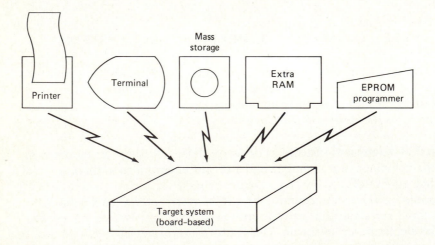

Figure 15.5 Using a target system as a development system.

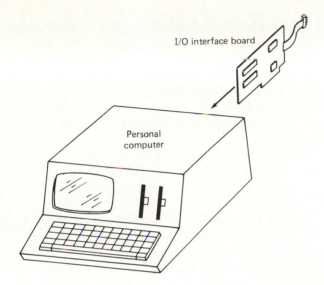

I/O interface board

Personal computer

Figure 15.6 Using a personal computer as a target system.

An example of such a system is the Intel iPDS system that can have both a standard operating system (CP/M) for word processing, data base management, etc., and a software-development operating system (ISIS) for microprocessor development applications.

Another approach is to use a well-known personal computer for which there is a significant amount of cross support for the types of microprocessors that you will be developing. These are also attractive because of the availability of project management, word processing, and other software that will run on these systems.

For microprocessor development, at least one vendor offers C-language support for the Motorola 6809 microprocessor, the DEC LSI-11, the Zilog Z80, and the Intel 8086/8 for any computer that runs one of the following operating systems: CP/M, PCDOS, MSDOS, RT-11, RSX-11, FLEX, UNIFLEX, or OS-9.

Similarly, cross support is available for assembly-language programming.

Dedicated microprocessor development systems Single-vendor microprocessor development systems are available from each of the major semiconductor manufacturers. Whenever a new processor is announced, the appropriate software for development will become available earlier on these development systems. However, once a user has a significant investment in this type of development equipment, it becomes an impediment to switching to another manufacturer's product. Of course, there are other problems with switching, such as the investment in software that may have already been developed around a particular microprocessor. This problem, however, can be minimized by the use of a high-level language that is supported for many microprocessors (such as C and Pascal).

Multivendor microprocessor development systems are available from test instrument manufacturers such as Tektronix or Hewlett-Packard. These companies and others support the development of microprocessors from many manufacturers. These systems are designed to provide sophisticated tools for microprocessor design. Some of the tools come as a side benefit of the UNIX operating system, which supports many tools for software development.

An additional advantage to be gained if an operating system is supported by many software vendors is that other useful software such as graphics packages can become available.

A final advantage of having a well-known operating system as the basis of the development system is that new users may already be familiar with the operating system resources through previous contact and thus their learning time will be shorter.

When a development system is purchased from an instrument supplier, the purchaser is not locked in to a single microprocessor vendor as is the case when a system is purchased from a semiconductor manufacturer. There can be disadvantages, however. The instrument companies will not be as quick to support a new microprocessor, but when one is designing at the board level, this is not a major factor, since it is not wise to design around a product that is not multiply sourced and such product support takes time.

Multiuser systems Software development systems can be either single-user or multiuser. Multiuser systems require more memory, larger mass memory, and faster processors than do single-user systems. A multiuser system actually requires a high-speed disk drive to be able to swap user programs in and out of the memory as the job load changes. Multiuser systems presently available based on stand-alone development systems typically handle up to eight users. Beyond that number a host-based development system is required such as a Digital Equipment Corp. VAX computer.

Downloading from a host-based development system In a host-based system the development software is operated on a mainframe computer alongside other tasks. There are a number of vendors of this kind of software. Examples are Hewlett-Packard, Tektronix, Language Resources, Boston Systems Office, and Human Computing Resources.

In either the host-based or stand-alone development system, the object code must be loaded into the target system and tested. There are several ways (see Fig. 15.7) that this can be accomplished:

1. A PROM programmer can be attached to the host and the code can be "burned" into the PROM.
2. The target system can be connected to the host and the program can be downloaded to the target under the control of a monitor EPROM.
3. An in-circuit emulator can be connected to the host and the code can be downloaded to the emulator memory through the emulator to the target or to a PROM programmer attached to the emulator.

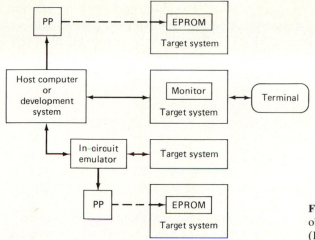

Figure 15.7 Four ways to transfer object code to a target system. (PP = PROM programmer.)

Some emulators can accept a large number of different microprocessors as well as provide facilities for hardware debugging such as logic analysis (specific event-triggered recording of electrical events on the bus and selected test points).

Other less-expensive and more-limited emulator systems are available to

- Examine and change target CPU registers
- Examine and change target memory or I/O ports
- Download programs
- Record program steps taken during execution
- Allow part of the program to be held in the emulator rather than in the target system
- Start, stop, and single-step the program

An example of a system that carries out these functions is the Microtek Lab's MICE-II emulator.

15.9 SELECTION OF A DEVELOPMENT SYSTEM

A very large number of factors can enter into the process of selecting a development system. Some of the factors involved are discussed in this section.

The size of the software project anticipated As we have discussed, the available solutions range from simple single-board systems to large mainframe computers. One of the major factors in determining the development system requirements is the size of the software system under development. If only a single project requiring only several kilobytes of software is anticipated, it is clearly unlikely that a multiuser development system will be required for its production.

Number of users When only a single user is required to be on the development system at one time, a variety of inexpensive single-user solutions are available depending on the size of the software under development. Multi-user systems require more sophisticated hardware and operating system support.

Native or cross development It is usually less expensive and easier to develop a software package on a development system that uses the same type of processor as the target system. Native compilers and assemblers are less expensive, and the code generated can be tested on the same processor as it is developed. Very often, however, a user does not wish to be locked into one manufacturer's product or product line. In this case, a user would either

- Purchase a general-purpose computer for which there is a wide variety of cross-support software available, such as a Digital Equipment Corp. VAX, LSI-11, or IBM personal computer
- Purchase a well-supported development system from an instrument manufacturer

In-house resources Resources can include the availability of computing facilities (VAX, IBM PC), project funding, and personnel experience with a particular operating system or development system.

Need for code optimality If code optimality is a factor, then aids such as optimizing compilers and software performance evaluation may be required.

Hardware-development requirements Hardware can include the availability of CAD packages for VLSI, gate arrays, board layout, and hardware simulators. It can also include the necessity for real-time emulators (no wait-states) and logic analysis.

8-bit, 16-bit, or 32-bit support As the word length increases, the resources of the development system must also increase. One reason for this is that the address space of the 16- and 32-bit microprocessors is much larger. The development of large programs on small machines becomes impractical.

SUMMARY

Software development in an engineering environment requires a disk-based operating system that is usually capable of supporting multiple users working on different program modules.

The use of a development system that is based on cross-compilers enables code to be developed for many different types of microprocessors on the same system. The switch to another microprocessor simply means the purchase of a new cross-compiler rather than a new development system. Personnel can also use the same operating system and editors, thereby reducing the learning time for the switch.

The code can be partially tested in the development system by running the code on either

1. A simulator program that uses the development processor to simulate the execution of target instructions
2. An emulator processor (of the same type as the target CPU) in the development system

Once the code is developed, it must be transferred to the target system. In a separate development system, this can be done by

1. Programming an EPROM and putting it in the target system.
2. Downloading the code serially to the target which uses a monitor program to load and start the program.
3. Downloading the code to an in-circuit emulator attached to both the target and the development system. The code is then run under the control of the emulator.

The target system can be the development system. If the development hardware and software are not required in the target system, they can be removed when the programs are complete.

A multiuser development system that is independent of the target system should provide a number of aids for code optimization, project management, accounting, shared databases, word processing, and system maintenance.

The development-system facilities most appropriate for a given project depend on the

1. Amount of software to be developed
2. Number of users of the development system
3. In-house resources
4. Need for optimality
5. Hardware-development capabilities

EXERCISES

15.1 List the principal hardware and software components required for the development of a major real-time software system. The project will require the full-time activity of three software engineers over a period of 3 months.

15.2 Name three types of editor programs and the advantages of each.

15.3 Describe the difference between a native compiler and a cross-compiler.

15.4 What is the ASCII name of the file used in the example in Sec. 15.5?

15.5 Translate the beginning (shown below) of the second S_1 record given in Sec. 15.5

```
''S11B1020''
```

15.6 Translate (*disassemble*) the machine-language portion of the S-record machine instructions that are listed below. The code appeared in the second S_1 record in the example of Sec. 15.5 as

```
''70022D40FFF8''
```

using the table of MC68000 instructions in App. C and the addressing mode descriptions in Chap. 7.

15.7 Describe several ways (before downloading to the target system) that you could debug a C program that has been written on DEC VAX computer and compiled using a cross-compiler for the Motorola MC68000 microprocessor. Assume that the program is still in VAX files.

15.8 After testing the program of Prob. 15.7 in the VAX, explain how the program can be transferred to the target and how it can be tested there.

15.9 What are the advantages of testing a program in the development system as opposed to the target system.

15.10 Why is testing the program in the target system important?

15.11 If you have a personal computer (e.g., an IBM PC) and you wish to develop software for an STD bus Motorola 6809 microprocessor, make a list of the types of software and hardware you should try to acquire.

REFERENCE

1. Williams, T., and J. Aseo: "Development Systems: Big Chips Require High-Level Tools," *Computer Design*, vol. 23, no. 1, 1984, pp. 149–164.

BIBLIOGRAPHY

The best sources of information on development system capabilities are the companies that supply the systems. There are basically three types of suppliers:

1. Instrument companies such as Gould, Hewlett-Packard, and Tektronix who supply systems for hardware and software development for microprocessors from many different manufacturers.
2. Microprocessor manufacturers such as Intel, National, Motorola, and Zilog who supply systems that support the development of hardware and software for their own microprocessor line.
3. Software houses such as Boston Systems Office, Human Computing Resources, and Language Resources that supply host-computer development software only for different kinds of microprocessors.
4. Emulator suppliers such as Microtek Labs that provide the control of the downloading and debugging process in the target.

There is a certain amount of crossover in these categories. For example, Tektronix also supplies an emulator that can be connected to a general-purpose host. They also supply host-resident software for VAXs. There are also some systems from the semiconductor manufacturers on which other vendor's development software could be mounted.

SIXTEEN
OPERATING SYSTEMS

An *operating system* is a collection of programs that controls and sequences the execution of the user programs on a computer system. It also provides the interface software between the user programs and the hardware of the computer system and allows these programs to communicate among themselves. User programs, or *applications programs*, are designed to carry out specific tasks which represent the end use of the computer. In a real-time system, they are the programs that perform the functions specified in the requirements document. The operating system, on the other hand, is an example of a *systems program* which is concerned with controlling the operation of the computer itself.

Operating systems are complex pieces of software. A complete discussion of the nature and design of operating systems would easily fill a book in itself. In this chapter we will restrict ourselves to a brief overview which introduces the basic concepts and provides some guidance on selecting commercially available operating systems. We have included a number of references at the end of the chapter to allow further reading.

16.1 INTRODUCTORY CONCEPTS AND TERMINOLOGY

16.1.1 Basic Elements of an Operating System

An operating system can usually be divided into three parts: a *command interpreter*, a *nucleus*, and a set of *I/O device drivers* (see Fig. 16.1).

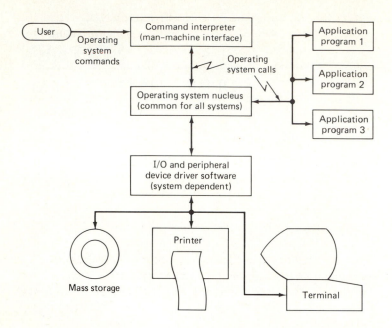

Figure 16.1 Basic operating-system organization.

Command interpreter The command interpreter is a program that allows a human user to give commands to the operating system. For example, the user might wish to command the operating system to retrieve an application program from disk storage, load it into memory, and then transfer execution to this program.

The command interpreter, along with applications programs, makes *system calls* to the operating system nucleus when it wishes to access an I/O device or the mass-storage system. A system call is basically a subroutine invocation that calls the operating-system nucleus with a parameter specifying what operating-system function is to be performed.

Nucleus The nucleus (also called the *kernel*) supplies a *logical interface* between the computer's hardware and the software running on the computer. As an example, a mass-storage system such as a disk drive is made to look like a file system to the programs running on the system. They refer to data on the mass-storage system by passing a file name to the nucleus which handles all the details of going from a file name to a particular storage area in a mass-storage device.

Ideally, the operating-system nucleus is the same on all computers which use the operating system regardless of the actual hardware configuration of the computer. Therefore, an application program can always perform I/O functions or access the mass-storage system if it makes use of the appropriate system calls. The applications program need not be altered to take account of the details of the interfaces to the I/O devices and mass-storage units on different computer systems, since the operating system handles the actual I/O and mass-storage access operations. Operating systems are sometimes called "software buses," since application programs can be "plugged" into any computer system which runs a version of the operating system.

In practice, the portability of application programs among computer systems using the same operating system may be limited by several factors. In some cases, several versions of an operating system exist, each of which has a slightly different nucleus. As a result, an application program which runs successfully on a computer using one version of an operating system may not run on another computer using a different version. Another problem, familiar to users of personal computers, is that the designers of some applications programs bypass the operating system and access system hardware directly in order to make use of some unique feature of a particular computer or to speed up execution of the program. As a result, the program will run only on computers which have a hardware configuration identical to that of the machine for which the program was originally written. A final problem is that many applications programs are sold only in machine-language form. There-fore, the program won't run on a computer using a different CPU even if the computer uses the same operating system.

I/O driver software   The final component of an operating system—the I/O device driver software—performs the actual control of hardware peripheral devices. As such, it is very much dependent on the nature of the devices interfaced to the computer system and on details such as I/O port addresses in the computer. Therefore, this portion of the operating system must be written, or at least modified, for every hardware configuration. The operating system nucleus uses the I/O driver software to access the hardware. As a result, the I/O driver software must be written to interface with the nucleus in a standard fashion. Much of the work in writing driver software for a particular I/O device is figuring out how to make the device conform to the standard device model used as a basis for the interface to the nucleus.

16.1.2 Operating-System Facilities

The facilities provided by the operating system depend on its intended area of application and on the type of computer system for which it is designed.

File systems   Operating systems used on computer systems with mass-storage devices, such as floppy disks or hard disks, are often called *disk operating systems* (DOS). They provide a *file system* to the user and the user's appli-cation programs. The file system hides the details of accessing tracks and sectors on the mass-storage media from the application program. Instead, the program (or the user) can make the operating system create a logical construct called a file, assign a name to the file, and store and retrieve data to and from the file.

The programmer's model for a file will vary depending on the operating system used. One model, which is becoming quite common, is simply a finite sequence of characters (bytes). The file size (i.e., number of characters contained in the file) increases (up to some maximum limit) as more data is added. The operating system maintains a *directory* which identifies and locates all files on the system. At a minimum, a directory entry contains the file's name and the information, such as sector and track numbers, required to find the file on the mass-storage device. A directory may also contain other information such as which users (on a multiuser system) are allowed to access the file.

Logical I/O system   Another common operating-system facility is a *logical I/O system*. Since application programs perform I/O functions by calling the operating system, they can be isolated from the physical details of the I/O device. Instead, the application program deals with a *logical device*. For example, a program may send its output to a logical device called the *console*. The actual I/O device which acts as the console could be a video terminal or it could be a printer-based terminal; it doesn't matter to the program using the "console," since the logical I/O system handles the actual interface to the physical I/O device. In fact, there is no reason why the console could not be a file on the mass-storage system, since the operating system can direct a stream of characters to the mass-storage system as easily as it can to a video terminal.

In some operating systems it is possible for the user to change the assignment between logical I/O devices and the physical devices. Thus the console could be changed from a video terminal to a file by a command to the operating system. This feature, called *I/O redirection*, is quite useful. For example, a user can first run an application program with its output directed to a video terminal in order to see the results. Then the program can be run again with the output redirected to a file so that the results can be stored. Of course, the same thing can be done without I/O redirection, but in that case, each application program must include routines to write to both the terminal and the file system and must include some means for the user to select where the output is to be sent.

Multitasking   Some operating systems give the appearance of allowing several programs to operate simultaneously on the computer. What actually happens is that each program is allowed to run for a short period of time in rotation so that the programs appear to be running simultaneously. These are called *multitasking*, or *multiprogramming*, operating systems. They have a *scheduler* program that controls the execution of programs on the system. Some schedulers simply rotate through the programs in a *round-robin* fashion, assigning each a brief *time-slice* in which to execute. Other schedulers have more complex algorithms to determine the order of task execution and the time each task will be allotted on the system.

Multitasking operating systems that service the requests of multiple human users are often called *multiuser*, or *time-shared*, operating systems. On these systems, some programs such as editors or the command interpreter operate interactively with the user. Other programs, called *background programs*, run without user interaction and are often invisible to users.

Memory and resource management Multitasking operating systems must have a means to allocate system *resources* such as memory and I/O devices among the programs operating on the system. For example, in a multiuser system, each user must have a private memory space and a private file system. When a user stops using the system, the memory manager in the operating system must reassign the user's memory to other programs on the system.

I/O devices must also be carefully managed. For example, if one program is sending data to a printer and is suspended to allow another program to run, the program that begins to run cannot get access to the printer, since it would corrupt the printout of the suspended program. One way to deal with this problem is to have a program in the operating system, sometimes called a *spooler*, which controls the printer. User programs which want to use the printer transfer the data to be printed to the spooler. The spooler handles the scheduling of the printing tasks and allows the other programs to continue executing without having to wait for the printer to become available

Message passing The ability to transfer data from one program to another, as in the case of a user program passing data to the spooler program, is an important facility in multitasking operating systems. This is particularly important in real-time applications where the control software may be divided up into a set of concurrently executing programs. These programs must be able to communicate with each other in order to transfer data and to maintain synchronism. The communications links between programs in a multitasking system have names such as *mailboxes*, *queues*, *buffers*, and *pipes*, depending on how they are implemented.

16.2 OPERATING SYSTEMS FOR MICROCOMPUTER DEVELOPMENT SYSTEMS

A microcomputer development system must have an operating system to support the application programs such as editors, compilers, and assemblers used by the designer in developing software. Since mass storage is a requirement for most development systems, the operating system is usually a disk operating system. The operating system usually supplies I/O drivers for video terminals and printers. It may also provide support for more specialized peripherals such as in-circuit emulators and EPROM programmers.

Operating-system facilities that are particularly useful for software development include the ability to perform I/O redirection and facilities to allow the user to alter the command interpreter program to suit the requirements of a particular development task. *Utility programs* to find differences between files, sort the contents of files, and search files for certain data patterns can be very useful when developing software. Naturally, application programs such as editors, compilers, assemblers, linker-loaders, and debuggers that run with the operating system must be available.

In many cases, the quantity and quality of the available application programs is the prime factor in determining the choice of an operating system. When choosing a development system, the designer should check which operating systems are available for use on the development system. Development systems which can use only a special operating system designed for that particular system should be avoided. No matter what the vendor promises, application software will probably be difficult to obtain and expensive. On the other hand, if a development system uses a version of a widely used operating system, market forces will normally ensure that a wide range of application programs are available at reasonable prices.

16.2.1 The CP/M Operating System

CP/M is a good example of an operating system suitable for smaller development projects. CP/M is a single-user, single-task, disk operating system that was developed by Digital Research Inc. for use with the Intel 8080 microprocessor and other similar microprocessors such as the Intel 8085 and the Zilog Z80. This operating system is used in many personal computers. Versions of CP/M have also been developed for the Intel 8086 (CP/M 86) and Motorola 68000 (CP/M 68K) 16-bit microprocessors. The MSDOS operating system from Microsoft Corp., which is used on the IBM personal computer and other personal computers, is also quite similar to CP/M.

CP/M is a small program that fits into about 16 kilobytes of memory. The command interpreter is called the console command processor (CCP). This operates much like a user program and can be replaced with a user written program if desired. It supports a limited set of commands which, among other functions, allow the user to load and execute programs from disk files, rename or erase disk files, and save files to disk. The latest version of CP/M (version 3.0) also supports I/O redirection.

The nucleus of the CP/M operating system is called the basic disk operating system (BDOS). It is responsible for all system management functions such as disk-file management, high-level (logical) I/O, and all other operating-system function calls available to user programs. The BDOS makes calls to the basic I/O system (BIOS) which contains the device drivers for the peripheral devices in the computer system. The BIOS portion of CP/M must be customized for every different hardware configuration.

CP/M supplies some simple utility programs such as a primitive text editor, a file-transfer program, an absolute assembler program for the 8080 microprocessor, and a simple loader and debug program. However, a vast quantity of application software is available for execution on computers using the CP/M operating system. These include high-quality screen-oriented editors, compilers for many high-level languages, assemblers for many microprocessors, and support software for development tools such as EPROM programmers and in-circuit emulators.

CP/M is rather processor-specific, so it supplies the best support for development of target systems based on the Intel 8080 family or the Zilog Z80, but it can be used to develop software for other microprocessors. Similarly, the MSDOS operating system is best suited to developing software for target systems based on the Intel 8086 family of 16-bit microprocessors but can be used to develop software for other microprocessors as well.

16.2.2 The Unix Operating System

The Unix operating system is a good example of a multiuser, multitasking operating system suitable for large development projects. It is a fairly large operating system that will not work with the 64K byte memory capabilities of standard 8-bit microcomputers. In addition, a hard-disk-based mass-storage system is normally required if the computer system is to make the best use of the capabilities of Unix. Unix does work well with the large memory spaces of modern 16- and 32-bit microcomputers. Versions of Unix are available for most of these as well as for many minicomputers.

The Unix command interpreter is called the *shell*. It provides a flexible interface between the user and the Unix nucleus. The commands provided are relatively simple, but they can be easily combined to suit the user's purpose. For example, the command

```
ls
```

will cause the nucleus to list the names of files contained in a directory on the user's terminal. The command

```
wc − w
```

will cause the nucleus to print the number of words in a file on the user's terminal. However, the two commands can be combined a follows

```
ls|wc − w
```

In this case, the wc command counts the number of words produced by the ls command. To accomplish this, the Unix nucleus creates a communications path called a *pipe* between the program that carries out the ls command and the program that carries out the wc command. The output of the ls command becomes the input data to the wc command. The result printed on the terminal will be the number of files in the directory. If this combination of commands is used frequently, it can be placed in a file with a name such as "filecount". If the user now types "filecount" into the terminal, the shell will execute the commands in the file and print the number of files in the directory to the terminal. The shell also allows I/O redirection. For example, the command

$$ls > temp1$$

will cause the file names produced by the ls command to be sent to a file named temp1 rather than to the user's terminal.

Most of the Unix nucleus is written in the C-programming language. As a result, C programs can access Unix facilities such as files or I/O device drivers through simple C function calls. The fact that Unix is primarily written in C also makes Unix very portable, since the amount of machine-specific assembly-language code in the operating system is fairly small.

Since Unix supports a number of users and large mass-storage devices, hundreds or thousands of files can exist on a single Unix system. Therefore, a single file directory for the entire system would not be adequate. Instead, Unix organizes the file system in a hierarchical fashion which supports multiple directories of files. Each user is assigned a separate directory. The user can create additional directories at lower levels in the file hierarchy in order to organize his or her own personal files.

Unix has been used extensively as an operating system to support software development. As a result, it has a large number of utility programs which are useful in program development. In addition, many application programs such as editors, compilers, and assemblers are available for use on Unix. Many manufacturers of multiuser microcomputer development systems (Tektronix and Hewlett-Packard are two examples) now offer versions of Unix for their development systems and provide support software for special-purpose peripherals such as in-circuit emulators.

16.3 REAL-TIME OPERATING SYSTEMS

Operating systems can also be useful in the target system. The complexity of the operating system required varies, depending on the nature of the application. A *monitor* is a simple operating system that provides simple I/O facilities and a few commands to control program execution and assist in debugging programs. A *real-time executive* is an operating system that includes real-time multitasking capability and a message facility between tasks. A full-fledged *real-time operating system* adds a disk file system to the capabilities provided by a real-time executive.

A real-time operating system is distinguished from normal multitasking operating systems by its ability to schedule tasks on the basis of external events which are signaled to the computer by interrupts. Operating systems which aren't intended for real-time applications usually don't give the user any control over how interrupt requests are handled and may actually disable interrupts for substantial periods while the operating-system nucleus is executing. As a result, response to interrupts may be slow or interrupt requests may even be missed entirely. Real-time operating systems are designed to provide fast response to interrupt requests. An *interrupt latency time* is often given in the specifications for a real-time operating system. This is the maximum amount of time it will take the operating system to recognize an interrupt request and begin servicing it.

Another distinguishing feature of real-time operating systems is that their command interpreters are usually rather simple and they don't include many utility programs. The goal of a real-time operating system is usually to serve the system it controls rather than human users.

The centralized multitasking capability offered by a real-time executive or a real-time operating system is not always required. In simple real-time systems the application program can often be configured as a combination of a main program which executes sequentially and a set of interrupt service routines which respond to external events. However, as the system gets more complex, it becomes more difficult to coordinate the interactions between the interrupt service routines and the main program. At some point, some well-organized means of coordination is required.

The multitasking capability of a real-time operating system provides a framework that allows the design of very complex real-time software which has well-defined and controlled interactions among its various components. In addition, the operating system supplies a number of prewritten and debugged software facilities such as interrupt handlers, data-transfer functions, real-time clocks, and I/O device drivers.

The design of a real-time operating system is best left to experts. However, some understanding of the concepts and design techniques used in real-time operating systems is useful for choosing and applying a commercially available real-time operating system. In addition, these design techniques, which are part of the discipline of *concurrent programming*, are also applicable to the design of simpler real-time systems which don't require a complete operating system. In particular, the problems introduced by interactions between interrupt service routines and other functions in a program are very similar to those that occur when separate tasks in a multitasking operating system interact.

16.3.1 Tasks and Task Scheduling

Definition of a task   At the heart of a real-time operating system is the concept of a *task*, or *process*. A task, or process, is an activity carried out by the computer. It consists of a program, data associated with the program, and computer resources such as memory space or I/O devices required to execute the program.

A memory map for a multitasking system is shown in Fig. 16.2. Each task and the operating system is assigned its own stack area in memory. This is used

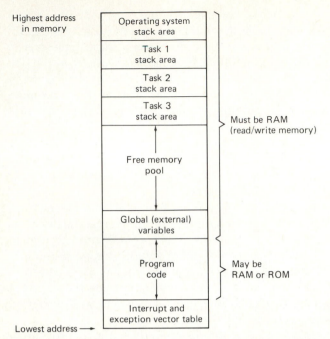

Highest address in memory

Operating system stack area

Task 1 stack area

Task 2 stack area

Task 3 stack area

Free memory pool

Global (external) variables

Program code

Interrupt and exception vector table

Lowest address →

Must be RAM (read/write memory)

May be RAM or ROM

Figure 16.2 Memory map for a multitasking system.

to store private data. A *free memory pool* is used by the operating system to create message channels or common data areas which allow tasks to exchange data. A certain number of global variables, accessible to all tasks and to the operating system, may also be required.

Two tasks can execute the same program but be distinct because they use different stack areas, message channels, and resources. For example, if a computer system had three identical A/D converters which provide input to the computer, three distinct tasks could be created to service the three devices. Each task will run the same program but will be assigned a different resource (A/D converter), a different stack area, and a different message channel to transfer the incoming data to other tasks. The tasks will run independently (*asynchronously*) depending on when their A/D converter has data available. Since tasks can share program code, it is very important to use only reentrant programs in a real-time multitasking system.

Task states Most real-time operating systems are designed so that a task can be in one of four states, as shown in the state transition diagram in Fig. 16.3. If a task is in the *running* state, the computer is executing that task. If the task is ready to be executed but not actually executing, it is in the *ready* state. A task that is unable to execute because it is waiting for an event or a resource is in the *suspended* state. The task is said to be *blocked*. For instance, if a task is an interrupt service routine, it must await the interrupt signal from the external device (an event) before it can execute. Similarly, a task which wishes to use a

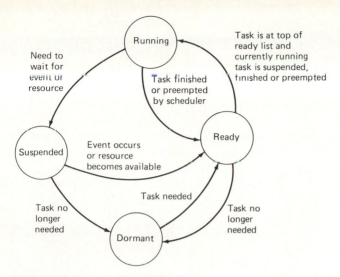

Figure 16.3 Task state transition diagram.

printer (a resource) may have to wait until another task has finished using the printer. Finally, if the task is not needed by the real-time system, it is in the *dormant* state.

Task scheduler The scheduler in a real-time operating system is responsible for controlling the transitions of the tasks among these states. The elements of a simple task scheduler are shown in Fig. 16.4. The scheduler program receives interrupt requests from the computer's interrupt system. In addition, it receives messages from the running task in the form of system calls to the operating

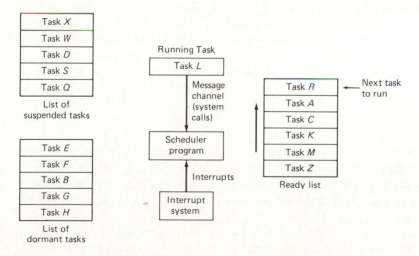

Figure 16.4 Elements of a task scheduler.

system. The scheduler maintains lists of ready, suspended, and dormant tasks. It also carries out the task switching function where the execution of the currently running task is stopped and a task on the ready list is started. A running task will continue running until one of the following events occurs:

1. The task makes a system call asking to be suspended until an event, such as an interrupt request, occurs.
2. The task makes a system call requesting a system resource which is unavailable. The task will then be placed in the list of suspended tasks.
3. The task makes a system call asking to be rescheduled, since it has completed its function. The task will then be placed in the ready list or the suspended list depending on whether it is to be restarted immediately or after some time delay.
4. The scheduler *preempts* the task to allow another task to execute. The task is then placed in the ready list.

When one of these events occurs, the scheduler must select the next task from the ready list and cause the computer to start executing it.

The scheduler's ability to control the execution of tasks is the key to the efficiency and speed of response of a real-time operating system. A task which is waiting for an event to occur or a resource to become available does not execute and therefore does not take up any CPU time. When the event occurs or the resource becomes available (usually signaled by an interrupt or a message from the running task), the scheduler allows the task to continue execution. The amount of time that elapses between the occurrence of an event and the execution of the task which was blocked on that event depends on the *scheduling strategy* used by the scheduler.

Scheduling strategies The actual scheduling method used to select the next task to run varies. In *round-robin scheduling* (Fig. 16.5), all tasks have equal priority and the ready list is configured as a simple first-in-first-out (FIFO) queue. Tasks enter the ready list at the end of the queue and tasks are selected for execution from the head of the queue. The scheduler gives each task an equal *time-slice* for execution before preempting it to allow the next task in the ready queue to execute. A task may, of course, cease executing before it is preempted at the end of the time-slice because of one of the reasons given above.

The length of the time-slices is usually determined by a real-time clock which interrupts the scheduler at regular intervals. When the scheduler receives a real-time-clock interrupt, it performs a *task switch*, stopping execution of the currently running task and starting execution of the next task in the ready list.

A major problem with this scheduling technique is that response to external events such as interrupts can be rather slow. If an interrupt service task is removed from the suspended list after an interrupt signal is received and

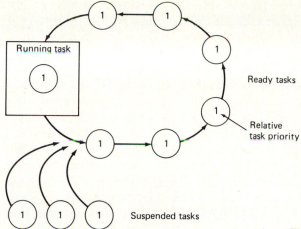

Ready tasks

Relative
task priority

Running task

Suspended tasks

Figure 16.5 Round-robin scheduling.

is added to the tail of the ready list, it must wait until all the tasks in front of it have received their time-slices before it can begin execution.

Priority-based preemptive scheduling (Fig. 16.6) avoids this problem. In this technique, tasks are assigned different priorities. The highest priority task that is not in the suspended, or dormant, state is always running. A task stops running only when it must wait for an event or a resource. The task with the next highest priority can then execute. Task priorities may be *static*, meaning that they are set when the operating system is initialized and cannot be changed thereafter; or *dynamic*, meaning that they can be changed by a system call while the system is running.

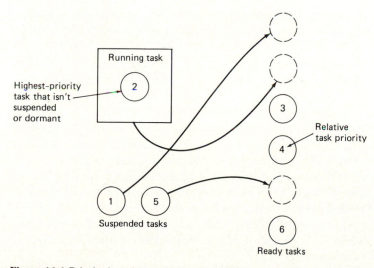

Running task

Highest-priority
task that isn't
suspended
or dormant

Relative
task priority

Suspended tasks

Ready tasks

Figure 16.6 Priority-based preemptive scheduling.

This scheduling technique guarantees very fast response to interrupts. For example, in Fig. 16.6, assume that the suspended task with a priority of 1 (highest priority) is an interrupt service routine waiting for an interrupt from its device. When the interrupt is received, the scheduler will place the task in the ready state. Since this task is now the highest-priority nonsuspended task, the scheduler will now preempt the currently running task (priority 2) and place the priority 1 task into the running state instead. A problem with this scheduling technique is that high-priority tasks that rarely become suspended may "hog" the CPU, preventing the lower-priority tasks from executing.

In practice, a mixture of round-robin and priority-based scheduling is often used. Interrupt service routines, which are usually short and are blocked waiting for an interrupt most of the time, are assigned a high priority. Other tasks are assigned equal priorities which are lower than the priorities assigned to the interrupt service routines. These low-priority tasks are scheduled in a round-robin fashion. However, the high-priority interrupt service routines are scheduled on a priority basis and do not have to wait in the ready queue. This ensures that all tasks have a chance to execute and that interrupt response will be quick.

The task control block (TCB) and task switching   In order to carry out its scheduling functions, the scheduler makes use of *task control blocks*, or *process descriptors*. Each task is assigned one of these blocks, which is simply an array of data about the task. An example of a task control block is shown in Fig. 16.7. The block contains data on the state of the task (i.e., dormant, suspended, ready, or running), the priority of the task, and events or resources for which the task is waiting. The starting address for the task's program code and the initial stack pointer value for the task's stack are also stored in the TCB. This information is used by the scheduler when a task is first activated and when a task is reactivated by moving it from the dormant to the ready state.

In addition, the task control block is used to store the task's *context*. The task context represents all the information that must be saved when the execution of a task is stopped and then restored when task execution is resumed. In most real-time operating systems, this information is the contents of the computer's registers just prior to the moment when execution of the task was stopped. Note that this is the same information (the machine state) which must be saved and restored when the system responds to an interrupt.

When a task switch occurs, the task context of the currently running task is stored in its task control block. Then the previously stored context of the task which is about to be executed is retrieved from its task control block and is loaded into the computer's registers. Execution is then transferred to this task. The task resumes execution at the point in its program which it reached the last time it was in the running state.

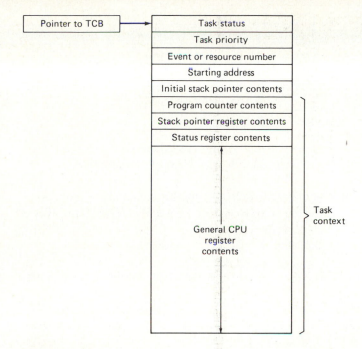

Figure 16.7 Task control block.

16.3.2 Task Synchronization and Data Transfer

There is no inherent synchronization between the execution of tasks; each task executes independently. However, in most real-time systems, tasks must work together and may also have to perform their functions at defined times. Some synchronization mechanisms must be provided.

Most real-time operating systems provide a complete real-time clock facility. A task can make a system call to suspend itself for a time interval or until a certain time of day occurs. The scheduler suspends the task and places it back on the ready list only when the requested time interval has elapsed or the requested time of day has arrived. Thus a task can ensure that it performs its functions at defined times.

The situation where two tasks must work in synchronism can be understood by considering the following example. A program for a robot-arm controller has among its tasks a task (task A) which controls the movement of the robot arm from point to point and another task (task B) which controls a tool attached to the arm. The two tasks must work sequentially. Task A moves the arm to a work position where task B operates the tool; task A then moves the arm to the next work position where task B again operates the tool.

A common technique used in real-time operating systems to synchronize two tasks such as task A and task B is to use a *semaphore* variable, s, and two

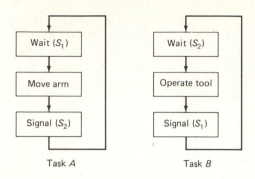

Figure 16.8 Synchronization of two tasks through the use of semaphores.

Task A   Task B

system calls, Wait(s) and Signal(s), which act on the semaphore. The operations performed by the operating system on receiving a Wait(s) system call are

If $s > 0$, then $s = s - 1$, else suspend the execution of the calling task.

The operation performed by the operating system on receiving a Signal(s) system call is

If a task has been suspended after making a Wait system call on this semaphore, then move it to the ready state, else $s = s + 1$.

In the case of our example, two semaphores, s_1 and s_2, shall be used. Semaphore s_1 is initialized to 1 and semaphore s_2 is initialized to 0. The two tasks make Wait( ) and Signal( ) system calls as shown in Fig. 16.8. Consider the situation immediately after initialization. Task A makes a Wait(s1) system call. Since s_1 has been initialized to 1, task A is allowed to continue execution and s_1 is decremented to 0. In the meantime, task B makes a Wait(s2) system call. Semaphore s_2 has been initialized to 0, so task B is suspended by the operating system. When task A has finished moving the arm, it makes a Signal(s2) system call which causes task B to be moved to the ready state. Task B can now execute. Task A, however, has looped back to the Wait(s1) system call. Semaphore s_1 is equal to 0 at this point, so task A is suspended until task B has completed its operations and makes the Signal(s1) system call. At this point the cycle begins again. The two tasks can be seen to be synchronized to each other via the semaphore mechanism.

Data transfer between tasks In many cases tasks must exchange data. The simplest way of doing this is to assign a common data area in the free memory pool where one task can write data and one or more tasks can read data. However, this simple approach is fraught with peril.

Consider our robot-arm controller software again. Task C sends X, Y, and Z coordinates which define the required end point position of the robot arm to task A which actually controls the movement of the arm. Task C writes the coordinates to three memory locations in a shared data area. These locations are read by task A. Say that task A has just read the X and Y coordinates when it is preempted by the scheduler. Task C executes before task A is allowed to resume execution and it writes new X, Y, and Z values to the data area. When task A resumes, it reads the new Z value. It now has received an incorrect coordinate consisting of the old X and Y values and the new Z value.

There are two separate problems here. Since the read-data operation can be interrupted by a write-data operation, the data can be corrupted. In order to avoid this, the acts of writing to the data area and reading from the data area must be made *mutually exclusive*. Sections of the program code which access the data area are called *critical sections*. They cannot be interrupted by any other task which also accesses the data. The second problem is that it is possible for the producer of data (task C) to overwrite the old data with new data before the consumer of data (task A) has read the old data. Alternatively, the consumer may read the same piece of data several times before the producer transfers the next item of data. This is referred to as the *producer-consumer problem*. Semaphores can be used to solve both the mutual-exclusion and producer-consumer problems associated with the user of common data areas to transfer data between tasks (see Prob. 16.2 at the end of this chapter). However, there is another method of data transfer which is often used in real-time operating systems. This is a *data channel* between tasks consisting of a *circular buffer*. These buffers have, as shown in Fig. 16.9, a *first-in-first-out*

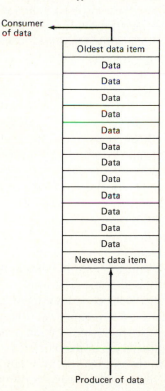

Consumer of data

| Oldest data item |
| Data |
| Data |
| Data |
| Data |
| Data |
| Data |
| Data |
| Data |
| Data |
| Data |
| Data |
| Newest data item |

Producer of data

Figure 16.9 Conceptual model of a circular buffer; first data inserted is first data removed.

(FIFO) characteristic. The buffer is like a data pipeline. Data is inserted in one end by the producer task and is removed at the other end by the consumer task.

The task producing data inserts a data item into a buffer by making a system call to the operating system. For example, the call might be

`Insert(data, channel)`

where the parameter "data" is the data item to be inserted and the parameter "channel" specifies the buffer to be used. The receiving task will make a system call such as

`Remove(dataptr, channel)`

where the parameter "dataptr" is a location in the task's private memory where the received item can be stored.

The actual operation of the circular buffer, which is handled by the operating system, can be understood with the aid of Fig. 16.10. The buffer consists of an array with a number (LENGTH) of data-storage locations. Two pointers, START and END, respectively, point to the first and last storage locations in the array. The pointer TAIL points to (i.e., contains the address of) the oldest data item stored in the array. This will be the next item removed from the buffer. The pointer HEAD points to the empty storage location just ahead of the newest data item in the buffer. This is the location where the next data item will be inserted in the buffer. Finally, the variable N contains the number of data items in the buffer.

When the buffer is initially empty, the pointers HEAD and TAIL are set equal to START. Data is added to the location pointed to by HEAD and HEAD is then incremented to point to the next empty location. Data is removed from the location pointed to by TAIL and TAIL is then incremented to point to the next location containing data. When HEAD or TAIL reach the end of the array (indicated by the END pointer), they are set equal to START again. This is why this arrangement is called a circular buffer: the HEAD and TAIL pointers move through the buffer in a circular fashion.

The variable N is used to detect buffer full and buffer empty conditions. N is incremented every time a data item is inserted and decremented every time a data item is removed. Before a data item is added, N is compared with LENGTH. If N equals LENGTH, the buffer is full and the data item cannot be inserted. Therefore, the task sending the data item is suspended until the receiving task removes some data from the buffer. Before a data item is removed, N is compared with 0. If N equals zero, the buffer is empty and there is no valid data to be removed. Therefore, the task receiving the data is suspended until the task sending data adds some more data to the buffer.

Data transfer via circular buffers deals with both the mutual-exclusion and producer-consumer problems. The task producing data cannot overwrite any

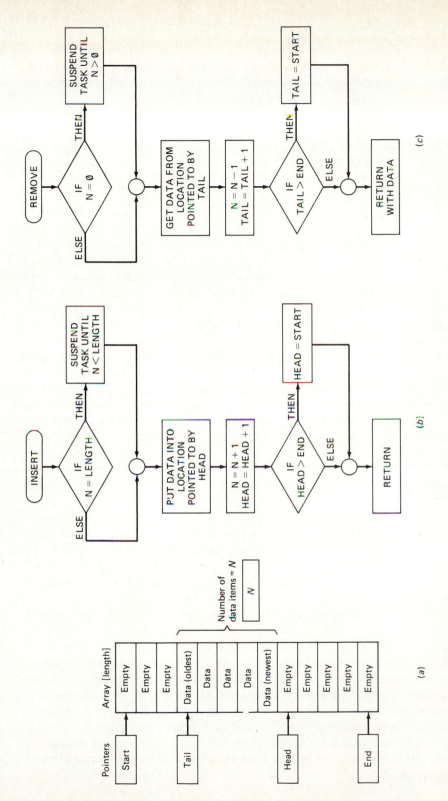

Figure 16.10 Circular buffers. (*a*) Data structures required. (*b*) Putting a data item into the buffer. (*c*) Getting a data item from the buffer.

493

data which has not been received by the task consuming the data. Similarly, the task consuming data cannot remove a data item from the buffer more than once and isn't allowed to remove data from an empty buffer.

16.4 FACTORS IN SELECTING A REAL-TIME OPERATING SYSTEM

In the past few years we have moved from a situation where a designer was frequently forced to design a real-time operating system because none was available for the particular microprocessor being used in the real-time system to a situation where there are often four or five competitive real-time operating systems commercially available for a particular microprocessor. They are available from manufacturers of microprocessors, such as Intel, Motorola, and National Semiconductor, and from independent software companies. The system designer's task has changed from designing operating systems to selecting and using operating systems.

The designer must first decide whether a full-fledged disk operating system is required or whether a simple executive will be sufficient. This will largely depend on the nature of the target computer and the real-time system. If the target computer has disk-based mass storage and it is necessary to use it to store data or programs, then a disk-based system is the obvious choice. An example would be a laboratory-data-acquisition system where large amounts of data have to be stored and the programs running on the system will change for every experiment that the system is used on. On the other hand, a simple executive that can be stored entirely in ROM (read-only memory) is the obvious choice for smaller dedicated systems where the program to be executed will not change.

There is another consideration, however. It may be possible to use a real-time disk operating system on the development system as well as the target system. Software can then be developed and debugged using the same operating system that will be used in the target system. This simplifies real-time software development considerably. If the real-time operating system can be stripped of its disk-handling functions and loaded into ROM, it may be a good choice even for dedicated systems because of the software-development advantages.

Once a particular class of operating system has been chosen, candidates from that class should be evaluated to see if they have the required facilities. We have discussed the facilities usually offered by real-time operating systems for task scheduling and for intertask synchronization and communication. The other important facility is the I/O device drivers supplied by the operating system. The system should support standard I/O devices such as video terminals and line printers. In addition, the operating system should be designed so that the user can add additional device drivers for special-purpose I/O devices easily.

The speed at which an operating system performs a task switch or responds to an interrupt may be critical in some real-time applications. In other applications, the amount of memory required by the operating-system program code and data structures may be important. Generally, as more features and flexibility are added to a real-time operating system, its response speed decreases and the memory space required increases. Therefore, a large number of operating-system features are not an unmixed blesssing. Some operating systems allow the user to remove unneeded facilities from the system before it is downloaded to the target computer. As a result, less memory is required and the operating system may be a bit faster.

Ease of use is the final factor that should be considered when choosing a real-time operating system. A real-time operating system will have some means by which the user links the application tasks and hardware-specific data such as I/O port addresses and special I/O device driver software to the operating system. This is known as *system configuration*. It is a time-consuming and error-prone operation at the best of times. The system configuration support offered by the real-time operating system should help rather than hinder this task. A *configuration program* that partially automates the task is a definite asset.

Good-quality documentation on the operating system is also vital to configuring and using the software. It is a good idea to examine the documentation supplied with a real-time operating system before purchasing it. If you cannot figure out how to use the operating system from the manual, the situation will not improve much after you have actually bought it.

Another factor affecting ease of use is whether there is a convenient interface between the operating system and the high-level language used to write the application program. Some real-time operating systems still force the user to write special-purpose assembly-language routines to allow a high-level language program to call operating-system functions. The vendor of the real-time operating system should provide *language interface libraries* of subroutines that allow system calls to be made directly from high-level programming languages such as C or Pascal.

SUMMARY

An operating system is a systems program that controls and sequences the execution of application programs on a computer system. Through the use of I/O device driver programs it provides a software interface between the application programs and the hardware I/O devices in a computer system.

In the development of real-time software, operating systems are used both in the development system and in the target system. Operating systems used on the development system are usually disk operating systems. Depending on the nature of the development system, they may support a single user or a number of simultaneous users. Probably the most important consideration in choosing an operating system for real-time software development is the quantity and quality of application programs such as editors, compilers, assemblers, and debuggers which are available for use with the operating system.

Operating systems designed for use on the target computer are called monitors, real-time executives, or real-time operating systems, depending on the facilities which they offer. They are distinguished from other operating systems by their ability to schedule tasks on the basis of external events signaled to the computer by interrupts. Rapid response to interrupts and an orientation toward the system being controlled rather than the human user are the hallmarks of a real-time operating sytem.

At the heart of a real-time operating system is a facility to schedule a number of independently executing tasks so that they appear to be running concurrently. In addition, facilities are provided to allow tasks to synchronize to a real-time clock or to each other. Data channels between tasks are also provided by the operating system. The programming techniques used to implement these facilities are applicable to simpler real-time programs as well.

EXERCISES

16.1 Problems identical to the ones that occur when tasks exchange data occur when an interrupt service routine transfers data to other functions in a program. Consider this case: An interrupt service routine responds to a real-time clock interrupt by updating four external (global) integer variables:

tenths
seconds
minutes
hours

These variables are read by other functions in the program when they wish to check the time. What problem can occur? Give an example. How can the problem be avoided?

16.2 Figure 16.11 shows the use of semaphores to solve the mutual-exclusion and producer-consumer problems associated with the use of a common data area to transfer data between tasks. Explain how the use of these semaphores solves these problems.

16.3 Semaphores can also be used by tasks to request exclusive use of resources such as printers. The task makes a Wait( ) call on a semaphore associated with a resource in order to request exclusive access to it. Once it has finished using the resource, it makes a Signal( ) call on the semaphore associated with the resource in order to release it and allow other tasks to use it. Figure 16.12 shows a situation in which two tasks send data to a common data area and to the printer. Both tasks use semaphores to gain exclusive access to the common data area and the printer. Describe the problem that can occur with the situation as shown in Fig. 16.12. (*Hint*: The problem is called *deadlock*.) What kinds of situations will lead to this problem? Suggest some ways of avoiding this problem.

16.4 Write C-language functions Insert(data) and Remove(dataptr) which insert and remove integer data from a circular buffer. The parameter "data" is the data to be inserted into the buffer. The parameter "dataptr" is a pointer to the variable where the data retrieved from the buffer is to be stored. If the buffer is already full and no data can be added, the Insert( ) function should return the constant FULL. If the buffer is empty and there is no data to remove, the Remove( )

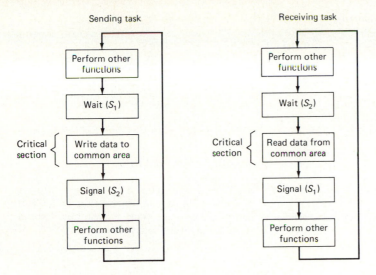

Figure 16.11 Use of semaphores in intertask data transfer.

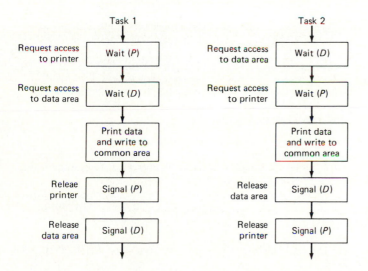

Figure 16.12 Deadlock.

function should return the constant EMPTY. Otherwise both functions should return the constant OK. Assume that the following definitions have been made:

```
#define FULL      0
#define EMPTY     0
#define OK        1
#define LENGTH    64
```

Assume also that the following external variable declarations have been made:

```
int cirbuff[LENGTH];    /*array for circular buffer*/
int *head, *tail;       /*pointers to head and tail of buffer*/
int n;                  /*number of integers in the buffer*/
```

Finally, assume that the following C statements are executed during system initialization:

```
head = tail = cirbuff;  /*pointers initialized to point to first loc-
                          ation in array*/
n = 0;                  /*no data in buffer initially*/
```

16.5 Circular buffers can be used to transfer data from an interrupt service routine to another function in the program. However, the portion of the function Remove(dataptr) (see Prob. 16.4) which adjusts *n* (the number of integers in the buffer) is a critical section. Explain why. Explain how you would ensure that the critical section of the function isn't interrupted.

BIBLIOGRAPHY

General texts on operating systems

Brinch Hansen, P.: *Operating System Principles*, Prentice-Hall, Englewood Cliffs, NJ, 1973.
Comer, D.: *Operating System Design*: *The XINU Approach*, Prentice-Hall, Englewood Cliffs, NJ, 1984.
Dahmke, M.: *Microcomputer Operating Systems*, McGraw-Hill, New York, 1982.
Kaisler, S.: *The Design of Operating Systems for Small Computers*, Wiley, New York, 1983.
Shaw, A. C.: *The Logical Design of Operating Systems*, Prentice-Hall, Englewood Cliffs, NJ, 1974.

Concurrent programming and real-time operating systems

Allworth, S. T.: *Introduction to Real-Time Software Design*, Springer-Verlag, New York, 1981.
Ben-Ari, M.: *Principles of Concurrent Programming*, Prentice-Hall, Englewood Cliffs, NJ, 1982.
Brinch Hansen, P.: *The Architecture of Concurrent Programs*, Prentice-Hall, Englewood Cliffs, NJ, 1977.
Holt, R. C., G. S. Graham, E. D. Ladzowska, and M. A. Scott: *Structured Concurrent Programming with Operating Systems Applications*, Addison-Wesley, Reading, MA, 1978.
Holt, R. C.: *Concurrent Euclid, the UNIX System, and TUNIS*, Addison-Wesley, Reading, MA, 1983.
Heath, W. S.: "A System Executive for Real-Time Microcomputer Programs," *IEEE Micro*, vol. 4, no. 5, June 1984, pp. 20–32.

BINARY NUMBERS AND CODES

A.1 BINARY NUMBERS

A.1.1 Unsigned Binary Numbers

In normal life we are used to dealing with numbers expressed in the decimal number system (base 10). However, computers perform their operations with electrical signals which normally have only two values—high or low, true or false, 1 or 0. As a result, a binary number system (base 2) is often used when dealing with the low-level details of computer operation. The two digits of the binary number system are represented by 0 and 1. Binary digits are commonly called bits. The binary number system is position-dependent, as is the decimal system, but each bit position corresponds to a power of 2 (i.e., $1, 2, 4, 8, 16, \ldots$) rather than a power of 10. Thus 1011 in binary equals 11 in decimal:

$$1 \times 2^3 + 0 \times 2^2 + 1 \times 2^1 + 1 \times 2^0 = 1 \times 10^1 + 1 \times 10^0$$

A decimal number can be converted to a binary number by an algorithm involving repeated division of the number by 2. The decimal number is divided by 2 and the remainder (1 or 0) becomes the least-significant bit of the binary number. The resulting quotient is divided by 2 again. The remainder of this division becomes the second bit in the binary number. The division process continues until the quotient is reduced to zero. As an example, consider the conversion of the decimal number 58 into binary form:

$$\frac{58}{2} = 29 \qquad \text{Remainder} = 0$$
$$\frac{29}{2} = 14 \qquad \text{Remainder} = 1$$
$$\frac{14}{2} = 7 \qquad \text{Remainder} = 0$$
$$\frac{7}{2} = 3 \qquad \text{Remainder} = 1$$
$$\frac{3}{2} = 1 \qquad \text{Remainder} = 1$$
$$\frac{1}{2} = 0 \qquad \text{Remainder} = 1$$

Therefore 58 decimal equals 111010 binary.

Conversion from binary to decimal can be performed by expressing the binary number in polynomial form and then evaluating the polynomial using decimal arithmetic. Consider the following example:

$$111010 = 1 \times 2^5 + 1 \times 2^4 + 1 \times 2^3 + 0 \times 2^2 + 1 \times 2^1 + 0 \times 2^0$$
$$= 1 \times 32 + 1 \times 16 + 1 \times 8 + 0 \times 4 + 1 \times 2 + 0 \times 1$$
$$= 32 + 16 + 8 + 2 = 58$$

In computer systems the number 0 (zero) is as good as any other. Numbering of sequential items such as memory addresses or bit positions commonly starts from 0. The registers and memory locations in a computer store binary *words* of a fixed length. A word of length n bits can represent positive integers from 0 to $2^n - 1$. Attempts to represent numbers larger than those which can be stored within a word will result in overflow. Normally the high-order bits of the number will be truncated (chopped off) because the register or memory location containing the number has no storage cells for them. Of course, larger numbers can be handled by using more than one memory location or register for storage of the number.

A.1.2 Representation of Signed Binary Numbers

So far we have discussed the representation of unsigned integers in the computer. In order to represent signed integers, one bit (normally the most-significant bit) becomes a *sign bit*. Usually, if the sign bit is a 0, the number is positive; and if the sign bit is a 1, the number is negative.

In the *sign-magnitude* representation of signed binary integers, the magnitude of the number is represented by the remaining (nonsign) bits in the word. For example,

$$+43 = +101011 = 0101011 \text{ (sign magnitude)}$$
$$\text{Sign bit}$$

$$-43 = -101011 = 1101011 \text{ (sign magnitude)}$$
$$\text{Sign bit}$$

Sign-magnitude representation is sometimes used by computer peripheral devices such as analog-to-digital converters. However, most computers use a different representation internally that allows arithmetic operations to be carried out more readily.

This representation is called *twos complement*. Positive binary numbers are represented in exactly the same fashion as are those for sign-magnitude representation. The representation for a negative number is created by first taking the representation for the equivalent positive number and *complementing* each bit in the number. To complement a bit, a 1 is changed to a 0, and vice versa. Now the result of the complementing operation is incremented by 1 to create the twos-complement representation of the negative integer. For instance, to create the twos-complement representation of the decimal number -27, the following steps are taken:

$$|-27| = +27 = 011011 \qquad (+27 \text{ in twos complement})$$
$$011011 \text{ complemented} = 100100$$
$$100100 + 1 = 100101 \qquad (-27 \text{ in twos complement})$$

In the twos-complement representation of signed numbers, the most-significant bit also distinguishes positive and negative numbers. Since one bit in the word is devoted to the sign bit, the range of signed integers that can be expressed in twos complement by a word containing n bits is from $+2^{n-1} - 1$ to -2^{n-1}.

Care must be taken when performing calculations with signed integers to avoid overflow, since overflow produces a result which appears to be of the opposite sign. For instance, if the sum of two positive twos-complement numbers exceeds the range of positive numbers that can be represented with a particular word length, the result will appear to be a negative number, since it will overflow into the sign bit. This can produce disastrous results in a real-time system.

A.2 BINARY-CODED DECIMAL NUMBERS

Some instruments that can be interfaced to a computer produce numerical data in a binary-coded decimal (BCD) format. In this format the binary word is divided into segments, called *fields*, of 4 bits. Each 4-bit field is used to represent one decimal digit of the number to be represented. Each decimal numeral is represented by its binary equivalent:

Decimal	BCD	Decimal	BCD
0	0000	5	0101
1	0001	6	0110
2	0010	7	0111
3	0011	8	1000
4	0100	9	1001

The binary numbers 1010 to 1111 are not used. Thus the decimal number 67 is represented in BCD form as follows:

$$6 \Rightarrow 0110$$
$$7 \Rightarrow 0111$$
$$67 = 01100111 \text{ (BCD)}$$

A.3 HEXADECIMAL AND OCTAL NUMBERS

Since binary numbers get very long even for fairly small numbers, two other number systems are commonly used to represent numbers in computer systems. The most common is the base-16 number system called *hexadecimal* (often abbreviated to hex). In this system, the first 10 numerals are represented by the standard arabic numerals 0 to 9. The remaining six numerals are represented by the letters A, B, C, D, E, and F. This explains why you see "odd" looking numbers like 3ED1 in the computer literature. Hexadecimal is used because it is very easy to convert binary numbers having multiples of 4 bits into hexadecimal notation. The binary number is simply divided into groups of 4 bits and each group is converted to the corresponding hexadecimal digit. For example,

$$1010110110101110 \Rightarrow 1010 \; 1101 \; 1010 \; 1110$$
$$1010_2 = A_{16}$$
$$1101_2 = D_{16}$$
$$1010_2 = A_{16}$$
$$1110_2 = E_{16}$$
$$1010110110101110_2 = ADAE_{16}$$

Most modern computer systems deal with data in multiples of 8 bits (8, 16, 32, 64), so the hexadecimal system is very convenient for representing this data. The basic 8-bit building block is commonly called a *byte*.

At one time the base-8 number system (octal) was also commonly used to represent numbers in computer sytems. In this case, the binary number is divided into groups of 3 bits:

$$10110011 \Rightarrow 010 \; 110 \; 011$$
$$010_2 = 2_8$$
$$110_2 = 6_8$$
$$011_2 = 3_8$$
$$10110011_2 = 263_8$$

The octal number system is best suited to computer systems where the normal data item is a multiple of 3 bits (e.g., 12, 24, or 36). For modern byte-oriented systems where the data word length is a multiple of 8 bits, the hexadecimal system is the more convenient choice.

Some equivalent number representations

Decimal	Binary	Octal	Hexadecimal
0	0000	0	0
1	0001	1	1
2	0010	2	2
3	0011	3	3
4	0100	4	4
5	0101	5	5
6	0110	6	6
7	0111	7	7
8	1000	10	8
9	1001	11	9
10	1010	12	A
11	1011	13	B
12	1100	14	C
13	1101	15	D
14	1110	16	E
15	1111	17	F

The C programming language, which is used in this book, accepts decimal, hexadecimal, and octal constants. Therefore, it is possible to use the most natural representation for a constant without having to perform tedious conversions. Memory or I/O port addresses are typically given in hexadecimal, for instance. In a C program these can be expressed directly in hexadecimal with no need to convert them to decimal notation. In the C language, constants in hexadecimal notation are written with a leading 0x or 0X (0 = zero) to indicate that they are hexadecimal. Octal constants are denoted by a leading 0 (zero). Other indicators of hexadecimal notation include a leading $ sign (Motorola format assembly language) or a trailing H (Intel format assembly language).

A.4 KBYTES AND MBYTES

Modern microcomputers usually have memory systems in which each location can store 1 byte of data. The size of a memory is therefore usually given in bytes. Since memories are often quite large, some new units have been coined to describe the storage capacities of memories and mass-storage devices such as disk drives. The units are based on decimal powers of 2. The *kilobyte* (Kbyte) is

$$2^{10} = 1024 \text{ bytes}$$

The *megabyte* (Mbyte) is

$$2^{20} = 1048576 \text{ bytes}$$

A.5 CHARACTER CODES

Besides dealing with numbers, computers must also handle the characters that are entered and displayed on a terminal or printed by a printer. These *alphanumeric* characters are represented by binary bit patterns or codes. The standard code used almost universally by microcomputers is called ASCII (pronounced "ask key"), which stands for American Standard Code for Information Interchange. Each character is represented by a 7-bit pattern and it is therefore possible to represent $2^7 - 1 = 127$ different characters. The ASCII character set includes the upper- and lowercase alphabet, the arabic numerals 0 to 9, common punctuation symbols, and *control characters* (see Table A.1). Control characters are used to control the operation of terminals and printers and include functions such as line feed and carriage return.

It is important to keep in mind the distinction between the ASCII code for arabic numerals and the actual numbers. If you want an ASCII terminal to display the digit 7, it must be sent the hexadecimal number 37 (the ASCII code for the numeral 7) not the hexadecimal number 7, which is the ASCII BEL character. The C programming language has functions in its library of standard functions which carry out the conversions between ASCII and binary representations of numbers.

A.6 LOGICAL OPERATIONS

The concept of the logical operators AND, OR, and NOT is familiar from everyday experience and from programming languages such as FORTRAN:

We will go if we have money AND it is NOT raining

IF (MONEY .AND. (.NOT. RAIN)) CALL ACTIVITY

Logical variables like MONEY and RAIN have only two possible values, true and false. A binary digit (bit) can also be considered as a logical variable since it can take on only two values, 1 and 0. It is possible to define a set of logical operations on bits:

$$
\begin{array}{lll}
1 \text{ AND } 1 = 1 & 1 \text{ OR } 1 = 1 & \text{NOT } 1 = 0 \\
1 \text{ AND } 0 = 0 & 1 \text{ OR } 0 = 1 & \text{NOT } 0 = 1 \\
0 \text{ AND } 1 = 0 & 0 \text{ OR } 1 = 1 & \\
0 \text{ AND } 0 = 0 & 0 \text{ OR } 0 = 0 & \\
\end{array}
$$

$$
\begin{array}{l}
1 \text{ XOR } 1 = 0 \\
1 \text{ XOR } 0 = 1 \\
0 \text{ XOR } 1 = 1 \\
0 \text{ XOR } 0 = 0 \\
\end{array}
$$

Table A.1

Graphic or control	ASCII code (hexadecimal)	Graphic or control	ASCII code (hexadecimal)	Graphic or control	ASCII code (hexadecimal)
NUL	00	+	2B	V	56
SOH	01	,	2C	W	57
STX	02	−	2D	X	58
ETX	03	.	2E	Y	59
EOT	04	/	2F	Z	5A
ENQ	05	0	30	[	5B
ACK	06	1	31	\	5C
BEL	07	2	32	]	5D
BS	08	3	33	∧	5E
HT	09	4	34	—	5F
LF	0A	5	35	`	60
VT	0B	6	36	a	61
FF	0C	7	37	b	62
CR	0D	8	38	c	63
SO	0E	9	39	d	64
SI	0F	:	3A	e	65
DLE	10	;	3B	f	66
DC1 (X-ON)	11	<	3C	g	67
DC2 (TAPE)	12	=	3D	h	68
DC3 (X-OFF)	13	>	3E	i	69
DC4 (~~TAPE~~)	14	?	3F	j	6A
NAK	15	@	40	k	6B
SYN	16	A	41	l	6C
ETB	17	B	42	m	6D
CAN	18	C	43	n	6E
EM	19	D	44	o	6F
SUB	1A	E	45	p	70
ESC	1B	F	46	q	71
FS	1C	G	47	r	72
GS	1D	H	48	s	73
RS	1E	I	49	t	74
US	1F	J	4A	u	75
SP	20	K	4B	v	76
!	21	L	4C	w	77
"	22	M	4D	x	78
#	23	N	4E	y	79
$	24	O	4F	z	7A
%	25	P	50	{	7B
&	26	Q	51	|	7C
'	27	R	52	}	7D
(	28	S	53	~	7E
)	29	T	54	Del (rub out)	7F
*	2A	U	55		

Note the two "OR-LIKE" functions, OR and XOR. OR is called the *inclusive-OR* function and is the function we normally associate with the concept of "oring". XOR is called the *exclusive-OR* funtion. The NOT function is often called the *complement*, or *ones complement*, operation.

These operations can be performed on a bit-by-bit basis on strings of bits. These *bitwise logical operations* are performed by carrying out the logical operations described above on each pair of corresponding bits. For example,

10101101 bitwise AND with 10001111

1 0 1 0 1 1 0 1
1 0 0 0 1 1 1 1
----
1 0 0 0 1 1 0 1

These operations are most often used to isolate, or *mask off*, a bit or bits within a word. For instance, if we wish to isolate the third bit in an 8-bit word, we would bitwise AND the word with a *mask* word containing of 0s in all bit positions but the third, which would contain a 1. From the definition of the AND operation we can see that ANDing a bit with a 0 will always result in a 0 while ANDing a bit with a 1 will result in the value of the bit.

1 0 1 1 0 1 1 0

AND

0 0 0 0 0 1 0 0
----
0 0 0 0 0 1 0 0

The bitwise logical operators can also be used to set the values of particular bits within a word. We could set the first 4 bits in an 8-bit word to 1s by ORing the word with a mask word consisting of 1s in the first 4-bit positions and 0s in the remaining positions as follows:

1 0 1 1 0 1 1 0

OR

0 0 0 0 1 1 1 1
----
1 0 1 1 1 1 1 1

EXERCISES

A.1. Determine the decimal equivalent of the following binary words based on each of the following assumptions: (1) The word represents an unsigned binary number; (2) The word represents a signed twos-complement number.

(*a*) 11011000.
(*b*) 11100001.
(*c*) 10010110.
(*d*) 01011111.
(*e*) 00011010.

A.2. Convert the following decimal numbers into binary twos-complement form.

(*a*) +111.
(*b*) −49.
(*c*) +73.
(*d*) −37.

A.3. Determine the decimal equivalents of the following binary-coded decimal (BCD) numbers.

(*a*) 10010011.
(*b*) 01101000.
(*c*) 01010111.

A.4. Determine the hexadecimal equivalents of the following unsigned binary numbers.

(*a*) 1011001111111010.
(*b*) 11000001.
(*c*) 11011110.
(*d*) 01011000.
(*e*) 00111100.

A.5. Convert the following hexadecimal numbers to unsigned binary form.

(*a*) BD.
(*b*) 8FE6.
(*c*) A7.
(*d*) 93C4.

A.6. How many bytes of data can be stored in a 256K-byte memory?

A.7. Perform bitwise AND, OR, and XOR operations on these two binary codes:

$$10111001$$
$$11010011$$

A.8. What logical operation and what mask word would you use to set the second and fourth bits in an 8-bit word to 0 while leaving the remaining bits unchanged?

BASIC LOGIC DEVICES

Information is stored and manipulated in computers in a *binary* representation. One of the main reasons for this is that "storage cells" can be constructed that can store binary information and arithmetic devices can be devised to manipulate binary information with predictable accuracy. The storage of a continuous (analog) signal is subject to degradation that depends on the length of time the data is stored in a cell (such as a capacitor "hold" circuit; Sec. 9.4). In addition to the difficulty involved in programming analog computers, analog electronics is subject to noise that increases with the number of calculations.

To enable digital calculations, analog quantities are converted to one of several possible equivalent binary representations in an analog-to-digital converter (Sec. 9.4).

Digital quantities can also be entered directly through a computer terminal (e.g., the number 1.2345). The terminal generates a binary-coded ASCII character for each alphabetic, numeric, or control key that is depressed. Base-10 digital numbers can then be converted to equivalent binary (base-2) quantities in software.

B.1 LOGIC FUNCTIONS

The computer operates on binary quantities with functions called *combinational and sequential logic functions* to produce other binary quantities. An example of a combinational logic function is the binary *decoder* (see Fig. B.1)

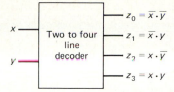

Figure B.1 Two-to-four line decoder.

that can be used on memory boards to *enable* the proper memory chips (see Sec. 8.3) and on memory-mapped interface boards to locate the interface registers in desired memory addresses (see Chap. 14). A two-line to four-line binary decoder implements the following function:

x	y	z_0	z_1	z_2	z_3
0	0	1	0	0	0
0	1	0	1	0	0
1	0	0	0	1	0
1	1	0	0	0	1

This table, known as a *truth table*, shows the outputs for all possible input combinations. We can see that each output is 1, or true, for only one combination of the input logic levels. If x and y are the two highest address lines in a 16-line memory system (64K bytes), then each of the z lines could be used to enable a unique 16K-byte block of memory.

If we call the 1s true and the 0s false, then we need a device that will make line z_3 true only if the address line x is true AND the line y is true. We can write this condition as

$$z_3 = x \cdot y$$

The dot represents the AND *operation*.

For z_2, the condition is that x must be true AND y must be false. When the desired state of a binary variable is false, it is either written with a bar over it or with an asterisk superscript. Thus

$$z_2 = x \cdot \bar{y}$$

Similarly,

$$z_1 = \bar{x} \cdot y$$
$$z_0 = \bar{x} \cdot \bar{y}$$

B.2 COMBINATIONAL LOGIC DEVICES

AND operation The device which is used to carry out the AND operation in the previous example is called an *AND gate*. For two inputs, an AND gate carries out the following function (Fig. B.2):

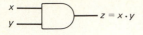

$z = x \cdot y$ **Figure B.2** AND gate.

x	y	z
0	0	0
0	1	0
1	0	0
1	1	1

Note that *both* x and y inputs must be true for the AND gate to produce a true output z.

Thus, an AND gate is all that is required to generate z_3 of the previous example.

NOT or inverter   In order to produce z_2, however, we must detect the false state of y and the true state of x. We can use an AND gate provided we *invert*, or *complement*, the observed state of the y line. This is done with an *inverter gate* (Fig. B.3). The inverter function is

$$z = \bar{x}$$

x	z
0	1
1	0

The small circle on the output side of the triangle can be interpreted as "take the complement."

The signal at the output of the inverter could then be labeled $\bar{x}$ and we will understand that it means that "x is true when this line is 0" (we might call it "x low-true").

With the inverter, the logic devices necessary to generate the z_2 function are shown in Fig. B.4. If you try another input combination such as $x = 0$ and $y = 1$ into the z_2 logic, the input to $z_2 = xy = 00$ and thus $z_2 = 0$, which is correct. The only xy combination that will make $z_2 = 1$ is $xy = 10$

Thus when $x = 1$ and $y = 0$ (that is, $xy = 10$), the input to the AND gate will become 11 and the output z_2 will be 1. The logic for the other functions z_0, z_1, and z_2 is also shown. We can see that whenever a variable is complemented (as in $\bar{x}$), an inverter is required before the AND gate.

OR operation   Another logic operation is the OR operation. Its logic symbol is shown in Fig. B.5. For two inputs the output is true if either the x OR the y

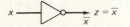

$z = \bar{x}$ **Figure B.3** NOT, or inverter, gate.

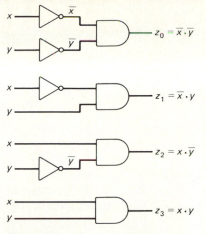

Figure B.4 Logic devices in a two-to-four line decoder.

Figure B.5 OR gate.

input is true. The OR operation on x and y is written

$$z = x + y$$

x	y	z
0	0	0
0	1	1
1	0	1
1	1	1

The circuit required to implement the function

$$z = x_1 + x_2 \cdot \bar{x}_3 = x_1 + (x_2 \cdot \bar{x}_3)$$

is shown in Fig. B.6.

Since AND has precedence over OR, the result of the AND is ORd with x_1.

NAND and NOR   A NAND gate is equivalent to an AND followed by a NOT. Its logic symbol is an AND gate with a circle (Fig. B.7). A NOR gate is equivalent to an OR followed by a NOT, as shown in Fig. B.7.

More than two inputs can be connected to all gates except the inverter. The output signal of all gates can be connected to other gate inputs.

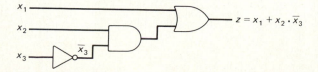

Figure B.6 Logic function with AND, NOT, and OR gates.

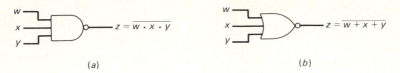

Figure B.7 Some other gates. (*a*) Three input NAND gate. (*b*) Three input NOR gate.

B.3 SEQUENTIAL LOGIC ELEMENTS

The previous circuits are examples of combinational logic in which there is no information storage. If an applied input logic level is changed, the output will change a very short time later (10 to 100 ns per gate later, depending on the technology).

A sequential circuit has a memory. The most basic element of a sequential circuit is a flip-flop. A flip-flop can be constructed from a pair of cross-coupled NOR gates, as shown in Fig. B.8.

If at a particular time the inputs R and S are assumed to be both 0, then the outputs can be either $Q = 0$, $\bar{Q} = 1$, or vice versa. Assume that $Q = 1$, $\bar{Q} = 0$ state. When $Q = 1$, the flip-flop is said to be *set*. If the R (reset) line becomes true (1) and stays there, we find that Q must go to 0, since the top NOR gate has an output of 0 when an input is 1. This, in turn, forces the other output to 1. The flip-flop is now called *reset*. If R then goes back to 0, the flip-flop stays reset.

Similarly, the flip-flop can be *set* by raising the S line. S must be 1 long enough for both gates to change state. If both R and S go to 1 together, both outputs will go to 0. The state will then be determined by the input line that stays high the longest.

D-type flip-flops (latches) Very often we want the flip-flop to change state only at a "clock time" or "strobe time" so that inputs may vary arbitrarily between

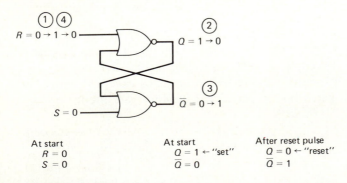

Figure B.8 *RS* flip-flop.

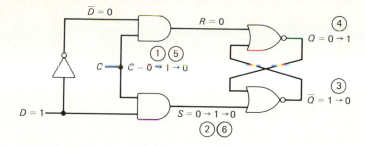

Figure B.9 *D* flip-flop or latch.

clocks. Also, we often have only one line whose state we would like to store at a particular time rather than having the state after a clock determined by both *R* and *S*.

A simple circuit for accomplishing these tasks is given in Fig. B.9. Assume that the input *D* is the data line on the CPU bus. The state of the flip-flop can be assumed arbitrarily. When the input *D* is stable (assume it is a 1), the strobe signal from the CPU will go to 1. Since both the strobe *C* and the input *D* are 1, then the *S* line is 1. Because of the inverter, *R* will be 0. Thus *Q* will be set and the strobe line can drop, making both *R* and *S* equal to 0 and locking out any further changes to *R* and *S* until the next strobe "opens" up the AND gates.

The logic in Fig. B.9 is called a *latch*, or *D*-type, flip-flop. The *D* stands for "delay," since the input on the *D* line *before* the clock transition appears on the *Q* output, *after* the clock.

Register If *N* latches are arranged side by side and used as a unit, as shown in Fig. B.10, the unit is called a *register*. In Chap. 9 the interfaces contain registers which are said to "latch" data from the system data bus at the occurrence of a "strobe" generated by the CPU's control lines. This situation is shown in Fig. B.10.

One reason that a latch is required is to capture data when it is valid on a data bus. Data is continually changing on a data bus depending on the machine

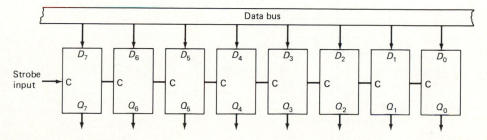

Figure B.10 8-bit register.

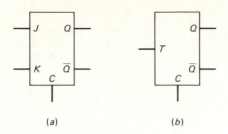

(a)   (b)

Figure B.11 Other types of flip-flops. (a) JK flip-flop. (b) T flip-flop.

cycle and clock cycle within it (see Sec. 2.3). When data is transferred over the data bus, it is valid for only one or two clock cycles (see the transfer of instruction data from memory to the CPU in Fig. 2.15, for example). Figure B.10 shows that each line of the data bus is connected to the input of a latch; and when the right combination of control signals occurs, the "strobe" is produced that causes the bus data to be "copied" into the register. Once the strobe has occurred, the data is stored in the register and can be read out by logic connected to the register outputs.

T and JK flip-flops   Other types of flip-flops also exist called T flip-flops and JK flip-flops. A T (trigger) flip-flop changes state at the clock time if $T = 1$; otherwise, for $T = 0$, the state does not change. A JK flip-flop acts like an RS flip-flop except that if both J and $K = 1$, the flip-flop will act like a trigger flip-flop and change state. These are shown in Fig. B.11.

B.4 LOGIC ELECTRONICS AND LIMITATIONS

To the board-level designer, the electrical characteristics of logic circuit inputs and outputs are more important than the internal details of the gate electronics. Most of the following discussion will focus on gate outputs that *drive* the system bus and gate inputs that *read* from the system bus. Figure B.12

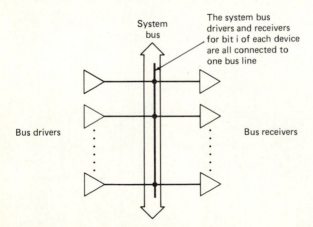

System bus

The system bus drivers and receivers for bit i of each device are all connected to one bus line

Bus drivers

Bus receivers

Figure B.12 Bus connections.

shows many outputs and many inputs connected to the bus. We will first look at input circuits. In this discussion we will use the term high to indicate a high-voltage state and low to indicate a low-voltage state. The use of these terms here does not imply any association with logical levels unless specified.

Gate input circuits   In a bus-based system there are many inputs attached to the bus. These constitute a load on the source gate's output that is driving the bus at a particular time. The type of load depends on the integrated circuit technology used in the system (see Chap. 8).

a   TTL   A TTL input circuit is shown in Fig. B.13. If we model as a switch the output of a gate that is driving this input, we can see that when the switch is open, the emitter current I is zero, since there is no potential difference to drive the current.

When the switch is closed, emitter current will flow out of the input lead. The input lead is said to *source* a current of 1.6 mA. This current is called the "input low current" on specification sheets and is given a negative sign to indicate that the current is out of the input lead. It is this current that is the principal loading of one gate upon a previous gate. For example, a gate that drives 10 inputs on a bus must be able to absorb (or *sink*) $10 \times 1.6 = 16$ mA when its output is low. The number of gate inputs that a gate output can drive is called its *fanout*.

b   MOS and CMOS   Both MOS and CMOS technologies have insulated gate inputs and thus there is no dc current flow regardless of the logic level of the gate that drives it. The principal loading factor here is capacitance. As the number of inputs attached to the output of a single gate is increased, the small capacitances of the gate-to-ground and conductors-to-

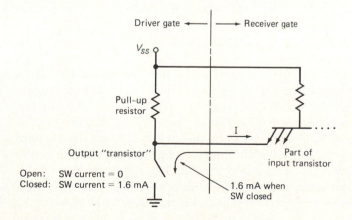

Figure B.13 Equivalent model of TTL input and output.

ground accumulate. Since each output device can be considered to be a switch in series with two resistors, when the switch is opened or closed, it takes time for the input voltage to the gate to charge up (or down) to the value required to affect the gate.

Gate output circuits

a *TTL* There are three main types of output circuits:

1. Standard TTL can be used on bus lines where there is a single driver gate per line such as a daisy chain (see Sec. 7.7).
2. TTL with open collector output is used for bus lines on which there can be several simultaneously active outputs such as for interrupt request lines, discussed in Sec. 7.7 (wired-OR applications).
3. Tristate is used for bus connections where there can be only one active gate driving the bus at a time, as described in Sec. 13.1 (e.g., data and address lines).

A TTL output circuit is shown in Fig. B.13. The output transistor acts like a switch. When the output is low, the output voltage is about 0.2 V; and when it is high, the output is about 3.4 V. Although a resistor is used in the circuit shown to pull the output high when the transistor is turned off, in practice, another transistor is used, called an *active pull-up* transistor which speeds up the switching. The two-transistor configuration is called a "totem-pole" output.

We can see that the output transistor must absorb all the current sourced by attached TTL inputs and this maximum current determines the fan-out capability.

The standard TTL gate cannot be used on a bus, since the output pull-up transistors of gates that are in the high state will then source current which will flow into the output transistor of gates that are in the low state and possibly destroy them.

Open-collector outputs An arrangement that *can* be used to drive a bus is the open-collector output circuit.

The open-collector arrangement is similar to that shown in Fig. B.13 except that the pull-up resistor is not in the gate, but a single pull-up resistor for all gates is attached to the bus line (see Fig. B.14). If several gates are attached to one bus line, then when all gates but the one currently driving the bus are driven to the high state, the bus can be pulled down by any driving gate and will be pulled up by the resistor when the gate's output transistor turns off.

If one or more output transistors switch to the low state, the bus will be pulled low. Then the connection acts like an OR circuit (called wired-OR) for a low-true association between physical levels and logical levels. That is,

$$\bar{z} = \bar{x} + \bar{y}$$

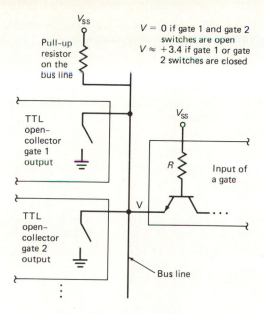

Figure B.14 Open-collector outputs.

A wired-OR connection is useful on the bus interrupt line to a processor. Open-collector interrupt request lines from several devices can be attached to a single interrupt line to the processor. If no interrupt is pending, all lines are high. When a processor wishes to interrupt the processor, it lowers its interrupt request line.

Tristate outputs Because of the previously mentioned slow pull up with the resistor in the open-collector output, the tristate configuration (Fig. B.15) is preferred when connecting to a bus. This is equivalent to isolating the output pin (attached to the bus) with two switches. When the top switch is closed, the output is high. When the bottom switch is closed, the output is low; and when both switches are open, the output is isolated or in a *high-impedance* state. The third, or high-impedance, state is exited by enabling a control line to the chip.

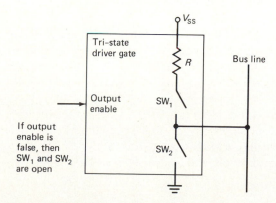

Figure B.15 Tristate output.

Tristate devices are used in bus systems by putting all outputs in high impedance except the one currently allowed to drive the bus.

b *MOS and CMOS*   These are not usually used to drive most buses, since the source transistors are of relatively high impedance, which would lead to a slow bus system. Standard MOS and CMOS output circuits are shown in Figs. 8.3 and 8.4. These are unsuitable for bus data and address line drivers for the same reasons that standard TTL cannot be used.

Tristate versions are available and are used in individual integrated circuits for handling component-level design and driving ultra-low-power bus systems. The tristate "switches" operate just as described for TTL.

MOTOROLA MC68000 FAMILY INSTRUCTION SET

MC68000L4•MC68000L6•MC68000L8•MC68000L10•MC68000L12

Mnemonic	Description	Operation	Condition Codes				
			X	N	Z	V	C
ABCD	Add Decimal with Extend	(Destination)$_{10}$ + (Source)$_{10}$ → Destination	*	U	*	U	*
ADD	Add Binary	(Destination) + (Source) → Destination	*	*	*	*	*
ADDA	Add Address	(Destination) + (Source) → Destination	–	–	–	–	–
ADDI	Add Immediate	(Destination) + Immediate Data → Destination	*	*	*	*	*
ADDQ	Add Quick	(Destination) + Immediate Data → Destination	*	*	*	*	*
ADDX	Add Extended	(Destination) + (Source) + X → Destination	*	*	*	*	*
AND	AND Logical	(Destination) Λ (Source) → Destination	–	*	*	0	0
ANDI	AND Immediate	(Destination) Λ Immediate Data → Destination	–	*	*	0	0
ASL, ASR	Arithmetic Shift	(Destination) Shifted by <count> → Destination	*	*	*	*	*
B$_{CC}$	Branch Conditionally	If $_{CC}$ then PC + d → PC	–	–	–	–	–
BCHG	Test a Bit and Change	~ (<bit number>) OF Destination → Z ~ (<bit number>) OF Destination → <bit number> OF Destination	–	–	*	–	–
BCLR	Test a Bit and Clear	~ (<bit number>) OF Destination → Z 0 → <bit number> OF Destination	–	–	*	–	–
BRA	Branch Always	PC + d → PC	–	–	–	–	–
BSET	Test a Bit and Set	~ (<bit number>) OF Destination → Z 1 → <bit number> OF Destination	–	–	*	–	–
BSR	Branch to Subroutine	PC → SP@ – ; PC + d → PC	–	–	–	–	–
BTST	Test a Bit	~ (<bit number>) OF Destination → Z	–	–	*	–	–
CHK	Check Register against Bounds	If Dn < 0 or Dn > (<ea>) then TRAP	–	*	U	U	U
CLR	Clear an Operand	0 → Destination	–	0	1	0	0
CMP	Compare	(Destination) – (Source)	–	*	*	*	*
CMPA	Compare Address	(Destination) – (Source)	–	*	*	*	*
CMPI	Compare Immediate	(Destination) – Immediate Data	–	*	*	*	*
CMPM	Compare Memory	(Destination) – (Source)	–	*	*	*	*
DB$_{CC}$	Test Condition, Decrement and Branch	If ~ $_{CC}$ then Dn – 1 → Dn; if Dn ≠ – 1 then PC + d → PC	–	–	–	–	–
DIVS	Signed Divide	(Destination)/(Source) → Destination	–	*	*	*	0
DIVU	Unsigned Divide	(Destination)/(Source) → Destination	–	*	*	*	0
EOR	Exclusive OR Logical	(Destination) ⊕ (Source) → Destination	–	*	*	0	0
EORI	Exclusive OR Immediate	(Destination) ⊕ Immediate Data → Destination	–	*	*	0	0
EXG	Exchange Register	Rx ↔ Ry	–	–	–	–	–
EXT	Sign Extend	(Destination) Sign-extended → Destination	–	*	*	0	0
JMP	Jump	Destination → PC	–	–	–	–	–
JSR	Jump to Subroutine	PC → SP@ – ; Destination → PC	–	–	–	–	–
LEA	Load Effective Address	Destination → An	–	–	–	–	–
LINK	Link and Allocate	An → SP@ – ; SP → An; SP + d → SP	–	–	–	–	–
LSL, LSR	Logical Shift	(Destination) Shifted by <count> → Destination	*	*	*	0	*
MOVE	Move Data from Source to Destination	(Source) → Destination	–	*	*	0	0
MOVE to CCR	Move to Condition Code	(Source) → CCR	*	*	*	*	*
MOVE to SR	Move to the Status Register	(Source) → SR	*	*	*	*	*

* affected   0 cleared   U defined
– unaffected   1 set

 MOTOROLA *Semiconductor Products Inc.*

MC68000L4•MC68000L6•MC68000L8•MC68000L10•MC68000L12

Mnemonic	Description	Operation	Condition Codes				
			X	N	Z	V	C
MOVE from SR	Move from the Status Register	SR → Destination	–	–	–	–	–
MOVE USP	Move User Stack Pointer	USP → An; An → USP	–	–	–	–	–
MOVEA	Move Address	(Source) → Destination	–	–	–	–	–
MOVEM	Move Multiple Registers	Registers → Destination (Source) → Registers	–	–	–	–	–
MOVEP	Move Peripheral Data	(Source) → Destination	–	–	–	–	–
MOVEQ	Move Quick	Immediate Data → Destination	–	*	*	0	0
MULS	Signed Multiply	(Destination)*(Source) → Destination	–	*	*	0	0
MULU	Unsigned Multiply	(Destination)*(Source) → Destination	–	*	*	0	0
NBCD	Negate Decimal with Extend	$0-(Destination)_{10}-X →$ Destination	*	U	*	U	*
NEG	Negate	0 – (Destination) → Destination	*	*	*	*	*
NEGX	Negate with Extend	0 – (Destination) – X → Destination	*	*	*	*	*
NOP	No Operation	–	–	–	–	–	–
NOT	Logical Complement	~ (Destination) → Destination	–	*	*	0	0
OR	Inclusive OR Logical	(Destination) v (Source) → Destination	–	*	*	0	0
ORI	Inclusive OR Immediate	(Destination) v Immediate Data → Destination	–	*	*	0	0
PEA	Push Effective Address	Destination → SP@ –	–	–	–	–	–
RESET	Reset External Devices	–	–	–	–	–	–
ROL, ROR	Rotate (Without Extend)	(Destination) Rotated by <count> → Destination	–	*	*	0	*
ROXL, ROXR	Rotate with Extend	(Destination) Rotated by <count> → Destination	*	*	*	0	*
RTE	Return from Exception	SP@ + → SR; SP@ + → PC	*	*	*	*	*
RTR	Return and Restore Condition Codes	SP@ + → CC; SP@ + → PC	*	*	*	*	*
RTS	Return from Subroutine	SP@ + → PC	–	–	–	–	–
SBCD	Subtract Decimal with Extend	$(Destination)_{10}-(Source)_{10}-X →$ Destination	*	U	*	U	*
S$_{CC}$	Set According to Condition	If CC then 1's → Destination else 0's → Destination	–	–	–	–	–
STOP	Load Status Register and Stop	Immediate Data → SR; STOP	*	*	*	*	*
SUB	Subtract Binary	(Destination) – (Source) → Destination	*	*	*	*	*
SUBA	Subtract Address	(Destination) – (Source) → Destination	–	–	–	–	–
SUBI	Subtract Immediate	(Destination) – Immediate Data → Destination	*	*	*	*	*
SUBQ	Subtract Quick	(Destination) – Immediate Data → Destination	*	*	*	*	*
SUBX	Subtract with Extend	(Destination) – (Source) – X → Destination	*	*	*	*	*
SWAP	Swap Register Halves	Register [31:16] ↔ Register [15:0]	–	*	*	0	0
TAS	Test and Set an Operand	(Destination) Tested → CC; 1 → [7] OF Destination	–	*	*	0	0
TRAP	Trap	PC → SSP@ – ; SR → SSP@ – ; (Vector) → PC	–	–	–	–	–
TRAPV	Trap on Overflow	If V then TRAP	–	–	–	–	–
TST	Test an Operand	(Destination) Tested → CC	–	*	*	0	0
UNLK	Unlink	An → SP; SP@ + → An	–	–	–	–	–

[ ] = bit number

* affected   0 cleared   U defined
– unaffected 1 set

Courtesy of Motorola Inc.

 MOTOROLA *Semiconductor Products Inc.*

INTEL 8086/8 FAMILY INSTRUCTION SET

DATA TRANSFER

MOV = Move:

	76543210	76543210	76543210	76543210
Register/memory to/from register	1 0 0 0 1 0 d w	mod reg r/m		
Immediate to register/memory	1 1 0 0 0 1 1 w	mod 0 0 0 r/m	data	data if w=1
Immediate to register	1 0 1 1 w reg	data	data if w=1	
Memory to accumulator	1 0 1 0 0 0 0 w	addr-low	addr-high	
Accumulator to memory	1 0 1 0 0 0 1 w	addr-low	addr-high	
Register/memory to segment register	1 0 0 0 1 1 1 0	mod 0 reg r/m		
Segment register to register/memory	1 0 0 0 1 1 0 0	mod 0 reg r/m		

PUSH = Push:

Register/memory	1 1 1 1 1 1 1 1	mod 1 1 0 r/m
Register	0 1 0 1 0 reg	
Segment register	0 0 0 reg 1 1 0	

POP = Pop:

Register/memory	1 0 0 0 1 1 1 1	mod 0 0 0 r/m
Register	0 1 0 1 1 reg	
Segment register	0 0 0 reg 1 1 1	

XCHG = Exchange:

Register/memory with register	1 0 0 0 0 1 1 w	mod reg r/m
Register with accumulator	1 0 0 1 0 reg	

IN/INW = Input to AL/AX from:

Fixed port	1 1 1 0 0 1 0 w	port
Variable port	1 1 1 0 1 1 0 w	

OUT/OUTW = Output from AL/AX to:

Fixed port	1 1 1 0 0 1 1 w	port
Variable port	1 1 1 0 1 1 1 w	
XLAT=Translate byte to AL	1 1 0 1 0 1 1 1	
LEA=Load EA to register	1 0 0 0 1 1 0 1	mod reg r/m
LDS=Load pointer to DS	1 1 0 0 0 1 0 1	mod reg r/m
LES=Load pointer to ES	1 1 0 0 0 1 0 0	mod reg r/m
LAHF=Load AH with flags	1 0 0 1 1 1 1 1	
SAHF=Store AH into flags	1 0 0 1 1 1 1 0	
PUSHF=Push flags	1 0 0 1 1 1 0 0	
POPF=Pop flags	1 0 0 1 1 1 0 1	

ARITHMETIC

ADD = Add:

Reg./memory with register to either	0 0 0 0 0 0 d w	mod reg r/m		
Immediate to register/memory	1 0 0 0 0 0 s w	mod 0 0 0 r/m	data	data if s:w=01
Immediate to accumulator	0 0 0 0 0 1 0 w	data	data if w=1	

ADC = Add with carry:

Reg./memory with register to either	0 0 0 1 0 0 d w	mod reg r/m		
Immediate to register/memory	1 0 0 0 0 0 s w	mod 0 1 0 r/m	data	data if s:w=01
Immediate to accumulator	0 0 0 1 0 1 0 w	data	data if w=1	

INC = Increment:

Register/memory	1 1 1 1 1 1 1 w	mod 0 0 0 r/m
Register	0 1 0 0 0 reg	
AAA=ASCII adjust for add	0 0 1 1 0 1 1 1	
DAA=Decimal adjust for add	0 0 1 0 0 1 1 1	

SUB = Subtract:

Reg./memory and register to either	0 0 1 0 1 0 d w	mod reg r/m		
Immediate from register/memory	1 0 0 0 0 0 s w	mod 1 0 1 r/m	data	data if s:w=01
Immediate from accumulator	0 0 1 0 1 1 0 w	data	data if w=1	

SBB = Subtract with borrow:

Reg./memory and register to either	0 0 0 1 1 0 d w	mod reg r/m		
Immediate from register/memory	1 0 0 0 0 0 s w	mod 0 1 1 r/m	data	data if s:w=01
Immediate from accumulator	0 0 0 1 1 1 0 w	data	data if w=1	

DEC = Decrement:

	76543210	76543210	76543210	76543210
Register/memory	1 1 1 1 1 1 1 w	mod 0 0 1 r/m		
Register	0 1 0 0 1 reg			
NEG=Change sign	1 1 1 1 0 1 1 w	mod 0 1 1 r/m		

CMP = Compare:

Register/memory and register	0 0 1 1 1 0 d w	mod reg r/m		
Immediate with register/memory	1 0 0 0 0 0 s w	mod 1 1 1 r/m	data	data if s:w=01
Immediate with accumulator	0 0 1 1 1 1 0 w	data	data if w=1	
AAS=ASCII adjust for subtract	0 0 1 1 1 1 1 1			
DAS=Decimal adjust for subtract	0 0 1 0 1 1 1 1			
MUL=Multiply (unsigned)	1 1 1 1 0 1 1 w	mod 1 0 0 r/m		
IMUL=Integer multiply (signed)	1 1 1 1 0 1 1 w	mod 1 0 1 r/m		
AAM=ASCII adjust for multiply	1 1 0 1 0 1 0 0	0 0 0 0 1 0 1 0		
DIV=Divide (unsigned)	1 1 1 1 0 1 1 w	mod 1 1 0 r/m		
IDIV=Integer divide (signed)	1 1 1 1 0 1 1 w	mod 1 1 1 r/m		
AAD=ASCII adjust for divide	1 1 0 1 0 1 0 1	0 0 0 0 1 0 1 0		
CBW=Convert byte to word	1 0 0 1 1 0 0 0			
CWD=Convert word to double word	1 0 0 1 1 0 0 1			

LOGIC

NOT=Invert	1 1 1 1 0 1 1 w	mod 0 1 0 r/m
SHL/SAL=Shift logical/arithmetic left	1 1 0 1 0 0 v w	mod 1 0 0 r/m
SHR=Shift logical right	1 1 0 1 0 0 v w	mod 1 0 1 r/m
SAR=Shift arithmetic right	1 1 0 1 0 0 v w	mod 1 1 1 r/m
ROL=Rotate left	1 1 0 1 0 0 v w	mod 0 0 0 r/m
ROR=Rotate right	1 1 0 1 0 0 v w	mod 0 0 1 r/m
RCL=Rotate through carry flag left	1 1 0 1 0 0 v w	mod 0 1 0 r/m
RCR=Rotate through carry right	1 1 0 1 0 0 v w	mod 0 1 1 r/m

AND = And:

Reg./memory and register to either	0 0 1 0 0 0 d w	mod reg r/m		
Immediate to register/memory	1 0 0 0 0 0 0 w	mod 1 0 0 r/m	data	data if w=1
Immediate to accumulator	0 0 1 0 0 1 0 w	data	data if w=1	

TEST = And function to flags, no result:

Register/memory and register	1 0 0 0 0 1 0 w	mod reg r/m		
Immediate data and register/memory	1 1 1 1 0 1 1 w	mod 0 0 0 r/m	data	data if w=1
Immediate data and accumulator	1 0 1 0 1 0 0 w	data	data if w=1	

OR = Or:

Reg./memory and register to either	0 0 0 0 1 0 d w	mod reg r/m		
Immediate to register/memory	1 0 0 0 0 0 0 w	mod 0 0 1 r/m	data	data if w=1
Immediate to accumulator	0 0 0 0 1 1 0 w	data	data if w=1	

XOR = Exclusive or:

Reg./memory and register to either	0 0 1 1 0 0 d w	mod reg r/m		
Immediate to register/memory	1 0 0 0 0 0 0 w	mod 1 1 0 r/m	data	data if w=1
Immediate to accumulator	0 0 1 1 0 1 0 w	data	data if w=1	

STRING MANIPULATION

REP=Repeat	1 1 1 1 0 0 1 z
MOVB/MOVW=Move byte/word	1 0 1 0 0 1 0 w
CMPB/CMPW=Compare byte/word	1 0 1 0 0 1 1 w
SCAB/SCAW=Scan byte/word	1 0 1 0 1 1 1 w
LODB/LODW=Load byte/wd to AL/AX	1 0 1 0 1 1 0 w
STOB/STOW=Stor byte/wd frm AL/A	1 0 1 0 1 0 1 w

CONTROL TRANSFER

CALL - Call:

	76543210	76543210	76543210
Direct within segment	11101000	disp-low	disp-high
Indirect within segment	11111111	mod 0 1 0 r/m	
Direct intersegment	10011010	offset-low	offset-high
		seg-low	seg-high
Indirect intersegment	11111111	mod 0 1 1 r/m	

JMP - Unconditional Jump:

	76543210	76543210	76543210
Direct within segment	11101001	disp-low	disp-high
Direct within segment-short	11101011	disp	
Indirect within segment	11111111	mod 1 0 0 r/m	
Direct intersegment	11101010	offset-low	offset-high
		seg-low	seg-high
Indirect intersegment	11111111	mod 1 0 1 r/m	

RET - Return from CALL:

	76543210	76543210	76543210
Within segment	11000011		
Within seg. adding immed to SP	11000010	data-low	data-high
Intersegment	11001011		
Intersegment. adding immediate to SP	11001010	data-low	data-high
JE/JZ–Jump on equal/zero	01110100	disp	
JL/JNGE–Jump on less/not greater or equal	01111100	disp	
JLE/JNG–Jump on less or equal/not greater	01111110	disp	
JB/JNAE–Jump on below/not above or equal	01110010	disp	
JBE/JNA–Jump on below or equal/not above	01110110	disp	
JP/JPE–Jump on parity/parity even	01111010	disp	
JO–Jump on overflow	01110000	disp	
JS–Jump on sign	01111000	disp	
JNE/JNZ–Jump on not equal/not zero	01110101	disp	
JNL/JGE–Jump on not less/greater or equal	01111101	disp	
JNLE/JG–Jump on not less or equal/greater	01111111	disp	

	76543210	76543210
JNB/JAE–Jump on not below/above or equal	01110011	disp
JNBE/JA–Jump on not below or equal/above	01110111	disp
JNP/JPO–Jump on not par/par odd	01111011	disp
JNO–Jump on not overflow	01110001	disp
JNS–Jump on not sign	01111001	disp
LOOP–Loop CX times	11100010	disp
LOOPZ/LOOPE–Loop while zero/equal	11100001	disp
LOOPNZ/LOOPNE–Loop while not zero/equal	11100000	disp
JCXZ–Jump on CX zero	11100011	disp

INT   Interrupt

	76543210	76543210
Type specified	11001101	type
Type 3	11001100	
INTO–Interrupt on overflow	11001110	
IRET–Interrupt return	11001111	

PROCESSOR CONTROL

	76543210	76543210
CLC–Clear carry	11111000	
CMC–Complement carry	11110101	
STC–Set carry	11111001	
CLD–Clear direction	11111100	
STD–Set direction	11111101	
CLI–Clear interrupt	11111010	
STI–Set interrupt	11111011	
HLT–Halt	11110100	
WAIT–Wait	10011011	
ESC–Escape (to external device)	11011 x	mod x r/m
LOCK–Bus lock prefix	11110000	

Footnotes:

AL = 8-bit accumulator
AX = 16-bit accumulator
CX = Count register
DS = Data segment
ES = Extra segment
Above/below refers to unsigned value.
Greater = more positive;
Less = less positive (more negative) signed values
if d = 1 then "to"; if d = 0 then "from"
if w = 1 then word instruction; if w = 0 then byte instruction

if s:w = 01 then 16 bits of immediate data form the operand.
if s:w = 11 then an immediate data byte is sign extended to
     form the 16-bit operand.
if v = 0 then "count" = 1; if v = 0 then "count" in (CL)
x = don't care.
if v = 0 then "count" = 1; if v = 1 then "count" in (CL) register.
z is used for string primitives for comparison with ZF FLAG.

SEGMENT OVERRIDE PREFIX

001 reg 110

if mod = 11 then r/m is treated as a REG field
if mod = 00 then DISP = 0*, disp-low and disp-high are absent
if mod = 01 then DISP = disp-low sign-extended to 16-bits, disp-high is absent
if mod = 10 then DISP = disp-high: disp-low

if r/m = 000 then EA = (BX) + (SI) + DISP
if r/m = 001 then EA = (BX) + (DI) + DISP
if r/m = 010 then EA = (BP) + (SI) + DISP
if r/m = 011 then EA = (BP) + (DI) + DISP
if r/m = 100 then EA = (SI) + DISP
if r/m = 101 then EA = (DI) + DISP
if r/m = 110 then EA = (BP) + DISP*
if r/m = 111 then EA = (BX) + DISP
DISP follows 2nd byte of instruction (before data if required)

*except if mod = 00 and r/m = 110 then EA = disp-high: disp-low.

REG is assigned according to the following table:

16-Bit (w = 1)		8-Bit (w = 0)		Segment	
000	AX	000	AL	00	ES
001	CX	001	CL	01	CS
010	DX	010	DL	10	SS
011	BX	011	BL	11	DS
100	SP	100	AH		
101	BP	101	CH		
110	SI	110	DH		
111	DI	111	BH		

Instructions which reference the flag register file as a 16-bit object use the symbol FLAGS to represent the file:

FLAGS = X:X:X:X:(OF):(DF):(IF):(TF):(SF):(ZF):X:(AF):X:(PF):X:(CF)

Mnemonics © Intel, 1978

Courtesy of Intel Corp.

THE C PROGRAMMING LANGUAGE

C is a popular systems programming language that is well-suited to real-time programming. It can be described as a "medium-level" language since it gives the programmer "low-level" facilities such as manipulation of individual bits within a word and direct access to memory locations and yet retains the high-level features of programming langauges such as Pascal or PL/1. C compilers are available for most 8-bit, 16-bit, and 32-bit microprocessors and for many minicomputers. As a result, C programs can be moved quite easily from one type of computer to another. For popular microprocessors, C compilers are available from several different vendors. This competitive situation results in high-quality compilers and reasonable prices.

The C language is used in programming examples in this book. This appendix provides an introduction to C that allows these programs to be understood. In addition, the features of C that make it particularly suitable as a real-time programming language are discussed. A complete description of the C language is not possible in the space available. Readers who wish to write and run C programs of realistic complexity should refer to one of the C texts listed at the end of the chapter and to the documentation for the C compiler they are using.

E.1 BASIC ELEMENTS OF A C PROGRAM

E.1.1 Functions

Figure E.1 illustrates the structure of a C program. The program is divided into *functions* which are similar to subroutines or procedures in other programming

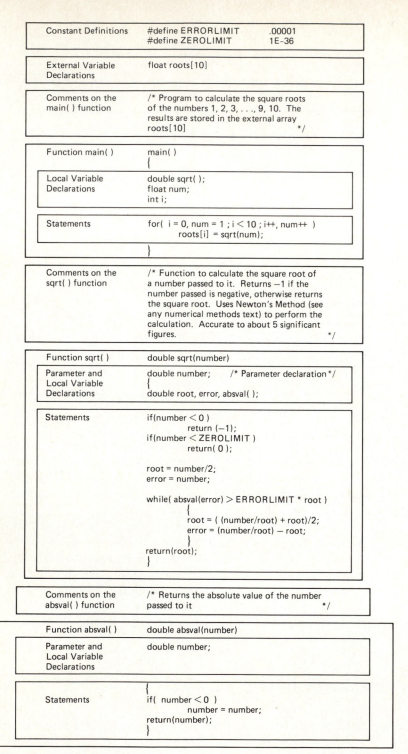

Constant Definitions	#define ERRORLIMIT .00001 #define ZEROLIMIT 1E–36
External Variable Declarations	float roots[10]
Comments on the main( ) function	/* Program to calculate the square roots of the numbers 1, 2, 3, . . ., 9, 10. The results are stored in the external array roots[10] */
Function main( )	main( ) {
Local Variable Declarations	double sqrt( ); float num; int i;

Statements	for(i = 0, num = 1 ; i < 10 ; i++, num++)
	        roots[i] = sqrt(num);
	
	}

Comments on the sqrt( ) function	/* Function to calculate the square root of a number passed to it. Returns −1 if the number passed is negative, otherwise returns the square root. Uses Newton's Method (see any numerical methods text) to perform the calculation. Accurate to about 5 significant figures. */
Function sqrt( )	double sqrt(number)

Parameter and Local Variable Declarations	double number; /* Parameter declaration*/
	{
	double root, error, absval( );

Statements	if(number < 0)
	        return (−1);
	if(number < ZEROLIMIT)
	        return(0);
	
	root = number/2;
	error = number;
	
	while(absval(error) > ERRORLIMIT * root)
	        {
	        root = ((number/root) + root)/2;
	        error = (number/root) − root;
	        }
	return(root);
	}

Comments on the absval( ) function	/* Returns the absolute value of the number passed to it */
Function absval( )	double absval(number)
Parameter and Local Variable Declarations	double number;

Statements	{
	if(number < 0)
	        number = number;
	return(number);
	}

Figure E.1 C-program structure.

524

languages. Execution of the program always starts with the function main( ). The function main( ) can call other functions in the program and these functions can, in turn, call other functions. A simple program may have only the function main( ) with no other functions.

E.1.2 Statements

The actual computation is carried out by program *statements* within the functions. Some statements carry out actions on data. For example, the statement

$$a = b + c;$$

puts the sum of the variables b and c into the variable a. Control statements implement the program control structures discussed in Sec. 4.7. For instance, the control statement

```
if(a==b)
        ⟨statement 1⟩
else
        ⟨statement 2⟩
```

is an embodiment of the IF-THEN-ELSE control structure. If the variable a is equal to the variable b, then statement 1 is executed; otherwise statement 2 is executed.

E.1.3 Blocks

Statements can be grouped into *blocks* that are treated as one statement by control statements. Braces { } are used to indicate the beginning and end of a block. For example, the control statement

```
if(a<b)
        {
        c = d + e;
        b = a - 1;
        e = a/b;
        }
else
        a = b;
```

executes the three statements between the braces if the variable a is less than the variable b. Note how indentation is used to format the control statement so that it is easier to read. C has few restrictions on the placement of statements, so the programmer is free to format the program to make it as readable as possible.

E.1.4 Constants

Constant definitions allow the programmer to assign names to numerical constants used in the program. For example, the constant definition

```
#define PI    3.14159
```

assigns the name PI to the the constant 3.14159. A portion of the compiler called the *preprocessor* will replace all occurrences of PI in the program with 3.14159.

E.1.5 Variables

Variable declarations tell the compiler how many variables there are in the program, the type of each variable (e.g., integer or real), and the portion of the program in which each variable is to be recognized. The portion of the program in which a variable is recognized is called the *scope* of a variable. Variables declared outside any functions are called *external* variables. They are recognized throughout the entire program; as a result, they are sometimes referred to as *global* variables. Variables declared within a function are recognized and valid only within that function; they are sometimes called *local* variables. A local variable declared in one function may have the same name as a local variable declared in another function, yet the two variables are recognized as being entirely distinct. The provision of local variables allows a programmer to develop a program module without worrying about possible conflicts with variables in other modules in the program.

E.1.6 Comments

Comments are used by the programmer to explain the function of the program. In C, any text enclosed by the symbols /* */ is considered to be a comment. For example, the line

```
/*This is a comment*/
```

will be ignored by the compiler but will be printed or displayed in program listings. At the very minimum, a comment should be inserted before every function in a C program, explaining what the function does and what interactions occur with other functions.

E.2 VARIABLE NAMES AND CONSTANTS

E.2.1 Identifiers

A C program is written using a character set consisting of the letters of the alphabet (upper- and lowercase), the digits 0 to 9, and some punctuation and special symbols. Names for variables, functions, and constants are called *identifiers*. They are made up of letters, digits, and the underscore symbol "＿". The first character must be a letter. The compiler reads only the first eight characters in an identifier, so identifiers for objects with overlapping scope must be different somewhere within the first eight characters. The other symbols accepted by C have special meaning to the compiler and must not be used in identifiers. A list of these reserved symbols is given in Table E.1.

Identifiers should be chosen so that they are descriptive of the meaning or function of the object they represent. Upper- and lowercase letters are distinct in C. Constant names are commonly written in uppercase in C, while variable names are commonly written in lowercase. Function names are sometimes written with the first letter in uppercase and the remaining letters in lowercase. Multiword identifiers can be written with the underscore symbol as follows:

```
Temperature_Sensor_Function( )
```

since spaces are not allowed within a name. Spaces or reserved symbols must be used between identifiers so that the compiler can properly recognize the individual identifiers. The start of a new line is considered to include a space. Otherwise, C makes few requirements on format. A statement could be written as

$$a = b + c - f*(g/h);$$

or as

$$a = b + c - f*(g/h);$$

The programmer is free to format the program so that it will be readable.

Identifiers must be chosen so that they are not the same as *keywords* in the C programming language. Keywords are predefined names reserved for the compiler. Table E.2 lists the keywords in C.

Table E.1 Reserved symbols in C

!	#	%	^	&
*	(	)	-	+
=	~	{	}	[
]	\	|	;	:
"	'	<	>	,
/	?	.		

Table E.2 C keywords

auto	break	case
char	continue	default
do	double	else
entry	extern	float
for	goto	if
int	long	register
return	short	sizeof
static	struct	switch
typedef	union	unsigned
while	void	enum

E.2.2 Numerical Constants

There are two types of numerical constants in C—integer constants and floating-point constants. An integer constant consists of a string of digits representing an integer. Integer constants can be expressed in decimal, octal, or hexadecimal form (see App. A for a discussion of hexadecimal and octal notation). If the string of digits begins with a zero, the constant is in octal form. If the string of digits begins with a zero followed by an x or X, the constant is in hexadecimal form. Otherwise the constant is in decimal form. Thus, the decimal number 12 could be expressed as follows:

12	Decimal notation
014	Octal notation
0xC	Hexadecimal notation

The ability to express constants in hexadecimal form is particularly useful in programs which directly control the computer hardware, since the addresses of memory locations and I/O ports are normally given in hexadecimal. In addition, data used in real-time applications often consists of patterns of bits; it is easier to convert these bit patterns to a hexadecimal number than to a decimal number (see App. A).

Floating-point constants represent real numbers. Floating-point constants are always in decimal notation. One format for floating-point constants consists of a string of digits with a decimal point in it. For example,

```
3.14159
-123.
.0023
```

are all floating-point constants. The other floating-point format is an exponential format consisting of a mantissa followed by an e or E followed by an exponent. The following real constants are expressed in two formats:

```
12E2         1200.
1.43e05      143000.
-.32E-01     -0.032
```

E.2.3 Character Constants

The characters displayed on a video display terminal or printed out by a printer are stored within the computer as numeric codes. The code used in C is the ASCII character code (see App. A). *Character constants* allow the numeric code for a particular character to be specified. Character constants are expressed as the character enclosed by single quotes ''. For instance, the character constant 'R' specifies the numeric code 82, which is the ASCII code for the character R. Note that the code for a character representing a digit is not equal to the numerical value of the digit. The ASCII code for the character constant '4' is 52, for example. If the number 52 is sent to a terminal, the character 4 will be displayed.

A *string constant* is a sequence of characters surrounded by double quotes. For example,

```
''This is a string constant''
```

E.3 DECLARATIONS AND VARIABLE TYPES

All variables used in a C program must be explicitly declared in a *declaration statement*. The declaration statement specifies the *storage class* and *type* of the variables listed in the statement. Declaration statements have the format

```
⟨storage class⟩⟨type⟩⟨variable name⟩,
  ⟨variable name⟩,...,⟨variable name⟩;
```

For example, the declaration statement

```
auto int var1, var2, var3;
```

declares the variables var1, var2, and var3 to be of type int (integer) with the auto (automatic) storage class.

The storage class of a variable gives the compiler information on the portion of the program in which the variable is to be recognized (the *scope* of the variable) and how it should be stored in memory. The storage class of a variable depends both on the storage-class specifier in the declaration statement and on the location in the program where the declaration occurs.

E.3.1 The auto Storage Class

Variables with the auto (automatic) storage class are local to the function in which they are declared. Thus, in the following simple program

```
main( )
{
auto int a, b, c;
a = 3;
b = 2;
c = a + b;
func( ); /*This calls the function func( )*/
}

func( )
{
auto int d, e, f;
d = 20;
e = 5;
f = d/e;
}
```

the variables *a, b, c* are known only within the function main( ) [i.e., their scope is the function main( )] and the variables *d, e, f* are known only in the function func( ).

Memory space is allocated to automatic variables when the function in which they are declared begins execution. This memory space is reused once the function completes execution. This results in efficient use of memory, since unused variables are not taking up memory space. However, it also means that automatic variables do not retain their value between successive invocations of the function.

Variables declared within a function are automatic by default, thus the declaration in the following program

```
main( )
{
int a, b, c;
a = 1;
b = 2;
c = b - a;
}
```

declares *a, b,* and *c* to be variables with automatic storage.

E.3.2 The static Storage Class

In some cases a local variable must retain its value between function invocations. The *static* storage class declares local variables that retain their value.
For example, in the function

```
counter( )
{
int a;
static int count;
a = 1;
count = count + a;
}
```

the variable *a* is an automatic variable (by default) while the variable count is a
static variable.

E.3.3 The extern Storage Class

External variables have a scope which consists of the entire program. External
variables are normally declared at the beginning of a program outside of any
functions. For example, in the program

```
extern int var;

main( )
{
int a, b;
a = 3;
b = 2;
var = a + b;
func( );  /*Calling the function func( )*/
}
func( )
{
int c;
c = var;
}
```

the external variable var is valid in both main( ) and func( ) while auto
variables *a* and *b* are valid only in main( ) and auto variable *c* is valid only in
func( ). Local variables with the same name as an external variable have
precedence within their scope. Variables declared outside any function are
external by default, so the extern storage class specifier is often omitted.

E.3.4 Integer Variable Types

The *type specification* in a declaration statement tells the compiler how much memory space to allocate to each variable and how to treat that variable during computations. Variables of type int are used to represent signed whole numbers (integers). The twos-complement form is used to represent signed numbers (see App. A). The range of integers that can be stored in an int variable depends on the particular C compiler and on the computer. Commonly, an int variable is stored in 2 bytes (16 bits) of memory. In that case, the number is 15 bits long with the sixteenth bit used to indicate the sign of the number. Thus the largest number that can be stored is $2^{15} - 1$, or 32,767. Larger positive integers can be stored if the qualifier *unsigned* is added to the int type specification, as in the following example:

```
unsigned int i, j, k;
```

An unsigned int is assumed to be positive, so the sign bit is not required. Larger signed whole numbers can be used if the variable is declared with the qualifier *long* added to the int type, as in

```
long int i, j, k;
```

The compiler normally allocates twice as many bytes to long int variables as to int variables.

E.3.5 Real Variable Types

Real, or floating-point, numbers are stored in variables of type *float*. A typical float variable in C has a mantissa which can be positive or negative and has about six significant digits. The exponent can range from about 10^{+38} to 10^{-38}. Applications requiring greater numerical precision are served by variables of type *double*. Double variables have about 14 significant digits. The range of exponents is the same as that for float variables.

Float and double variables should be used with care in real-time applications. Most microcomputers and minicomputers perform calculations on floating-point numbers much more slowly than on integers. If an int variable can be used, it should be used in preference to a float or double variable.

In general, the size of integer and real variables is machine- and compiler-dependent. The compiler documentation should always be used to determine the actual length of these variables in a particular system.

E.3.6 Character Type

The type specifier *char* is used for variables which store character codes. In the ASCII character code used by C, a character is represented by a 7-bit number. Therefore, a variable of type char is normally assigned 1 byte (8 bits) of memory space, since this is the smallest unit that can be assigned. A char variable can also be used to store small whole numbers, but whether they are treated as signed or unsigned depends on the particular compiler. The *unsigned char* declaration can be used to ensure that the char variable will be treated as an unsigned number.

E.3.7 Arrays

Arrays of the previous variable types can be declared in C. For example, the declaration

```
char c[10];
```

declares c to be an array of 10 char variables. Multidimension arrays can also be declared, as in the following example:

```
int mat[2][5];
```

The name mat is declared to be an array of two rows by five columns of int variables. Elements in an array are numbered from zero. Thus the first element in the array c, declared above, is referred to as c[0] and the last element is referred to as c[9].

E.4 OPERATORS AND EXPRESSIONS

Operators are the symbols in a programming language that cause the program to perform some action on data. For instance, the + operator causes the program to add two variables. An *expression* consists of operators and the variables or constants they act upon. As an example,

```
a + b
```

is an expression. Expressions have a *value* which is the result obtained when the expression is *evaluated*. The expression

```
c − d
```

evaluates (when the program is executed) to a value which is the difference between the value of *c* and the value of *d*.

E.4.1 Arithmetic Operators

C has the standard arithmetic operators found in most programming languages. They are listed in Table E.3. If the standard division operator (/) is applied to

Table E.3 Arithmetic operators

+	Addition
−	Subtraction
*	Multiplication
/	Division

integers, a truncated quotient is produced; for example, $5/3 = 1$. A *modulus* operator ($\%$) is available that produces the remainder in an integer division; for instance, $5\%3 = 2$.

E.4.2 Assignment Operators

The *assignment* operator ($=$) causes the value of the expression on the right of the operator to be stored in the variable on the left. As in other programming languages, the assignment operator is not the same as the *equality operator* ($==$ in C) which checks whether two expressions are equal. The assignment

$$\mathrm{prod} = 4 * 3$$

stores 12 (the value of the expression $4 * 3$) in the variable prod. The assignment operator can be combined with other operators in the following fashion:

$$x \langle \mathrm{operator} \rangle = y$$

This has the same result as

$$x = x \langle \mathrm{operator} \rangle y$$

For example,

$$w + = 7$$

has the same effect as

$$w = w + 7$$

but using the first expression often gives the compiler a chance to produce more efficient object code.

E.4.3 Increment and Decrement Operators

C has two operators which increment or decrement variables. The increment operator ($++$) increments a variable by 1, while the decrement operator ($--$) decrements a variable by 1. For example, the expressions

$$++a$$

$$--b$$

increment variable *a* and decrement variable *b*. These operators may be used before or after the variable; the position affects when the variable is used in the remainder of an expression. For instance, if *b* is 4, the expression

$$w = --b$$

sets *w* to 3 since *b* is decremented before its value is used. On the other hand, the expression

$$w = b--$$

sets *w* to 4 since *b* is only decremented after its value is used in the assignment. The use of the increment and decrement operators often gives the compiler an opportunity to generate more efficient object code than would be the case if an expression such as

$$a = a + 1$$

were used.

E.4.4 Bitwise Logical Operators

A feature of C which is particularly useful in real-time applications is its set of *bitwise logical* operators. These carry out the boolean logic operations AND, OR, NOT, and exclusive-OR (see App. A). These operators deal with integers on a bit-by-bit basis. For instance, the NOT operator ($\sim$) converts every bit in an integer which is a 1 to a 0 and every bit which is a 0 to a 1. For example, if

$$k = 0x3F87 \qquad (0011111110000111)$$

then $\sim k$ is

$$0xC078 \qquad (1100000001111000)$$

The effects of the other bitwise logical operators, AND (&), OR (|), and exclusive OR (^), are shown in Table E.4.

The bitwise AND of two numbers has a 1 in every bit position where both numbers have a 1 and has a 0 in every position where one or both the numbers have a 0. The bitwise OR of two numbers has a 1 in every bit position where either number has a 1 and a 0 in every bit position where both numbers have a 0. The bitwise exclusive-OR of two numbers has a 1 in every bit position where

Table E.4 Examples of bitwise logical operations

j = 0xAF3	equals	1010011011110011
k = 0x740E	equals	0111010000001110
j & k	evaluates to	0010010000000010
j | k	evaluates to	1111011011111111
j ^ k	evaluates to	1101001011111101

the two numbers have opposite bit values and a 0 in every bit position where they have the same bit value.

C also incorporates two operators which shift the bits in an integer left or right. The expression

$$i << j$$

shifts the bits in *i* left by *j* bit positions. The expression

$$i >> 3$$

shifts the bits in *i* right by 3 bit positions. As an example, if *i* is 0x09 (0000000000001001), then the preceding expression evaluates to (0000000000000001). Zeros are shifted in to replace the bits shifted out.

Programs involved in the control of computer hardware often have a need to deal with individual bits within an integer. As an example, peripheral devices interfaced to a computer often contain a *status register* which can be read by the computer. The organization of a status register for a hypothetical device is shown in Fig. E.2. The states of the individual bits within the register correspond to conditions within the device. For instance, the state of bit D0 indicates whether the device is ready to receive data from the computer. If the bit is a 1, then data can be sent; if it is a 0, then the device is not ready to receive. A program which transfers data to the device from the computer must monitor the status of bit D0 in order to send data only when the peripheral device can accept the data.

A C program can use the bitwise logical operators to *mask off* the desired bit in the status register so that it can be tested. If the contents of the status

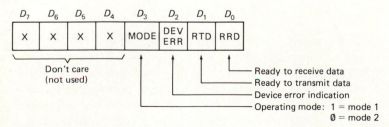

Figure E.2 Contents of a status register.

register have been read by the program and stored in the variable statusreg, then the expression

$$rrd = statusreg \& 0x01$$

will set all but the lowest-order bit in the variable rrd to 0. The lowest order bit of rrd will equal the value of the lowest order bit in statusreg, which is the bit to be tested. The variable rrd can now be compared with zero to test whether the peripheral device is ready to receive data.

E.4.5 Relational Operators

Relational operators test to see if the relation specified by the operator is true. An expression evaluates to 1 if the relation is true, otherwise it evaluates to 0. The relational operators are listed in Table E.5.

E.4.6 Logical Connective Operators

The tests performed by the relational operators can be combined by *logical connectives*. The three logical connective operators are && (AND connective), || (OR connective), and ! (NOT connective). The expression

$$(a == 3) \&\& !(b < 4)$$

evaluates to 1 (true) if *a* is equal to 3 AND *b* is NOT less than 4.

E.4.7 Precedence of Operators

The example above illustrates the use of parentheses ( ) to ensure the proper order of evaluation. Operators in C have levels of *precedence* which determine the order of evaluation in expressions involving multiple operators. Table E.6 lists the precedence of the C operators we have described so far.

Table E.5 Relational operators

Operator	Sample expression	Test performed
>	a > b	Is a greater than b
>=	a >= b	Is a greater than or equal to b
<	a < b	Is a less than b
<=	a <= b	Is a less than or equal to b
==	a == b	Is a equal to b
!=	a != b	Is a not equal to b

Table E.6 Precedence of C operators

Operator type	Operators	
Primary	( ) [ ]	Highest precedence
Unary	! ~ ++ -- -	
Arithmetic	* / %	
Arithmetic	+ -	
Shift	<< >>	
Relational	< <= > >=	
Relational	== !=	
Bitwise logical	&	
Bitwise logical	^	
Bitwise logical	|	
Logical connective	&&	
Logical connective	||	
Assignment	= += -= &= |= etc.	Lowest precedence

When an expression involves operators with different levels of precedence, the subexpression with the highest precedence operator is evaluated first. Thus in the expression

$$b+c*d$$

the subexpression $c*d$ is evaluated first.

C normally evaluates expressions involving operators of equal precedence (those on the same line in Table E.6) from left to right. Assignment operators are evaluated right to left so that expressions such as

$$a = b = c = d = 0$$

can be written. The unary operators are also evaluated from right to left. Thus,

$$-x++$$

is evaluated as

$$-(x++)$$

rather than

$$(-x)++$$

E.4.8 Expressions with Mixed-Variable Types

C allows expressions in which the variables are of a different type. The compiler automatically performs type conversions so that all the variables are of the same type. Table E.7 shows how variables are *promoted* in expressions.

Table E.7 Promotion of variables

double		(preferred type)
	float	
long		(preferred type)
unsigned		(preferred type)
int		(preferred type)
	char	

Nonpreferred variable types (char and float) are promoted to the next preferred type above them in the table. If there is more than one type of variable remaining in the expression, all variables are then promoted to the highest remaining type. Thus an expression involving an int and a float variable will result in both variables being promoted to type double, since the float variable will first be promoted to the preferred type double, and then the int variable will be promoted to the highest remaining type in the expression, which is now double.

When conversions take place across assignment operators, the value of the right side is converted to the type of the variable on the left. Thus

```
float x, y;
int i;
i=x+y;
```

results in the sum of *x* and *y*, which is of type double, being converted to type int (by truncating the fractional portion of the sum) and stored in variable *i*. Note that this could produce undefined results if the sum of *x* and *y* is greater than the largest integer that can be represented by an int variable. Normally, conversions "upward," say, from int to long int, are safe in that they do not destroy any information. An exception is the conversion from int to unsigned int, which produces undefined results if the value of the int variable is negative. Conversions "downward," say, from int to char, require truncation, normally of high-order bits, and thus may destroy some information.

E.5 STATEMENTS AND CONTROL STRUCTURES

A simple statement in C has the form

```
x=a+b;
```

The semicolon acts as a statement terminator and is required. *Compound statements*, or blocks, can be created by using braces { } to group simple statements together, as in

```
{
x = a + b;
j = i + 4;
c = d * a;
}
```

Compound statements are treated like a single statement by the control structures described in this section.

E.5.1 The if else Statement

The C language has the structured control constructs discussed in Chap. 4. The IF-THEN-ELSE structure is represented by

```
if(⟨expression⟩)
        ⟨statement 1⟩
else
        ⟨statement 2⟩
```

If ⟨expression⟩ evaluates to a nonzero value (i.e., true) then ⟨statement 1⟩ is executed; otherwise ⟨statement 2⟩ is executed. The else portion of the statement is optional and may be omitted if no action is to be performed if ⟨expression⟩ equals zero.

Multiple if else statements can be grouped to form a multiway decision structure. The structure

```
if(⟨expression 1⟩)
        ⟨statement 1⟩
else if(⟨expression 2⟩)
        ⟨statement 2⟩
else if(⟨expression 3⟩)
        ⟨statement 3⟩
else
        ⟨statement 4⟩
```

causes the expressions to be evaluated in sequence. If an expression evaluates to true (i.e., nonzero), then the statement associated with it is executed and execution transfers to the first statement after the structure. The statement associated with the else at the end of the structure is the default if no other statement is executed. It can be omitted if no default is required. The following gives an example of this structure:

```
if(watertemp<30)
        {
        status = COOL;
        heater = ON;
        }
else if(watertemp<40)
        {
        status = NOMINAL;
        heater = OFF;
        }
else

        {
        status = HOT;
        heater = OFF;
        alarm = ON;
        }
```

E.5.2 The switch Statement

Another multiway branching construct is supplied by the switch statement:

```
switch(⟨integer expression⟩)
        {
        case⟨constant⟩:
                ⟨statements⟩
        case⟨constant⟩:
                ⟨statements⟩
        case⟨constant⟩:
        . . . .
        default:
                ⟨statements⟩
        }
```

The expression is evaluated and then compared with the integer or character constant in each case. If any case matches, execution begins at that point. The statements associated with default are executed if no match is found. Again, default is optional and may be omitted.

Once the statements associated with the matching case are executed, program execution proceeds to the statements in the next case of the switch structure. If this is not desired, the use of the *break* statement will cause execution to transfer to the first statement following the structure.

In the following example, names in capitals such as INITIAL_STATE and CC1_CHARGE_STATE represent integer constants (defined at the beginning of the program) which are compared with the integer variable nextstate.

```
switch(nextstate)
    {
    case INITIAL_STATE:
            initialize( );
            break;

    case WAIT_STATE:
            wait( );
            break;

    case CC1_CHARGE_STATE:
            cc1_Charge( );
            break;

    case CVC_CHARGE_STATE:
            cvc_Charge( );
            break;

    case CC2_CHARGE_STATE:
            cc2_Charge( );
    break:

    case MALFUNCTION_STATE:
            malfunction( );
            break;

    default:
            malfunction( );
    }
```

The example illustrates the use of the switch construct as a sequencer for a finite-state machine. In this example, the state machine represents the battery-charger controller discussed in Chap. 4 (see Fig. 4.4). The variable nextstate contains the number of the state to be entered by the finite-state machine. Each case corresponds to one of the states in the machine. The first statement in each case calls a function which performs the actions required by that state and determines the next state. The function stores the number corresponding to the next state in the variable nextstate so that the next loop through the switch statement will cause this state to be entered.

E.5.3 The while and do while Statements

The while statement allows a statement or a block of statements to be executed repetitively. It has the format

```
while(⟨expression⟩)
    ⟨statement⟩
```

If the expression evaluates to true (nonzero), the statement is executed, and the process repeats. If the expression becomes false (zero), the statement is not executed and execution transfers to the next statement in the program. The statement

```
while(a<10)
    {
    if(b==a)
    ++a;
    else
    b=a;
    }
```

illustrates the use of the while statement and also demonstrates that one control construct (if else) can be *nested* inside another.

The do while statement has the format

```
do
    ⟨statement⟩
while(⟨expression⟩);
```

In this case the statement is executed first and then the expression is evaluated. If the expression is true, the loop is repeated; otherwise execution transfers to the next statement in the program.

E.5.4 The for Statement

The for statement is another technique for the repetitive execution of program statements. Its format is

```
for(⟨expression 1⟩; ⟨expression 2⟩; ⟨expression 3⟩)
    ⟨statement⟩
```

The operation of the for statement is equivalent to

```
⟨expression 1⟩;
while(⟨expression 2⟩);
    {
    ⟨statement⟩
    ⟨expression 3⟩;
    }
```

Normally ⟨expression 1⟩ is used to initialize a loop index, ⟨expression 2⟩ tests the loop index, and ⟨expression 3⟩ modifies the loop index. For example,

```
for(i = 0; i < 10; i++ )
    c[i] = 0;
```

forms a loop that sets the elements of the array c[10] to zero. The use of the C *comma* operator allows more than one variable to be initialized and modified in the for statement. In the following statement,

```
for(i = 0, j = 1; i < 10; i++, j *= 2)
    c[i] = j;
```

the variable *j* is initialized to 1 and then repeatedly multiplied by 2 so that the elements of the array c[10] are set to the first 10 powers of 2.

E.6 FUNCTIONS

A function is a defined block of variable declarations and statements that can be invoked (caused to execute) by a single statement. The statement that invokes a function in C is the *function call*, which has the format

```
⟨function name⟩(⟨argument list⟩);
```

For example,

```
distance(x1, y1, x2, y2);
```

invokes the function distance( ) and passes the variables x1, y1, x2, y2 to it. The variables passed to a function are called the function's *arguments*, or *parameters*. The parentheses after the function name must be used even if no arguments are passed to the function.

E.6.1 Passing Arguments

In C, arguments are said to be *passed by value*. This means that copies are made of the variables to be passed and the copies are transferred to the function. Thus the called function cannot affect the variables themselves. This, plus the use of local variables, gives the programmer the opportunity to design program modules which do not interact with other modules in unexpected ways.

A function can return a value to the function that called it with the statement

```
return(⟨expression⟩);
```

The value of the expression is returned to the calling function. When a function appears in an expression, its value is the value returned by the function. Thus, in the expression

$$sqrt(x) < 4$$

the function sqrt(x) is invoked and the returned value is then compared with 4.

Functions are assumed to return values of type int by default. If another type of value is to be returned, then the function must be declared in the calling function. For example, in the program

```
main( )
{
double x, y, cube( );
x = 2.345;
y = cube(x);
}
```

the function cube( ) returns a double value and so it is declared in the function main( ) which calls cube( ).

E.6.2 Function Definition

A function *definition* consists of the declarations and statements that make up the function. It has the format given in Table E.8.

The arguments declared in the function definition are called *formal arguments*. The arguments passed to the function in a function call must be the same type as the formal arguments but need not have the same names.

Functions which return values of type int or do not return a value at all, need not have their type explicitly declared in the function definition. Thus, the definition

Table E.8 Format of a function definition

Format	Example
⟨type⟩⟨name⟩(⟨argument list⟩) ⟨argument declarations⟩ {	double distance(a1, b1, a2, b2) double a1, b1, a2, b2; {

⟨local variable declarations⟩
⟨statements⟩
	double temp1, temp2, sqrt( );
	temp1 = (a1 − a2) * (a1 − a2);
	temp2 = (b1 − b2) * (b1 − b2);
	return(sqrt(temp1 + temp2));

}	}

```
donothing( )
{
}
```

is a valid function definition.

Function definitions must occur outside any other function definitions. Nested function definitions are not allowed in C. However, nesting of function calls is perfectly legal in C. In fact, a function can even call itself.

In Chap. 5 we discussed the need for reentrant programs in real-time systems that allow interrupts from external sources. If a function in a C program uses only auto variables, it will be reentrant. Any functions in a C program that are used by the interrupt handler should be made reentrant.

E.7 POINTERS

A *pointer* in C is a variable which contains the address of the memory location in which another variable is stored. If the pointer variable ptr contains the address of the variable c, then ptr is said to "point to" c. Figure E.3 illustrates the relationship between the pointer variable and the variable it points to. Some variables require more than one memory location for storage, in which case the pointer to the variable contains the address of the first location used to store the variable.

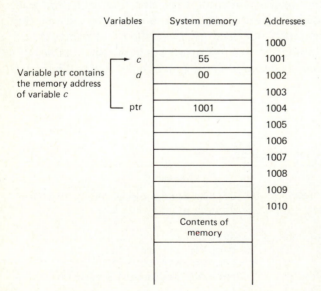

Figure E.3 Relationship between a pointer and the variable it points to.

E.7.1 Pointer Declarations

A pointer variable must be declared like any other variable. The pointer variable ptr might be declared as follows:

```
char *ptr;
```

The type declaration in a pointer declaration refers to the type of the variable that the pointer points to, not to the type of the pointer itself. Thus, the pointer ptr is declared to point at variables of type char. The asterisk * is a signal to the compiler that this is a pointer declaration rather than the declaration of a normal variable.

E.7.2 Pointer Initialization

After a pointer is declared, it has no initial value. The pointer must be assigned some value. If we wish to point at a specific memory location for some reason, a constant can be assigned to the pointer variable. For example,

```
ptr = 1000;
```

The variable ptr will contain the number 1000 and can be said to point to memory location 1000. Usually a pointer is used to point to a variable, in which case the address of the variable can be assigned to the pointer by the *unary &* operator. For instance,

```
ptr = &c;
```

will assign the address of the variable *c* (1001) to the pointer variable ptr.

One other method of initializing a pointer is specific to character pointers. String constants such as

```
''String constant''
```

are stored in memory as an array of character constants. The first memory location contains the character constant 'S' (ASCII code 83), the second location, the character constant 't' (ASCII code 116), and so on. The array is terminated by an ASCII NUL character (00) so that programs can find the end of a string constant. A character pointer to the first character in the string constant can be initialized as follows:

```
ptr = ''String constant'';
```

This statement causes the compiler to allocate storage for the string constant, initialize the storage locations to the character constants in the string, and assign the address of the first storage location to the variable ptr.

E.7.3 Pointer Indirection

The contents of a variable can be accessed through the use of a pointer to the variable. This process is called *indirection*, since the contents are accessed indirectly by first getting the address of the variable from the pointer and then using the address to get the contents of the variable. In C the unary indirection operator * is used for this purpose. In the following statements,

```
char c, d;
char * ptr;
c = 55;
ptr = &c;
d = *ptr;
```

1. Variables *c* and *d* are declared to be of type char.
2. Variable ptr is declared to be a pointer to char variables.
3. Variable *c* is assigned the value 55.
4. Variable ptr is assigned the address of variable *c*. The situation in memory is now as is shown in Fig. E.3.
5. The indirection operator causes the constants of variable ptr (1001) to be used as an address to access a memory location. The contents of this memory location (55) are assigned to variable *d*. The situation is now as shown in Fig. E.4.

Because the address of variable *c* has been assigned to the variable ptr, the contents of variable *c* are transferred to the variable *d*. Or, in other words, the variable *c* has been assigned to the variable *d*. Thus, the statement

```
d = c;
```

is equivalent to

```
ptr = &c;
d = *ptr;
```

The two uses of the asterisk with pointer variables is sometimes confusing. In a pointer declaration such as

```
char *ptr;
```

the asterisk simply acts as a signal to the compiler that this is a pointer

Variables	System memory	Addresses
		1000
c	55	1001
d	55	1002
		1003
ptr	1001	1004
		1005
		1006
		1007
		1008
		1009
		1010
		1011

Figure E.4 Memory contents after execution of d = *ptr.

declaration. In a statement such as

$$d = *ptr;$$

the asterisk is an operator which causes the compiler to assign the contents of the variable pointed to by ptr to d rather than assigning the contents of ptr to d as it would with the statement

$$d = ptr;$$

The indirection operator can be used in expressions other than assignments. For example, if ptr points to c, then

$$++*ptr$$

is the same as

$$++c$$

and

$$b + *ptr$$

is equivalent to

$$b + c$$

One common use of pointers is in cases where a function must change a local variable in another function. Simply passing the variable to the function will not work, since only the contents (value) of the variable are passed to the function; the function does not know the address of the variable. On the other hand, if a pointer to the local variable is passed to the function, the function can use the pointer to get access to the variable via the indirection operator. For example, the function

```
shuffle(a, b, c, d)
int *a, *b, *c, *d;
{
int temp;
temp = *a;
*a = *d;
*d = temp;
temp = *b;
*b = *c;
*c = temp;
}
```

uses the four pointers a, b, c, d to swap variables local to another function. The function shuffle( ) might be invoked as follows:

```
shuffle(&w, &x, &y, &z);
```

Note that a unary & operator is applied to the variables w, x, y, and z, so that pointers to the variables are created which are passed to the function shuffle( ).

E.7.4 Pointer Arithmetic

The arithmetic operations allowed on pointers are limited to those that make sense for variables that represent memory addresses. The allowed operations are

Addition of an integer to a pointer
Subtraction of an integer from a pointer
Incrementation or decrementation of a pointer
Comparisons of two pointers using the relational operators
Subtraction of two pointers which point to variables of the same type

On most machines, a pointer can be assigned to an unsigned int variable, and vice versa, without changing it. This allows "illegal" operations to be carried out on pointers when required.

C scales the pointer operations so that the differing amount of memory space taken up by different types of variables is taken into account. For example, the operation

```
ptr++
```

will increment the pointer ptr by 1 if it points to char variables which take up only one memory location. On the other hand, if ptr points to long int variables which take up four memory locations, it will be incremented by 4.

Pointer arithmetic operations are most often used to access elements in an array or other regular data structure. C stores array elements in contiguous addresses in memory with the first element stored at the lowest address. If a pointer is created to this first element by a statement such as

```
ptr = &a[0]:
```

then the expression

```
*(ptr + i)
```

refers to the array element a[i]. In fact, it is not necessary to create a separate pointer to the first element in an array. In C, the name of an array is a pointer to the first element in the array. Thus, if an array is declared as

```
int a[10];
```

the variable *a* is a pointer to element a[0]. The third element in the array could be accessed either by the expression

```
a[2]
```

or by the expression

```
*(a + 2)
```

Using pointers for array operations may sometimes be more efficient than using array indexes. For example, the operation

```
i = 0;
j = ARRAYSIZE - 1;
while(j >= i)
        {
        temp = a[i];
        a[i] = a[j]
        a[j] = temp;
        i++;
        j--;
        }
```

which reverses the order of the elements in an array, uses array indexes to access the elements of the array. The two index variables, i and j, are used in a calculation to determine the addresses of array elements a[i] and a[j]. The calculation adds the index variable to the pointer a, which points to the first element in the array a[ ]. By contrast, the operation

```
b = a + ARRAYSIZE - 1;
while(b >= a)
        {
        temp = *a;
        *a = *b;
        *b = temp;
        a++;
        b--;
        }
```

performs the same function but uses pointers to access the array elements directly. No address calculations involving index variables are necessary, thereby saving two additions per iteration. We should point out, however, that some C compilers are quite clever. They will recognize situations where array access can be made more efficient and will generate the appropriate code. Therefore, there may be no penalty to the use of array indexes.

When an array is passed to a function, only the pointer to the array is actually passed. This is an exception to the normal C practice of passing variables by value. As a result, the function can directly access and alter the array elements.

E.8 INPUT/OUTPUT

The C language has no built-in statements to perform input or output operations. Instead, functions must be written to send or receive data from an external device. Most C compilers come with prewritten functions that handle communications with terminals and printers. These functions are contained in a function library of the type described in Chap. 5. If a C program calls a function which is not defined within the program, the linker portion of the compiler will search the function library for its definition. If the definition is found, it will be incorporated into the program.

Since the input/output functions contained in the library are not an intrinsic part of the C language, they can be changed to suit the requirements of particular computers and additional functions can easily be added to cope with special-purpose peripheral devices. This flexibility adds to the portability of C and adds to its appeal in applications involving the direct control of hardware.

E.8.1 Memory Mapped I/O

In order to write functions that communicate with or control an external device, the programmer must have some means to access the registers in the device. In computer systems which have memory-mapped I/O systems, these registers appear to be memory locations. Therefore, pointers can be assigned to the addresses of the registers and the contents of the registers can be accessed by indirection. For example, if we have an external device with a status register at address 0x4000, the statements

```
char *statusreg;
statusreg = 0x4000;
```

define statusreg as a pointer to a single byte of memory (char) and initialize it to the address of the status register. Now the contents of the status register can be read by statements such as

```
status = *statusreg;
```

The pointer must be declared to point to a variable that occupies only one memory location. On most microcomputer systems, one memory location contains 1 byte (8 bits) of data. Therefore pointers to char variables should be used. If a pointer to a variable occupying more than one memory location is used, the results will be unpredictable.

E.8.2 I/O to Separate I/O Address Spaces

Pointers cannot be used to access I/O devices in computer systems which have a separate address space for I/O devices. C compilers for machines with this type of I/O system almost invariably have library functions which allow direct access to the I/O system. For instance, most C compilers for the 8080/Z80 family of microcomputers have functions called input( ) or inp( ) and output( ) or outp( ) which provide access to the microcomputer's I/O system. The function input( ) is passed an argument specifying the address of the I/O port to be accessed and it returns the contents of the external device register at that port address. The function output( ) is passed the address of the I/O port to be accessed and the data to be sent to the external device register; it then sends the data to the register.

EXERCISES

E.1. Which of the following are legal identifiers in C? For the illegal identifiers, explain why they are illegal.

(*a*) register	(*e*) FORTY_FIVE_SECONDS	(*i*) id_code
(*b*) fault!	(*f*) a	(*j*) queen3
(*c*) area	(*g*) 3prime	(*k*) X452
(*d*) string.var	(*h*) #5	(*l*) ON/OFF

E.2. Convert the following bit patterns into hexadecimal constants acceptable to a C compiler.

(*a*) 01010010
(*b*) 0111101011100011
(*c*) 1100101100001111
(*d*) 01110001

E.3. List the variables in the following C program and give their storage class.

```
#define PI 3.14159
int radius;
float area;
main( )
{
int diameter;
float circumference;
radius = 2;
diameter = radius * 2;
circumference = diameter * PI;
area = radius * radius * PI;
}
```

E.4. If unsigned int variables are stored in 16 bits of memory and long int are stored in 32 bits of memory, what are the largest positive integers that can be stored in each type (in decimal notation)?

E.5. In the following C program, what will be the final value of each variable after the program is executed?

```
main( )
{
int a, b, c, d, e, f;
a = 10;
b = 3;
c = a%b;
b += a;
d = a - b/c;
d *= a;
e = d++;
f = --d;
a = f - d*c + e/b;
}
```

E.6. In the following C program, what is the final value of each variable (in hexadecimal)?

```
#define MASK1      0xCF03
#define ALL_ON     0xFFFF
#define MASK2      0xAAAA
#define ALL_OFF    0x0000
```

```
main( )
{
int status, pattern, switches, lamps;
status = pattern = ALL_ON;
switches = lamps = ALL_OFF;
status &= MASK1;
switches |= MASK2;
pattern = MASK2;
lamps = switches & pattern;
status = ~status;
switches = lamps & status | pattern;
}
```

E.7. In the following C program, what is the final value of each variable (in decimal notation)?

```
main( )
{
int k, m, n, o, p, q;
k = 3;
m = k << 1;
n = k << 2;
o = k << 3;
p = o >> 1;
q = o >> 2;
}
```

What arithmetic expression is equivalent to shifting a positive integer *j* left by *n* positions? What arithmetic expression is equivalent to shifting a positive integer *j* right by *n* positions?

E.8. If a signed integer is shifted right, two possible results can occur, depending on the particular C compiler used. Some C compilers will cause the computer to perform an *arithmetic shift* (when the program is executed) in which a 1 is shifted into the leftmost bit position if the previous bit in that position was also a 1. A zero will be shifted in otherwise. Other compilers will cause the computer to perform a *logical shift* in which a 0 is always shifted into the leftmost bit position. Write a C function called checkshift( ) which determines whether an arithmetic shift or a logical shift is performed for a signed integer. The function returns a 1 if an arithmetic shift is performed and a 0 if a logical shift is performed. All C compilers will cause the computer to perform a logical shift when the integer is unsigned.

E.9. What are the values of the variables in the following C program after the program has executed?

```
main( )
{
int a, b, c;
float x, y, z;
a = 100;
b = 150;
x = 2.0;
c = a * x / b;
y = b / a;
z = b / a * x;
}
```

E.10. What are the values of the variables in the following C program after the program has executed?

```
main( )
{
int h, i, j, k, m, n, p;
h = 6;
i = 7;
j = 3;
k = (h & i) * j;
m = k * i;
n = m++;
if (k < (h * j) || m == n)
        j = 5;
else
        j = 2;
if (i > (k / j) && m > 10)
        p = 0;
else
        p = 1;
}
```

E.11 A native assembler for a microcomputer whose memory is organized into locations containing 1 byte produces an object module having the following characteristics:

1. The first two bytes in the module contain the 16-bit starting address of the machine-language program. The low-order (less-significant) 8 bits are stored in the first byte (i.e., the location with the lower address).
2. The next two bytes contain a 16-bit number equal to the number of bytes in the machine-language program. Again the low-order 8 bits are stored in the first byte.
3. The remainder of the module contains the machine-langauge program.
4. The object module is stored in memory in contiguous locations with ascending addresses.

Write a C function called loader( ) which places the machine-language program into the proper locations in memory. The function is passed a pointer to a character which points to the memory location where the first byte of the object module is stored.

BIBLIOGRAPHY

C-language texts

Gehani, N.: *C: An Advanced Introduction*, Computer Science Press, Rockville, MD, 1985.
Hancock, L., and M. Krieger: *The C Primer*, McGraw-Hill, New York, 1982.
Kernighan, B., and D. Ritchie: *The C Programming Language*, Prentice-Hall, Englewood Cliffs, NJ, 1978.
Kochan, S. G.: *Programming in C*, Hayden, Rochelle Park, NJ, 1983.
Plum, T.: *Learning to Program in C*, Plum Hall Inc., Cardiff, NJ, 1983.
Purdum, J.: *C Programming Guide*, Que Corp., Indianapolis, IN, 1983.

Other C references

Byte: "Theme Issue on the C Programming Language," August 1983.
Feuer, A. R.: *The C Puzzle Book*, Prentice-Hall, Englewood Cliffs, NJ, 1982.
Zahn, C. T.: *C Notes: A Guide to the C Programming Language*, Yourdon, New York, 1979.